BIOLOGICAL PRINCIPLES WITH HUMAN PERSPECTIVES

TO THE STUDENT: A Study Guide for the textbook is available through your college bookstore under the title Study Guide to accompany BIOLOGICAL PRINCIPLES WITH HUMAN PERSPECTIVES, 2nd Edition by Gideon E. Nelson. The Study Guide can help you with course material by acting as a tutorial, review and study aid. If the Study Guide is not in stock, ask the bookstore manager to order a copy for you.

2nd edition

BIOLOGICAL PRINCIPLES WITH HUMAN PERSPECTIVES

Gideon E. Nelson
University of South Florida

John Wiley & Sons
New York   Chichester   Brisbane   Toronto   Singapore

Production Supervised by Linda Indig
Book and cover designed by Madelyn Lesure
Photo Research by Kathy Bendo
Cover Photo: Micrograph of human skull tissue;
　　　　　Russ Kinne, 1972/Photo Researchers, Inc.
Manuscript editor was Pam Landau under the supervision of Deborah Herbert

Library of Congress Cataloging in Publication Data:

Nelson, Gideon E.
　　Biological principles with human perspectives.

　　Includes indexes.
　　1. Biology.　2. Human biology.　I. Title.

QH308.2.N44 1984　　　　599.9　　　　83-10343
ISBN 0-471-86277-0

Printed in the United States of America

10 9 8 7 6 5 4 3 2 1

preface

This textbook is intended for use in a one-term, introductory biology course for general college students. Considerable effort has been taken to make the discussion understandable, informative, and interesting.

The goal of the book is to provide students with a background of knowledge with which they can better evaluate and interpret various kinds of biological information. Test-tube babies, monoclonal antibodies for treating cancer, and the perils of acid rain make exciting news, but individuals need a background of basic biological knowledge to assess the accuracy or significance of such information.

The book is planned around three objectives. One of these is to present major topics in the field of biology that also affect people's everyday lives. To implement this goal, human examples are used extensively to demonstrate basic biological principles or concepts. Students are usually more interested, and thus more motivated, to learn principles that apply to themselves as well as to other organisms. After all, human beings are among the most important and interesting biological creatures on earth. In addition, many biological concepts can be as well explained with human examples as with frogs or flatworms.

Another objective is to make the book as usable for students as possible. Each chapter begins with an abbreviated chapter outline and a few basic questions to provide readers with a preview of the chapter. Within each chapter, careful attention has been given to the reading level, the explanation of complex concepts, and to making certain that there is a minimal use of scientific terminology. An extensive glossary is provided at the end of the book. The important basic concepts in each chapter are summarized in boldfaced type following the sections that relate to them. The purpose is to identify clearly the concepts for the reader, and to relate them immediately to the appropriate body of text material. Each chapter concludes with a list of key terms and a set of study questions that help students review the chapter. A separate study guide with chapter outlines and quizzes is also available from the publisher.

Instructors may wish to know how this edition differs from the previous one. First, all material retained from the previous edition has been closely examined and brought up to date with new information as necessary. Second, many of the recommendations made by reviewers for changes in text or illustrations have been incorporated when appropriate. Finally, a series of new topics was added to various chapters. These include bioethics, tissue types, biological rhythms, hybridoma cells and monoclonal antibodies, the human life cycle, altruism, and acid rain.

I thank the many reviewers who provided invaluable aid in improving the manuscript for this edition. They are: Drs. Stewart Swihart, Marvin Alvarez, Deborah Nickerson, and Gertrude Hinsch of the University of South Florida; Raymond R. White, City College of San Francisco; Kenneth D. Bergman, Keene State College; Martin S. Rochford, Fullerton College; William L. Cavenee, Central Arizona College; Barbara E. Bowman, Wichita State University; Allen F. Reid, State University College of Arts and Science; Charles L. Ralph, Colorado State University; Barry O. Ochs, Shippensburg State College; Mary Ann Bianchi, State of Connecticut, Manchester Community College. I also acknowledge the unsung heroes of this and countless other books from John Wiley & Sons, namely the many wonderfully helpful individuals who are involved in the intricacies of editing, producing, illustrating, and marketing textbooks.

To my wife, Nancy Anne, I dedicate this edition as a small token of appreciation for her enormous patience, continuous and enthusiastic support of my endeavors, and generous affection.

Gideon E. Nelson

contents

BIOLOGICAL PRINCIPLES WITH HUMAN PERSPECTIVES

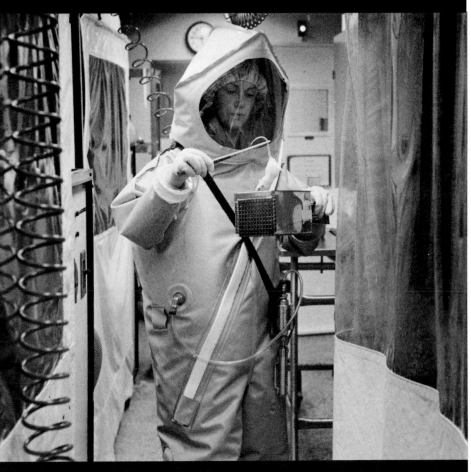

A research laboratory for tropical diseases at the Atlanta Disease Control Center.

BIOLOGY: AN INQUIRY INTO LIFE

chapter 1

questions to think about

In what ways did biology have its roots in early human being's interactions with nature?

How would you describe the scope of the field of modern biology?

In what ways does contemporary biology interact and influence our lives today?

Why has the scientific method been so successful in helping us understand the world around us?

What are the roles of basic and applied research in science?

How is ethics becoming a part of biomedical science?

Equipped with his five senses, man explores the universe around him and calls the adventure science.

Edwin Powell Hubble

The above quotation is an especially apt one because science—the systematic gathering of knowledge obtained by observation—is truely an exciting adventure. Our five senses: vision, smell, equilibrium, hearing, and touch, are our basic tools, and devices such as microscopes or centrifuges become extensions of these basic tools. With them mankind has made astounding discoveries about the world around him. So far as anyone can tell, we will never exhaust this exciting potential for advancing our knowledge. Moreover, what seems incredible now often becomes commonplace knowledge in the future.

One branch of science, *biology*, specializes in attempting to understand and define the phenomenon called life. To accomplish this, biologists like other scientists use the scientific method as described later in the chapter. The field of biology has grown so fast during this century that it has become divided into numerous subfields. In wondering how all of this came about, we can take a very general look at how the field of biology evolved.

Most of the meager evidence that we have about the cultures of prehistoric human beings indicates their sense of close kinship with nature and their special interest in other living creatures. Paintings by our prehistoric ancestors on the walls of caves in France and Spain depict wild animals (Figure 1-1). Nearly all the art objects of ancient peoples, paintings and carvings alike, were based on biological subjects: animals, human beings, or more fanciful creatures. An anthropologist would say that this interest is not at all unusual, considering how closely primitive peoples are involved with their environment.

EARLY HUMANS AND NATURE

Picture yourself, for a moment, as a member of a small tribe living about 10,000 years ago. Your home, possibly a cave, was provided by nature. Even if your group

Figure 1-1 *Paintings by prehistoric humans on the wall of a cave in France.*

Figure 1-2 *Members of a Neandertal group as they may have lived thousands of years ago. Their facial features and body forms were probably similar to ours.*

lived in manmade shelters of some sort, the shelters would be constructed of plant materials, or perhaps of animal hides. As protection from the elements, the group would have to make clothing from animal skins (Figure 1-2). Most of your waking hours would be spent locating food: picking berries, grubbing for edible roots, and occasionally hunting small animals with simple weapons such as rocks, clubs, and spears. Occasionally, a member of the tribe might be fortunate enough to happen on the remains of a kill made by a large predator and succeed in driving away the other carrion feeders long enough to grab his share.

Because the human body is so defenseless (lacking even a decent coat of fur) you are almost helpless against many perils of the environment. Only two features enable you to survive from day to day: a relatively large body size, which makes you one of the larger predators, and cunning in avoiding danger.

As a hunter-gatherer life is rigorous and life spans are short. Phenomena such as lightning, rainstorms, changing seasons, and the variation in forms of plant and animal life, being such integral parts of nature, would undoubtedly awe and mystify your tribe. This awe of natural events as well as the mysteries of birth and death frequently became the basis for various tribal rituals and religions. Thus early humankind evolved in an intimate association with nature, a bond that persisted for many thousands of years.

Domestication of Plants and Animals

Long ago prehistoric man began to domesticate a number of wild plants and animals for his own use. This not only provided more abundant food but also allowed more people to live on a smaller plot of ground. We tend to forget that all of our present-day pets, livestock, and food plants were taken from the wild and developed into the forms we know today. Table 1-1 lists some of our domestic animals and approximately how long they have been in this form.

As centuries passed and human cultures evolved and blossomed, humans began to organize their knowledge of nature into the broad field of *natural history.* One aspect of early natural history concerned the use of plants for drugs and medicines. The early herbalists sometimes overworked their imaginations in this respect. For instance, it was widely believed that a plant or part of a plant that resembled an internal

organ would cure ailments of that organ. Thus, an extract made from a heart-shaped leaf might be prescribed for a person suffering from heart problems. Nevertheless, the overall contributions of these early observers provided the rudiments of our present knowledge of drugs and their uses.

NATURAL HISTORY AND MODERN BIOLOGY

Out of the broad field of natural history—that is, man's interest and curiosity about animals and plants—eventually grew a more specialized study, the systematic gathering of data about living things in nature and how they function, the branch of science we now call *biology.* Thus, in a general sense, biology had its earliest roots in our primitive ancestors' interactions with nature. We have systematized the study of life into a formal science; nevertheless, nearly everyone still feels some kind of personal kinship with nature. Urban dwellers try to capture a little of it by growing plants and keeping pets; many of us get closer to nature by going camping and hiking, and some adventuresome souls even go backpacking in wilderness areas (Figure 1-3). Backpacking is so popular in fact that three million individuals now hike on the beautiful and rugged Appalachian trail each year.

Even with our love of nature we seem to retain considerable awe and fear of nature. The camper, hiker, and backpacker all need special clothing and other gear as protection against the rigors of the environment. And, in the fashion of our ancient ancestors, modern campers like to build campfires at night and huddle around them. The fire not only provides warmth, but more importantly, also creates a sense of companionship and a feeling of unity with nature that is difficult to put into words. It is not surprising that the remains of countless ancient campfires provide the best evidences we have of where ancient man lived and how he spread across continents.

All in all, then, it is not surprising to find so many individuals interested in biology in one respect or another. Numerous college students take courses in biology because they find it satisfying to learn more about themselves and the world of living things around them. We are, after all, an influential part of that world. Only

TABLE 1-1. The Domestication Time for a Variety of Animals (After Dobzhansky, 1955)

Species	Domestication	Number of Varieties
Pigeon	Prehistoric	140
Donkey	Prehistoric	15
Guinea pig	Prehistoric	25
Dog	10,000–8000 B.C.	200
Cattle	6000–2000 B.C.	60
Pig	5000–2000 B.C.	35
Chicken	3000 B.C.	125
Horse	3000–2000 B.C.	60
Cat	2000 B.C.	25
Duck	1000 B.C.	30
Canary	1500 A.D.	20

Source: Table modified from paper by K. F. Dyer, Evolution Observed— Some Examples of Evolution Occurring in Historical Times. *Journal of Biological Education,* Volume 2, pp. 371–338, 1968.

Figure 1-3   *A backpacker in the Grand Teton National Park, Wyoming.*

by knowing more about it can we know more about ourselves. Also, as we shall see throughout this book, discoveries in the field of biology have improved the quality of human life immensely and will likely continue to do so in the future.

Basic Concepts

Biology is a branch of science that studies life and its processes by means of systematic observations.

Human cultures evolved in an intimate association with nature.

BIOLOGY TODAY

In recent times, the discipline of biology has expanded so rapidly into such a variety of fields that the inclusive term *life sciences* is sometimes used to encompass the diverse specialties of medicine, dentistry, nutrition, pharmacology, botany, biochemistry, physiology, zoology, and numerous other disciplines. Their common thread is that they all deal with living matter of some sort. The amount of knowledge in the life sciences is so vast and complex that it is already far beyond the grasp of any single human mind (Figure 1-4). As a result, a professional biologist is, by necessity, a specialist in some small aspect of biology. It is these individuals who

Figure 1-4   *The accumulation of new knowledge in the Life Sciences is so rapid that this periodical consists of only the tables of contents from numerous scientific journals. (Reprinted with permission of the Institute for Scientific Information® from Current Contents ®/Life Sciences. © 1983 by ISI.)*

continue to add new bits of knowledge to biology by searching for answers to the myriad questions in their particular spheres of interest. Often it seems that they contribute bits of almost trivial information: the number of birds nesting in an Appalachian woodland, the chemical constituents of plant sap, or the reproductive processes of an obscure bacterium, but all these unrelated pieces of information have a place, like parts of a jigsaw puzzle, in a much larger picture. Sooner or later someone puts the pieces together and our understanding of the natural world takes a major step forward.

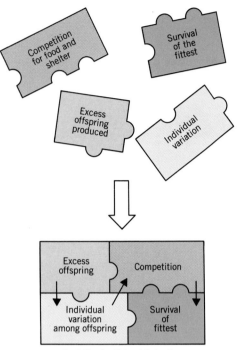

Evolution by Natural Selection

Figure 1-5 *Charles Darwin brought together seemingly unrelated ideas and observations to form his theory of evolution by natural selection. Advances in scientific knowledge frequently occur in this manner.*

These steps can sometimes be immense. Charles Darwin united numerous preexisting ideas, as well as his own data and thoughts, into a whole new concept of how populations of living things evolve in relation to their environment (Figure 1-5). This concept at first shook the foundations of biological science. In more recent times, James Watson and Francis Crick made an inspired guess about the structure of the DNA molecule, based on existing information from a variety of sources (including their own studies), that revolutionized many aspects of modern biology. Such people are the heroes, the immortals in the history of biology. We should not forget, however, that all scientific immortals use the findings and ideas of others to a considerable extent in making their own discoveries.

Some of the complexity of modern biology also arises from the necessity to integrate knowledge from other scientific fields. Data and theories from mathematics, physics, chemistry, and geology are used in various biological fields. Even astronomy provides data for biologists concerned with the age of the earth, space travel, and life on other planets.

The discoveries of organic matter in stony meteorites and organic compounds in outer space broaden the possible answers to such questions about the origin of life. Contemporary biology has become increasingly intertwined with the social and behavioral sciences in matters of human population growth and control, famine, drug use, aggression and antisocial behavior, and the use and abuse of natural resources (Figure 1-6). Hopefully, the future will see the successful application of biological theories to the solution of some of these serious problems.

Basic Concept

Modern biology encompasses many different specialities, uses knowledge from other scientific fields, and is becoming greatly involved in human social problems.

OBTAINING NEW BIOLOGICAL KNOWLEDGE

The discipline of biology, like all sciences, would quickly stagnate if new knowledge were not continually being added to it. This knowledge, as mentioned earlier, grows out of the studies conducted by specialists in many different areas of biology. The investigators all work within the broad framework of the scientific method, a procedure formulated back in the seventeenth century by Francis Bacon, a philosopher. This method is not a single standardized technique that must be followed rigorously, nor do all scientists use it in the same way. Rather, it is a practical and reliable way to approach problems and to obtain reliable solutions to them.

Scientists usually concentrate their research efforts in areas that are of special interest to them, and in which they usually have specialized training. The goal of the researcher is to make a significant contribution to biological knowledge or perhaps even to challenge an existing theory. A crucial beginning step is to pose significant questions for the investigation to answer. Some investigators may begin by proposing a *hypothesis,* that is, as an unproven conclusion that challenges

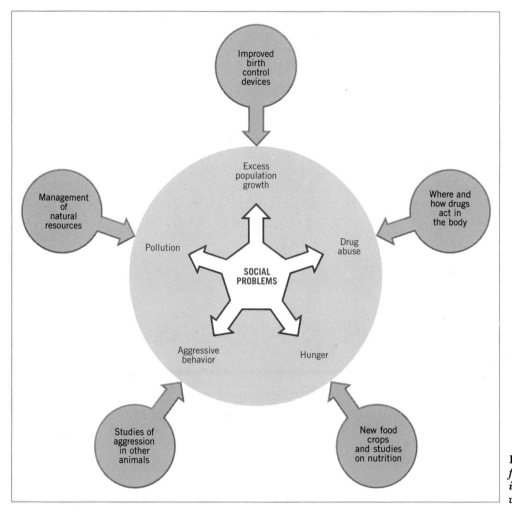

Figure 1-6 *How knowledge from various biological fields is used to help alleviate various social problems.*

or extends an older conclusion and thus provides a basis for further investigation and experimentation. This posing of questions and alternate hypotheses is one of the most important parts of the study because they guide the course of the research. Here, as in the design of experiments, the interpretation of data, or the formulation of hypotheses, scientists often utilize a certain amount of *intuition* — hunches about how to proceed next and what kinds of questions to ask. Intuition is not usually described as a part of formal scientific methodology, yet many successful scientists declare that intuition is important in their scientific work. However, intuition as well as hypotheses and data interpretation must be tested continually and changed

frequently as new facts are discovered. This constant evaluation procedure is the heart of the *scientific method.*

After formulating some initial questions or hypotheses and planning how to conduct his study, the researcher proceeds to obtain evidence known as *data.* Data are accumulated by making observations of some sort; these may be obtained from nature or in the laboratory. Data gathering is usually aided by use of equipment such as cameras, microscopes, electronic equipment, and various devices for making chemical and physical measurements (Figure 1-7).

A valuable aspect of data gathering is the use of experimentation to verify or disprove an hypothesis.

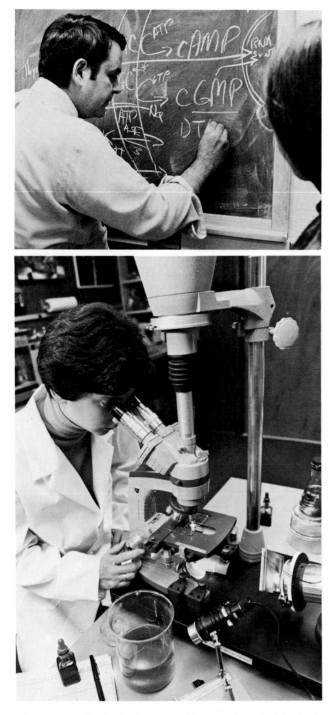

Figure 1-7 *Individuals engaged in a variety of biological studies and research activities.*

An *experiment* is a procedure designed to discover which factors in a given situation affect one another, that is, to determine what kinds of cause-and-effect relationships exist under the conditions of the experiment. Two general rules govern experimentation. The experiment should be conducted in such a manner that every *variable* — that is, every significant factor that might affect the outcome — is under the control of the experimenter. The ideal way to do this is to run a series of tests (experiments) that are all alike except in one specific factor, which is allowed to vary. In this way the most significant aspects of the cause-and-effect relationship are identified.

The second general rule is that experiments be of such a nature that they can be repeated by other scientists, yielding similar results. *Repeatability* is an extremely important part of scientific methodology, for it is the major basis for accepting or rejecting many hypotheses. To take just one example: A scientist reported a few years ago that he had cultured living bacteria from the inside of a meteorite by using a special technique. If true, this would have been an outstanding discovery, because it would indicate a means by which life was being spread throughout the universe. However, other scientists, using the described procedure, were unable to obtain any forms of life from inside similar meteorites. The "outstanding" discovery was unacceptable to most scientists because it failed the test of repeatability.

If a series of experiments produces results that are consistent with a new hypothesis, then the hypothesis can be said to have reached a level of greater certainty. That is, there is now a higher probability that the hypothesis correctly summarizes or interprets a set of facts. Some scientists term this type of hypothesis a *theory,* although there is no sharp distinction between the two (Figure 1-8).

At this point the investigator has contributed two kinds of knowledge to science — the facts (data) that have been gathered, and the hypotheses or theories that were derived from them. Theories are useful in two ways: First, they link sets of data. Second, and sometimes more importantly, they make predictions that may lead to additional lines of investigation (Figure 1-9).

A theory that has been repeatedly verified and appears to have wide application in biology may assume

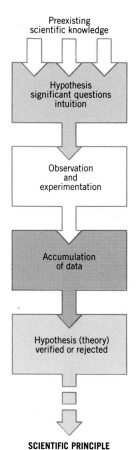

Figure 1-8 *One possible sequence of steps in the scientific method.*

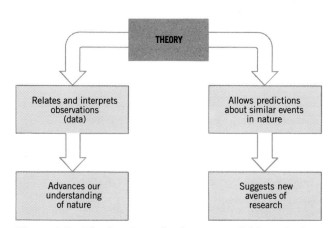

Figure 1-9 *The functions of a theory or valid hypothesis.*

the status of a *biological principle.* Such principles are sometimes called *scientific laws,* although this term tends to make them sound too awesome and permanent. A principle is a statement that applies, with a high degree of probability, to a range of events. "Living matter is made of cells or cell products" is often stated as a biological principle in many books. This is a sound and useful principle in the field of biology. It is essential to remember, nonetheless, that even the most accepted principles are manmade and subject to change if contradictory observations emerge in the future.

In addition to gathering facts, hypothesizing, and testing hypotheses, individuals attempt at intervals to summarize the research efforts of other biologists in order to present the current status of progress in various fields. This type of summary usually appears as a long review paper or as a monograph in the scientific literature. Textbooks such as this one represent another attempt to bring together basic information and principles from various fields of study.

The systematic accumulation of observations as described above is termed *research.* Scientific research is expensive and time consuming, and it requires the efforts of highly trained individuals. Yet no more efficient way has been found to obtain new knowledge about nature.

Basic Concepts
New biological knowledge is obtained by studies conducted within the framework of the scientific method.

The scientific method involves observing, hypothesizing, experimenting, constant *verification,* and formulating principles.

Basic and Applied Research
In biology as in other sciences, one often encounters a distinction between *basic* and *applied research.* In general, basic research includes those investigations that are not directed at immediate practical applications. For example, a marine ecologist might spend a period of years working out the dynamics of a salt marsh: the kinds of organisms living in the marsh, how they obtain energy, how they survive, and so on, with no particular thought about the practical uses of these findings (Fig-

ure 1-10). The value of this type of research is its contribution to man's understanding of the world.

Most research of the basic type is conducted by individuals associated with academic institutions. In fact, one of the major attractions of a university teaching career is the opportunity to engage in research and other scholarly activities of one's own choosing.

Applied research is directed toward the solution of a problem that is of immediate concern to someone or, in some cases, toward applying a new scientific discovery to practical uses. This makes it a part of the applied sciences, or technology, that dominate so much of contemporary life. But the underlying ideas for these applied endeavors depend heavily on basic research findings, which means that the two types of research activities are intimately related. Thus a fisheries biologist employed by a seafood company might study the data derived from the salt marsh study described above and use them for designing a "shrimp farm" for the commercial production of shrimp. In fact, a series of events similar to these actually occurred in a southeastern coastal area a few years ago.

Controversy sometimes arises over the relative importance of basic and applied research: both are expensive to maintain, and both often depend on public tax dollars for financial support.

Members of Congress are frequently asked to approve large appropriations of money for research in the basic sciences. It is difficult to explain to these individuals, who for the most part are nonscientists, or to the general public, why this expensive process is so important, and why scientists cannot predict the ultimate value or applications of their efforts in basic research. No one can make this kind of prediction, because this judgment can only be made in retrospect. Some results of basic research eventually find great use and some do not. When researchers discovered that certain strains of bacteria exchange hereditary material, no one thought it was particularly important or that it would have any practical application. Now, however, this knowledge is the basis of revolutionary techniques that may enable humans to correct some of their own faulty genes. Consider the great significance of our present knowledge of the atom, or of antibiotics, or of DNA. The initial knowledge of these topics, gained from basic research studies, was not recognized as very important, and years passed

Figure 1-10   *An extensive salt marsh on the coast of Georgia. The vegetation is* Spartina alterniflora, *a grass that is adapted to living in sea water.*

before their values were appreciated. Thus it is absolutely necessary, in the opinions of most scientists, to continue supporting and encouraging basic research endeavors in order to advance man's understanding of himself and nature.

Basic Concepts

Utilization of the scientific method to obtain knowledge is called research.

Research may be basic or applied. Both are vital to the advancement of scientific knowledge.

BIOETHICS

The life sciences have long used observation and experimentation to discover new knowledge about the universe. The implementation of the process is presumed to be neither influenced by value judgments and beliefs of the investigator, nor was it the responsibility of the scientist to tell society how to use new scientific knowledge. Now this attitude is being challenged! This is especially true in the health-related fields, where so many decisions are based on scientific discoveries, and which directly affect society and individual lives.

Some decisions are easily made, such as whether or not to perform a new life-saving surgical operation or to administer a disease-preventing drug. Other decisions, however, are not so clear-cut. What about decisions involving abortion, using humans in experiments, genetic engineering, and euthanasia? Should mechanical life-support devices be removed from a terminally ill patient who is suffering greatly and expresses the wish to die? Who should make this choice —patient, family members, or physicians? As you can see, in all of these cases, choices are not easy to make and, indeed, will depend mostly on the value systems (beliefs) of the people who are involved. The scientific method is not helpful here because it does not deal with *value judgments.* Nonetheless, situations keep cropping up in the scientific and biomedical world where value judgments are critical and must be made by someone. Controversy often results when the value judgments are not universally accepted—and they seldom are.

As a result of this dilemma, and the desire to make rational value-judgment decisions, scientists have turned to the field of ethics. *Ethics,* a branch of the discipline of philosophy, deals with what is *right* or *good* in relation to human conduct. Ethics attempts to relate what people do (their morality) and what they *ought* to do in various situations. In this sense, we can speak of making *moral-ethical* decisions. Whether or not to perform an abortion is often a moral-ethical decision. Making choices such as these always involves the values (beliefs) held by the individual or the group making the decision.

At this point you may be wondering, "Is there really any rational way to solve perplexing moral-ethical

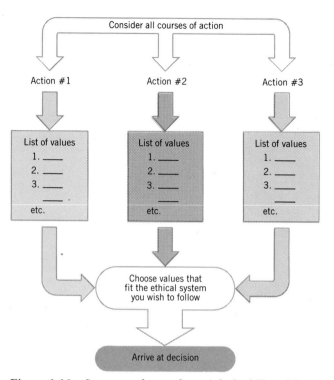

Figure 1-11 *Sequence of steps that might be followed in making moral-ethical decisions.*

problems?" One approach that has been suggested is to consider all possible courses of action (solutions) that are available, list the values associated with the courses of action from the most desirable to the least desirable, and then to make your decision on the basis of this analysis (Figure 1-11). In our example of the terminally ill patient described above, the choices are to remove the life-support devices or not to remove them. If removed, the patient dies and is spared further pain and suffering. If they are not removed, the patient continues to live and suffer. The choice here is whether to place a higher value on a human life or to end the suffering of an individual who is going to die in a short time anyway.

Some individuals are uncomfortable with ethical decision-making processes because there seems to be no predetermined or ultimate right or wrong. Indeed, one major school of ethical thought asserts that the goodness of an action is determined by the results of the action. If an action benefits an individual and the group around him or her, then it is a "good" action.

This is called the *Consequentialist Theory*, and is widely followed in contemporary society. Another opposing school of thought, the *Deontological Theory*, states that good or right actions are based on predetermined principles and not on the consequences of the act itself.

Consider the principle that human life is sacred. No actions would then ever be taken to endanger another human being. One would never unplug a life-support system from even a terminally ill patient because this would violate the principle. Which theory should one follow? Is one theory "better" than the other? In the actual process of making moral-ethical decisions, individuals often find that in some situations, the consequentialist concept works best (the greatest good or happiness for the most people). In other situations the application of a universal principle as advocated by the deontological theory may be a better solution.

The application of ethics to biomedical problems is referred to as *bioethics.* It is a rapidly growing movement as shown by the number of hospitals that have bioethical committees on their staffs. Some biologists urge that bioethics be expanded to cover all life on our planet. Throughout the text bioethical problems will be presented from various areas of biology. As they are brought up, I hope that you will find it challenging and interesting to arrive at your own ethical decisions concerning them. Indeed, one goal of the bioethical movement is to keep people sufficiently informed so that they can participate in the many moral-ethical decisions that arise almost daily.

Basic Concept

Bioethics is the application of ethical theories to value-oriented problems that arise in biomedical fields.

KEY TERMS

applied research	data	intuition	scientific method
basic research	deontological theory	life sciences	theory
bioethics	domestication	moral-ethical problem	value judgment
biological principle	ethics	natural history	variable
biology	experiment	repeatability	verification
consequentialist theory	hypothesis	scientific law	

SELF-TEST QUESTIONS

1. Define the terms *science* and *biology.*
2. In what ways, in addition to vision, are we equipped to observe nature?
3. How did prehistoric human cultures express their close kinship with nature?
4. What two features probably enabled prehistoric human beings to survive the many perils around them?
5. What is the origin of our present-day domesticated animals and food plants?
6. The scientific endeavor called biology had its roots in an earlier field called what?
7. Which of our present-day activities reflect our continuing interest in nature?
8. What types of fields are included under the life sciences?
9. Why is it said that a professional biologist is "by necessity" a specialist of some sort?
10. Briefly tell the relationship of contemporary biology to:
    (a) other scientific fields such as chemistry and astronomy; and
    (b) the social and behavioral sciences.
11. In the sciences, all investigators use a general technique called what?

12. What are two commonly used beginning steps in the scientific method?

13. Define each of the following terms:
    (a) hypothesis;
    (b) intuition;
    (c) observation;
    (d) experimentation;
    (e) theory;
    (f) scientific law

14. To be of greatest value, an experiment should always follow what two rules?

15. What is the value of the "test of repeatability" in science?

16. Give a broad definition of the term *research*.

17. Describe the difference between applied and basic research.

18. Why is it important to society to promote both basic and applied research?

19. Why does controversy often exist over the relative importance of these two kinds of research?

20. (a) Define the term *bioethics*.
    (b) What is meant by a moral-ethical dilemma?
    (c) Give an example of a moral-ethical decision that might have to be made.

chapter 2

*Dr. Heinz Meng with a
trained hawk.*

THE HUMAN
ORGANISM AS
A BIOLOGICAL
ENTITY

chapter 2

questions to think about

What structural features do human beings share with other forms of life?

What basic functional features do humans share with other forms of life?

Why is man classified as a vertebrate, a mammal, and a primate?

In what ways is man a distinctive rather than a unique organism?

How is man a unique organism?

No knowledge can be more satisfactory to a man than that of his own frame, its parts, their functions, and actions.

Thomas Jefferson,
Letter to Thomas Cooper, 1814

Just exactly what are human beings? Are we simply additional units in a world of complex biological entities? Or are we unique creatures among all the organisms inhabiting the world? Are we like other animals but blessed by a divine touch? Diverse ideas about the nature of man have been proposed and debated for centuries. The more we learn about ourselves, the more cautious we become about making dogmatic statements concerning who and what we are. Being human, perhaps we are too self-centered ever to see ourselves objectively. On the other hand, to be human is to be curious, and curiosity about the relation of human organisms to the rest of the biological world is too important and fascinating for us to ignore.

In this chapter we take a broad look at the human organism as a biological organism in a very broad sense. The first part of this chapter examines man's unity (similarity) with other forms of life. The rest of the chapter considers features that make us distinct and unique in the living world. After reading this presentation, perhaps you will be stimulated to formulate your own answers to the profound question, "What is a human being?"

UNITY WITH OTHER FORMS OF LIFE

Human being are biological organisms and, as such, share a number of structural and functional features with many other forms of life. Few people would dispute this general idea. Our anatomy corresponds bone for bone, organ for organ, and system for system with that of numerous other animals. There are differences, of course, between the makeup of a human being and that of a horse or some other kind of vertebrate, but the differences are more those of proportions and special adaptations than of fundamental qualities. As shown in Figure 2-1 even the skeleton of a man and that of a bird correspond closely despite the many special adaptations the bird has for flight. The same kinds of similarities are found with respect to our physiology. The way we move, digest food, use energy,

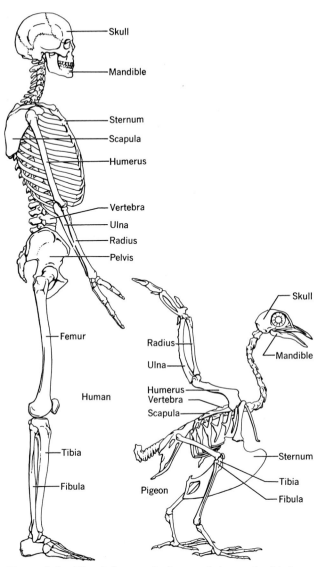

Figure 2-1 *The skeletons of a human being and a bird. The many similarities in their skeletal structures emphasize the close kinship between humans and other animals.*

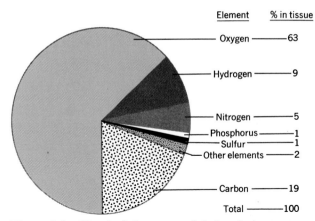

Figure 2-2 *Chemical elements and their relative proportions by weight in human tissues. Note that the most abundant elements are oxygen, carbon, hydrogen, and nitrogen. A similar pattern is found in the tissues of all animals.*

and reproduce are similar to those of many other animals. Biologists interpret these many similarities in structure and function to mean that human beings share a strong kinship with other members of the animal kingdom.

This biological kinship may seem too obvious to discuss, yet how deep does it really go? To find out, we must look at some very basic biological characteristics found in a wide range of organisms and determine whether these are also shared with human beings.

Elements and Compounds

If samples of human tissues are analyzed chemically, between 20 and 30 elements (fundamental units of matter) are found. The most abundant of these are always oxygen, carbon, hydrogen, and nitrogen. In fact, if we add phosphorus and sulfur to the list (Figure 2-2), we have the major elements that comprise the principle substances of which all living things are composed: namely, proteins, carbohydrates, lipids, and nucleic acids.

Our tissue analysis would also reveal a group of additional elements that are present in smaller amounts but are, nevertheless, essential to life. These include sodium, chlorine, calcium, potassium, and magnesium. Still another group, the *trace elements,* are found in exceedingly small amounts. Iron, iodine, and zinc are examples. Although present in only minuscule amounts, the trace elements have a variety of functions that are essential to life. Iron, for example, forms an integral part of hemoglobin molecules that transport oxygen in the blood. Table 2-1 lists some of the important elements in animal tissue and their general functions. Finally, our tissue analysis might show trace elements whose functions in the body have never been established. These substances may be contaminants

TABLE 2-1.   Some Important Elements in Animal Tissues and Their General Functions

Element	Some General Functions
Oxygen	Helps make up all organic compounds and water; an important reactant in basic biochemical reactions such as cellular respiration
Carbon	Forms the chainlike "backbone" of organic compounds; a versatile element that takes part in many chemical reactions in living matter
Hydrogen	Essential part of organic compounds and water; an essential element in many energy transfer reactions in living matter
Nitrogen	Essential in the structure of proteins and nucleic acids (genetic material)
Phosphorus	Necessary in composition of bones and teeth, plays key role in energy transfer reactions in cells; a component of nucleic acids; an important ion in body fluids
Sulfur	Important component of some proteins such as hair and other cellular compounds; a basic ion, as sulfate, in body fluids
Sodium	Most abundant ion in tissue fluids outside of cells; major role in regulating water distribution in the body; important in bones
Chlorine	A major ion, as chloride, in tissue fluids and within cells; major role in regulating water distribution in the body
Calcium	Necessary for bone and teeth formation; basic ion in body fluids
Potassium	Major ion within cells
Magnesium	Major ion within cells; essential for functioning of certain enzymes
Iron	Essential in oxygen transport in hemoglobin and in many enzymes
Zinc	Required by many enzymes for successful functioning; required in the hormone insulin
Iodine	Necessary part of the thyroid hormones

from the environment. Examples are mercury, gold, aluminum, and lead.

If we compare the elements found in human tissues with those in tissues from other animals, two major points emerge. First, the variety and amounts of elements found in the human body are similar to those found in other animals. Second, there are no elements found regularly only in human tissues and not in other forms of life.

It is worthy of note that all of the biologically essential elements exist in varying amounts in the soil, water, and atmosphere of our planet. To be made available for use by human organisms and other life, most of these elements must be extracted from the environment by plants and then transferred to animals. As the animals and plants die, the elements are returned to the earth to be recycled throughout the living world. These complex mineral cycles, involving interactions between the living and the nonliving parts of the environment, are essential to all life. Three of the most important—the carbon, nitrogen, and water cycles—are summarized in Figure 2-3. More details about these cycles are provided in Chapter 19. In the absence of these cycles, the elements in the environment would not be available in the chemical forms required, and life could not exist.

Human creatures, like other forms of life, depend on these cycles for their existence. One of the risks that is heightened by human activities (such as adding pollutants to the environment) is that important *mineral cycles* might be damaged. For example, pesticides accumulating in the soil in an area might deplete the population of nitrogen-fixing bacteria living there. These microscopic organisms convert atmospheric nitrogen, which is unusable by most forms of life, into nitrogen compounds that plants can use. Animals in turn obtain their nitrogen from these plants. Hence a chain reaction might be started that would jeopardize all life because a small but essential link was broken in the nitrogen cycle. This is a theoretical example and probably has not happened anywhere; nevertheless, the risk is a real one because all forms of life are interdependent.

Our description of the elements found in living tissues is misleading in one respect: Most elements are not found in their simple elemental form. Some exist in the body as electrically charged particles called *ions.* Some important ions are hydrogen (H^+), sodium (Na^+), calcium (Ca^{++}), chloride (Cl^-), potassium (K^+), and magnesium (Mg^{++}). Ions have many important functional roles in the body, such as regulating the passage of materials through cell membranes, aiding the trans-

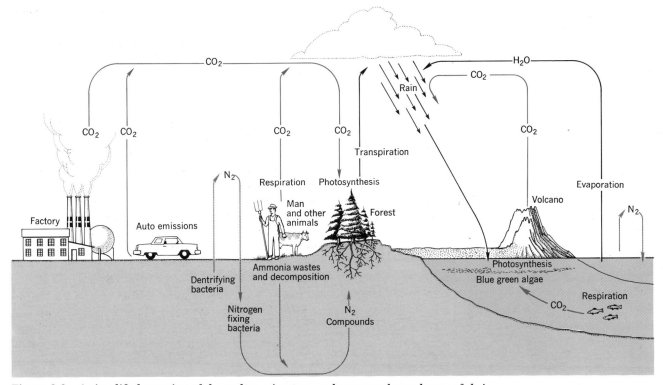

Figure 2-3   *A simplified overview of the carbon, nitrogen, and water cycles and some of their interactions. The existence of all life depends on the functioning of these and many additional mineral cycles.*

mission of impulses along nerve fibers, and maintaining alkalinity or acidity in body fluids. Ions also react with each other to form mineral substances such as the hard portions of bones and teeth.

Most elements in the body are found combined with additional elements to form a large variety of chemical *compounds.* These range from relatively simple compounds such as water to complex substances made of carbon chains such as carbohydrates and proteins. The elements composing a compound are held together by energy forces called *chemical* or *energy bonds.*

Of the smaller compounds, water stands out as the most abundant as well as the most versatile. It surrounds all cells, is a solvent for virtually all compounds, takes part in many biochemical reactions, makes the formation of ions possible (ionization), and provides a transporting medium in the body.

The most remarkable compounds found in living matter are the carbon-containing organic substances called proteins, carbohydrates, lipids, and nucleic acids. They are often termed *macromolecules* (large molecules) in reference to the thousands of elements of which they are made (a molecule is the smallest particle of an element or compound that retains the properties of that substance). These organic compounds are produced only by living cells and have special functions described in subsequent chapters. The relative proportions of organic materials, water, and minerals (nonorganic substances) are shown in Figure 2-4). An important point here is that the same kinds of organic materials are found in all forms of life. Moreover, they have essentially the same functions in all living creatures.

Although chemistry is not emphasized in this text, it is important to keep in mind that all living things are made of interacting chemical materials, and all of the activities and processes of life itself result from chemical reactions. Thus, some basic knowledge of chemistry will make your understanding of biology much richer and more complete.

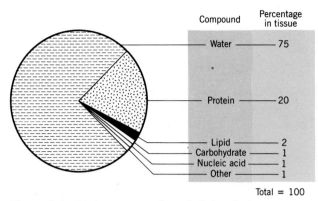

Figure 2-4 *Major compounds and their relative proportions by weight in tissues. The percentages vary somewhat in different tissues.*

Basic Concepts

Living matter, including human tissues, shows the following chemical features:

1. **Carbon, hydrogen, oxygen, and nitrogen are the most abundant elements in it.**

2. **Water is the most abundant and versatile compound present.**

3. **Proteins, lipids, carbohydrates, and nucleic acids make up the organic matter of all living things.**

Cells and Organs

The elements, ions, and compounds described above do not all exist in an unorganized mass. Rather, they are contained in highly organized units called *cells.* Cells are the basic structural units of all forms of life and hence are of great interest to biologists. Even tiny, microscopic forms of life, such as protozoans and algae, consist of what appear to be specialized single cells. Sometimes these organisms are referred to as unicellular (one-celled) life.

In the higher forms of life, cells are grouped to form *tissues*—groups of cells usually highly specialized, that perform specific functions. Muscle is such a tissue, consisting of cells capable of contracting and thus performing work in the body. Groups of associated tissues constitute *organs* (Figure 2-5). The stomach, for exam-

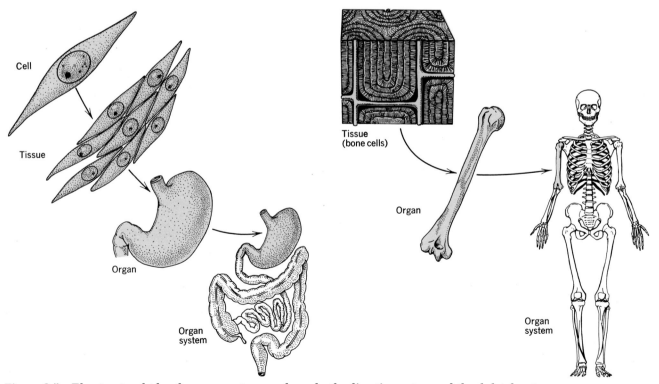

Figure 2-5 *The structural plan for organ systems as shown by the digestive system and the skeletal system.*

ple is an organ made of layers of smooth muscle tissue, but also includes additional tissues such as blood vessels, glands, and nerves. An organ functions as an integrated unit even though it is made of diverse kinds of tissues; hence its function is different from that of any one of its component tissues. In people as in other animals, the stomach or any other organ is part of an even larger *organ system,* the digestive system in this instance. Again, an organ system has functions of a greater diversity than that of its constituent organ parts. Finally, the *organism* itself consists of a variety of organ systems, all with different specialized functions, and all coordinated for maintaining the individual.

Table 2-2 lists and briefly describes the organ systems found in higher animals.

This organization into cells, tissues, organs, and organ systems is the general structural plan for most animals, including human beings. It demonstrates again the basic similarity among multicellular animals.

Anatomical similarities and differences allow biologists to classify organisms into various groups. For example, we find that man is classified as a *vertebrate,* due to various anatomical features common to all such animals, including the possession of a backbone (vertebral column), a concentration of specialized sensory devices in the head area (smell, taste, vision, hearing,

TABLE 2-2. The Major Organ Systems and Their Functions

System	*Description*	*General Functions*
Integumentary	Skin and associated structures such as claws, nails, and hair	Covers and protects the body; contains sense organs; helps regulate body temperature
Muscular	Muscles of the skeleton, heart, and internal organs	Moves the skeleton, pumps blood; moves materials through internal organs
Skeletal	Bones and joints	Forms a framework for the body; protects organs; forms blood cells
Cardiovascular	Heart, blood vessels, and blood	Distributes materials around the body; protects against disease; heals injuries; helps regulate body temperature
Lymphatic	Small vessels, nodes, glands, and lymph	Protects against diseases; forms blood cells; transports lipids
Respiratory	Lungs and associated passageways	Brings oxygen into the body takes carbon dioxide away
Nervous	Brain, spinal cord, nerves, and sense organs	Receives stimuli; stores knowledge; activates the muscular system; regulates many activities in the body
Endocrine	Glands and tissues that produce hormones	Regulates many activities in the body
Urinary	Kidneys, bladder, and connecting tubes	Regulates composition of the blood; collects and eliminates urine
Reproductive	Testes, ovaries, and associated organs	Produces reproductive cells; reproduces new organisms
Digestive	Mouth, esophagus, stomach, intestines, and associated organs such as the liver and pancreas	Ingests, digests, and absorbs food

and equilibrium, for example), as well as the presence of a protective case (cranium) around the brain. Vertebrates also have other unique features, including thyroid and pituitary hormonal glands, a gallbladder, pancreas, spleen, and a special type of liver.

The category of vertebrates is a very broad grouping including a considerable variety of organisms: lampreys, fish, amphibians, reptiles, and birds, as well as the mammals. Despite the diverseness of their body forms, all vertebrates share these same features. Biologists interpret this similarity as evidence of their evolutionary kinship. We human beings are vertebrates because the basic anatomical plan of our body is like that of other members of this category.

Among the vertebrates, human animals are members of the Class Mammalia because we possess hair on our bodies, suckle our newborn with milk from specialized glands (mammaries), are warm-blooded, and have a variety of other specialized features typical of mammals as a whole. Table 2-3 lists some additional mammalian features.

Basic Concepts

Cells are the basic structural units in all forms of life.

The structural plan of all organisms is based on cells, tissues, organs, and organ systems.

Human beings are classed as vertebrates and as mammals by virtue of their structural characteristics.

TABLE 2-3.   Characteristics Found in All Mammals

1. Young nourished by milk from female's mammary glands
2. Hair on body during some stage of life cycle
3. Body temperature regulated internally
4. Seven cervical (neck) vertebrae
5. Teeth in bony sockets
6. Five toes
7. Well-developed brain
8. Thorax and abdomen separated by a muscular diaphragm
9. Genitalia external to body
10. Internal fertilization
11. Four-chambered heart
12. Aortic arch curves to the left and descends along the midline of the body
13. Mature red blood cells lack nuclei
14. Three middle ear bones

Basic Physiology

If we agree that human beings are constructed of cells, organs, and systems like those of other animals, does it follow that the functioning (physiology) of these parts is also similar? One answer to this is to consider a few functional events that are basic to other animals and then determine whether these events are also important in human physiology.

One of the basic needs of organisms is chemical energy. Every activity undertaken by a cell, an organ, or the organism itself expends energy. Moreover, this energy must be in a particular chemical form to be usable. This special source of energy is a compound, *adenosine triphosphate* (abbreviated as *ATP*), found in all cells. When a cell requires energy (as it does continuously), it always obtains it from ATP molecules. This reaction changes ATP into another chemical, *adenosine diphosphate* (*ADP*). If we envision this energy expenditure on the scale of a mammal made of trillions of cells, then the number of ATP molecules needed to keep the mammal's body functioning at any moment is truly astonishing. ATP molecules are essential in the maintenance of all life forms; human beings included.

The ATP supply must somehow be replenished continuously or else cells cease to function. This renewal process occurs in specialized parts of the cell where the chemical energy from the breakdown of small organic compounds is used to convert ADP molecules into ATP molecules. This is called the *ADP-ATP cycle,* a cycle being a continuous event (Figure 2-6).

How widespread are the events of cellular respiration and ATP energy use? Evidently *all* forms of life, from bacteria through human beings, depend on these same basic chemical events for their existence. There are variations from species to species, of course, but they are not nearly as remarkable as the general uniformity that one finds in these processes in all living organisms.

A second basic physiological consideration is the source of the chemical compounds used in cellular respiration. The immediate source is the cell itself, which contains a certain amount of stored nutrients in the form of carbohydrates and fats. These substances can be converted into smaller molecules and used in the cellular respiration process. Eventually, however, the cell must obtain more of these substances from its environment. That is, the organism must take them in,

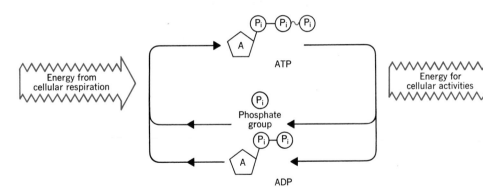

Figure 2-6 *The ATP-ADP cycle. The energy source for all cellular activities is derived from ATP molecules. This ATP supply is continually replenished with an input of energy from cellular respiration. All forms of life use this method for obtaining and using energy.*

process them, and make them available to its cells. This involves the events of feeding, digestion, absorption, and transport to the cells. Organisms show a great diversity in the means of obtaining food, but the events of digestion, absorption, and transport are similar in a wide range of animals.

Complex organisms obviously require some method of coordinating and controlling the diverse activities conducted by their organs and systems. This is accomplished through the hormonal, nervous, and excretory systems. Each of these systems forms the substance of chapters later in the text; there is no need to discuss them in detail here. These systems in human organisms are similar functionally to those in other mammals. (The human brain is an extremely important exception, possessing capabilities for language, intelligence, and creativity that are unknown in the rest of the animal world.)

It seems clear then that many of the basic features of human physiology are like those of other mammals. Indeed, most aspects of human biology support our belief in a close kinship between human beings and other forms of life.

Basic Concept

Human beings are physiologically like other organisms in the use of ATP for energy needs, in cellular respiration, in digesting and utilizing nutrients, and in the types of control systems they utilize.

UNITY WITH DIVERSITY

Man as a Primate

Few people would argue with a view of kinship between man and other animals. However, we humans possess a variety of distinctive features as well. We need to examine these differences in order to round out our perspective of man as a biological entity.

As noted previously, we are classed as mammals because our anatomical and physiological features are more similar to other mammals than to any other vertebrate group. This in itself is a type of distinctness, in that it separates mammals from all other vertebrates. We even like to think that mammals are the most advanced or sophisticated of animals, in an evolutionary sense, although their numbers and variety of species are small compared with groups such as insects.

Among the mammals, human organisms are in a major subcategory called the *Order Primates,* along with the apes, monkeys, lemurs, tarsiers, and such exotic creatures as galagos, pottos, lorises and aye-ayes (Figure 2-7). Primates as a group are distinguished by a number of features. The most important are an advanced brain structure (especially large cerebral hemispheres); eyes directed forward in such a manner as to provide three-dimensional stereoscopic vision: excellent eyesight (large visual areas in the brain); poor sense of smell as compared with many other mammals; flattened nails in place of claws; prehensile (grasping) hands and feet; an opposable thumb (which can be placed opposite the fingertips); and a well-developed collarbone. All of these features are viewed as adaptations related to tree dwelling. The problem with this categorization is that primates are such a diverse group that there are some exceptions for almost every one of these features. A few primates have claws (galagos, tree shrews, and some lemurs). Some are tree dwellers and some are terrestrial; their diets vary considerably; and a few are bipedal ("two-footed") like man, but most are not. The primates also range in size from quite tiny,

(a)

(c)

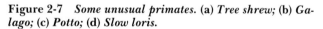

(b)

(d)

Figure 2-7 *Some unusual primates.* **(a)** *Tree shrew;* **(b)** *Galago;* **(c)** *Potto;* **(d)** *Slow loris.*

mouse-sized lemurs to the largest primates, gorillas and man. Also, some features such as stereoscopic vision are possessed by many nonprimates. Consequently, we cannot point to any single trait and say that it distinguishes a primate from other mammals.

Despite all of this seeming vagueness, it is definite that man fits all of the requirements to be classed as a primate. Again, like being a mammal, this in itself sets us apart from most other life. Because humans, apes, and monkeys share so many anatomical features, they are usually placed together in a suborder of the primates, the *Anthropoidea.* Loosely translated this name means, "in the form of a man." Human beings are anatomically most similar to the great apes (or vice versa). However, this similarity does *not* mean, as it is

often interpreted by laymen, that man is a descendant of the apes. The apes of today are as "modern" as we are, evolutionarily speaking, and hence they cannot possibly be our ancestors. Neither of two contempo-

Figure 2-8 *A majestic female gorilla (left), and a chimpanzee (right) nursing her year-old baby.*

rary organisms of *any* kind can be the ancestor of the other. It is unfortunate that the term "ape-man" has become synonymous with prehistoric man for so many people. There is no fossil evidence to indicate that early man had much resemblance at all to an ape—a matter that we will discuss later in the chapter on evolution. This erroneous impression has also been reinforced by movies, television, and cartoons depicting ancient man as some sort of half-ape. Surely our ancient ancestors deserve a far nobler treatment in the eyes of the public.

Man's closest relatives in the animal world are the great apes: gibbons, orangutans, chimpanzees, and gorillas (Figure 2-8). Still, we are sufficiently different, in the opinions of anthropologists and paleontologists, to be placed in our own distinct family of the great apes, the *Family Hominidae,* which is shared only with our prehistoric ancestors.

The distinguishing features that separate us from the great apes include the amount of gray matter in the cerebral cortex of the brain, greatly reduced jaws and small front teeth, bipedality, distinctive culture, and unique sexual behavior. Perhaps our uniqueness is summarized by our scientific name *Homo sapiens,* a Latin name meaning "wise man." Some fellow *Homo sapiens* might debate the aptness of this name, in view of the behavior of some, but compared with other

primates a human being is certainly an intellectual giant.

Basic Concepts

Based on anatomical features, man is classified in the Order Primates, Suborder Anthropoidea, and Family Hominidae.

Man's nearest living relatives are the great apes.

Man's Distinctive Features

Of the distinctive features that help set us off from the rest of the animal world, perhaps the most outstanding consists of certain structural and functional features of our brain. The volume of the human brain is about 1500 cubic centimeters (cc) and it weighs approximately 1400 grams (about 3 pounds). By comparison a male gorilla, three times the weight of an average man, has a cranial capacity of only 650 cc. The weight of the human brain is surpassed only by those of large whales (9000 grams) and elephants (5000 grams). Thus humans have a very large and dense brain in proportion to their body size.

The unusually large cerebral hemispheres of the human brain (Figure 2-9) are covered with a layer of gray matter (*cerebral cortex*) thicker than that found in any other animal's brain. This *associative cortex,* as it is

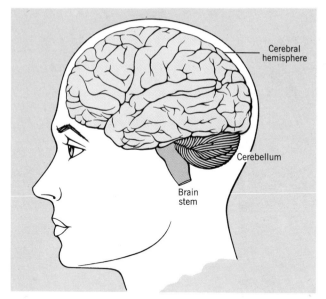

Figure 2-9 *Side view of the brain showing the large size of the cerebral hemispheres. The associative cortex forms the outer layer of this portion of the brain.*

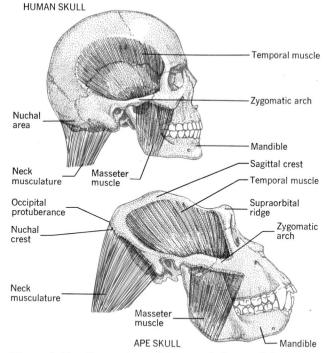

Figure 2-10 *Comparison of the rounded, smooth human skull with the ridged skull of a gorilla. The prominent nuchal crest of the gorilla skull serves as the attachment site for the gorilla's massive neck muscles.*

often termed, is evidently responsible for our superior intellect. Precisely how this feature makes us the intellectual giants that we are is not really known. Some theories about it are presented in a later chapter on the nervous system. One of the frontiers of biological research involves attempts to understand how knowledge is stored in the cortex and retrieved at a later time by the individual.

To accommodate and protect our large, dense mass of brain cells, human beings have an unusually rounded and smooth cranium. This globular shape is even more emphasized because humans do not have a muzzle like most of the other primates. Our muzzle reduction apparently accompanied the evolution of the large cranium. Our jaws are receding and filled with small teeth, the cheekbones are rounded, ears are flattened against the side of the skull, and our forehead is high.

The cranium is balanced on the end of an upright backbone and hence requires only small neck muscles to hold it erect. By contrast, an animal that moves on all fours requires rather massive muscles to support the head. The large bony ridges on their skulls provide attachment sites for muscles (Figure 2-10). Consequently their skulls do not have the smooth, rounded contours that ours have.

Our most fundamental adaptation, many anthropologists believe, is bipedal (two-footed) locomotion. In fact, our striding gait (way of walking) is unique among animals. On the other hand, *bipedalism* itself is not uncommon among vertebrates: some lizards run on their hind limbs, many dinosaurs stood partially upright, and birds are entirely bipedal. Among mammals, kangaroos, some rodents, and nonhuman primates also use an upright posture to varying degrees. Nevertheless, the anatomy and form of the human body reflect so many adaptations to this mode of locomotion and posture that it is a major contributor to our distinctness.

Human limbs and feet demonstrate some of the adaptations to bipedalism. Our legs are longer, larger, and more muscular than our upper limbs—obviously well adapted to supporting and moving our body. The thigh is the site of two massive groups of muscles, the hamstrings on the back side and the quadriceps muscles on the front side (Figure 2-11). Together these muscles flex and extend the lower leg on the hingelike

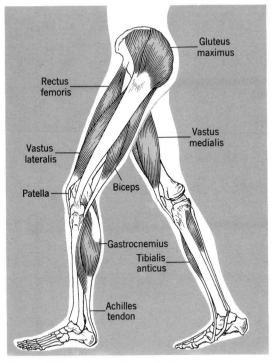

Figure 2-11 *Human leg and thigh musculature. The massive quadriceps muscles on the front of the thigh and hamstring muscles on the back of the thigh extend and flex the lower leg in man's unusual striding gait.*

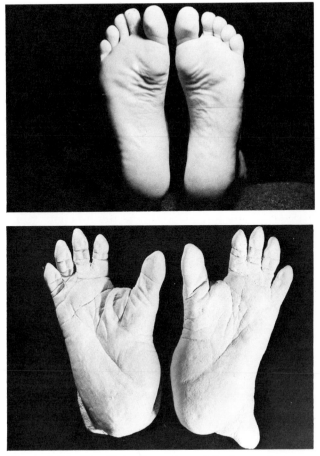

Figure 2-12 *The distinctive nature of the human foot compared with the foot of an ape. The human foot is well adapted for walking and supporting the body.*

knee joint when we walk or run. The human foot has been rather drastically reshaped and is unlike that of any other primate. The toes are quite short and have lost their grasping ability. The unique outline of a human footprint indicates that the weight-bearing parts of the foot are the heel, outside edge of the foot, ball, and toes, especially the first toe. Figure 2-12 shows how different a human foot appears compared with that of an ape.

The human leg and foot appear to be highly specialized for walking and body support, but not as well adapted for running. Despite our interest in track events and jogging, even highly trained runners cannot attain 30 miles per hour, nor is their endurance very respectable compared with that of many other mammals. On the other hand, our limb and foot structure is superbly adapted for long-distance travel with a striding gait, an activity that prehistoric man must have used extensively in seeking food, exploring his surroundings, and extending his range into new lands.

Fossil findings indicate that our ancestors spread into the major continents rather rapidly in this manner.

The apparent absence of hair or fur on the human body is frequently pointed out as an unusual trait. In fact, the expression "naked ape" was the title of Desmond Morris's widely read book on human behavior. In reality, much of the body is fairly densely covered with hair follicles and tiny hairs. If all of these follicles should suddenly produce inch-long hair, then we would be as furry as our pet dog or cat.

Only on the scalp, armpits, and pubic area does human hair become noticeably long. The adaptive function of having tufts of hair on these areas of the body is obscure. Perhaps the tufts are ornamental, like the mane of a lion, and functioned as sex attractants for primitive man. Pubic hair and armpit hair grow only at the onset of sexual maturation—a highly visual sign that sexual maturity has taken place. Even today, scalp and facial hair are extensively groomed as ornamental objects, often designed to appeal to the opposite sex. Regardless of the speculation, the pattern of human hair distribution on the body is unusual and notable.

Human skin is richly endowed with sweat glands, more so than in most mammals, and these function as part of the body's cooling system. In addition there are sebaceous (oil) glands associated with the hair follicles in many parts of the body. The *sebum* or oil produced by these glands is a complex mixture of waxes, fatty acids, cholesterol, and dead cells. Its function is not known. When the glands become plugged or infected, the results are blackheads or pimples (acne). In the armpits, the mixture of sweat and sebum is decomposed by bacteria to produce a pungent, unmistakable body odor. Because this body odor is so characteristic of people, one wonders that advantage, if any, it might have conferred on our ancient ancestors. Many other mammals produce distinct odors, often from special scent glands, that usually relate to territorial behavior and to finding mates. It boggles the mind a bit, especially in our almost compulsively clean culture, to think of an odoriferous armpit as having a useful function. Perhaps body odor and human skin oil was a warning that discouraged certain predators from attacking humans.

Probably directly related to man's scanty hair covering, human skin contains pigment bodies (melanocytes) that help protect it from sunlight and also impart color to the skin. When light-skinned people are exposed to sunlight, their melanocytes produce granules of the dark pigment, melanin. This makes the skin darker. Melanin absorbs some of the ultraviolet wavelengths of sunlight that are so injurious to living cells.

Darker-skinned people contain more melanin than do lighter-skinned ones, but categorizing people as "white" or "black" is fallacious biologically. All human beings have pigmented skin, normally, and the only truly white people are albinos, who have an abnormal genetic condition. Hence skin color is a poor measure of racial differences among peoples, even though it is still used all too often.

The differences in the body forms of sexually mature human males and females are pronounced and much greater than that found in other primates. The large size of the breasts of the female and the penis of the male are examples of these features. In addition, many aspects of human sexual behavior are unique and will be discussed in the chapter on reproduction.

These are a few of the many special adaptations that typify human beings. Subsequent chapters will describe additional ones. For a more extensive but highly readable account of man's physical characteristics, the reader is referred to a well-written paperback, *Man*, by Richard J. Harrison and William Montagna (Appleton-Century-Crofts, 1969).

Basic Concept

Distinctive anatomical features of *Homo sapiens* include his unusually large cerebrum and thick cerebral cortex, smooth rounded cranium, numerous adaptations associated with bipedalism, small amounts of body hair, pigmented skin, numerous sweat and oil glands, and differences in male and female body form.

Man's Unique Features

In the process of discussing distinctive human attributes in the previous section we unavoidably included some of his unique features because the two are closely related. These included the nature of the cerebral cortex, some of the adaptations associated with bipedalism and an upright stance, and differences related to reproductive organs and behavior. Although these are distinct and notable, they only "set the stage" so to speak for the astounding cultural achievements that eventually accompanied or were made possible by these features. These achievements fall under two broad categories: tool making and language development.

Recent fossil evidence indicates that our probable prehuman ancestors were bipedal nearly four million years ago, considerably predating the evolution of his brain. A major consequence of becoming bipedal was to free the upper limbs and hands because this allowed prehumans to become tool makers and users. At first they just sharpened the edges of large pebbles, perhaps for skinning animals or cutting off chunks of flesh. Later, a host of useful implements evolved including spear points, arrowheads, and stone axes. Although the improvement in these primitive tools came slowly, it was destined to have a drastic influence on human culture and history. In a sense this was the beginning of our *technological evolution,* a process that continues

Figure 2-13 *The personal computer with its numerous applications in science and the business world provides an example of our technological evolution.*

Figure 2-14 *Woodpecker finch using a twig to obtain insects from holes in a tree trunk.*

today and perhaps has no limits in time (Figure 2-13). By contrast, no other organisms have made such a breakthrough. A few animals use simple tools, such as a woodpecker finch using a twig to dislodge insects from crevices in bark (Figure 2-14) or a sea otter crushing sea urchins against a rock held against its chest, but none have gotten beyond this crude level of achievement.

The next great advance in human uniqueness came with the unusually rapid evolution of the brain, especially the cerebral hemispheres. This initiated a time of more sophisticated tool manufacture, building of shelters, food sharing, and a distinctive social behavior. Roughly ten thousand years ago our ancestors began an era of *social* and *cultural evolution* which became the dominant influence in further human development. In this time interval our anatomy has not changed measurably (judging from fossil skeletal remains), but our cultural and social advancement has been rapid as well as immense in scope. Our future biological evolution by natural selection may likely be inconsequential because we can already manipulate our heredity (and thus our future evolution) to some extent with modern medical science. By contrast, our social evolution, an outgrowth of our "braininess," may still be in its infancy.

As a brainy social creature, the achievements of *Homo sapiens* are staggering compared with the rest of the animal world. The most outstanding is the possession of a language. This is often noted as the most diagnostic single trait of man. Other animals communicate in simple ways to one another, but only man uses a system of complex, abstract verbalizations to express his physical, emotional, and intellectual state. In addition, we record our language in various forms for other humans to use.

Included among our complex social interactions are the abilities for extensive learning, creativity in numerous art forms, a compassion for other forms of life, the performance of unselfish actions, ethics that attempt to distinguish between good and evil, codes of laws to protect society and individuals, and imaginative and brilliant achievements in science and technology.

For the sake of objectivity it should be noted that human beings frequently engage in actions that are not so admirable and often are puzzling. For example, all too often we are greedy, cruel to forms of life including our own kind, engage in war against our own species, take part in antisocial acts, consume drugs known to be harmful, and seem capable of all the evil that a complex brain can conjure—in short, far more self-destructive than any other animal. It is puzzling why these behaviors exist at all because many of them are biologically self-defeating. It has been suggested that these acts are the unavoidable side effects of the evolution of our "super brain." Perhaps they are spinoffs of a too-rapid social evolution. We can only fervently hope that greater knowledge of human biology will someday provide better methods of coping with such destructive and wasteful behavior.

Basic Concepts

Unique characteristics of *Homo sapiens* include an enlarged cerebrum, certain structural features related to bipedalism, reproductive anatomy, and cultural features such as tool making and language.

Bipedalism allowed man to become a tool maker and to participate in a technological evolution.

Man's large brain facilitated the development of his complex social structure.

KEY TERMS

adenosine diphosphate	chemical bond	ion	organ system
adenosine triphosphate	Class Mammalia	macromolecule	physiology
ADP-ATP cycle	compound	molecule	technological evolution
anatomy	cultural evolution	mineral cycle	tissue
bipedalism	element	Order Primates	trace-elements
cell	Family Hominidae	organ	vertebrate
cerebral cortex	*Homo sapiens*	organic substance	

SELF-TEST QUESTIONS

1. Define (a) the term *element,* and (b) list the four most abundant elements found in living tissues.
2. Name two *trace* elements found in living matter.
3. How do the variety and amounts of elements in human tissues compare with those found in other animals?
4. By what basic mechanism do the chemical elements move from the nonliving part of the environment into living things?
5. Name three mineral cycles that are especially important to the living world.
6. Define (a) the term *ion* and give several examples found in living tissues, and (b) some functions of ions in the body.
7. Name the most abundant compound in the body and describe its functions.
8. List the four major categories of organic compounds in the body.
9. What is the basic structural unit for all forms of life?
10. Starting with cells, what is the basic structural plan found in all higher animals?
11. List the major features that categorize human beings as (a) vertebrates, and (b) mammals.
12. What is the special energy source for all living cells?
13. Explain the relationship between cellular respiration and ATP in the body.
14. In terms of cellular respiration, why is it necessary for an animal to take in food materials?
15. What three major systems are used by complex organisms to coordinate and control their diverse activities?
16. How do human organ systems compare with those of other animals from the standpoint of function?
17. List the features that collectively distinguish primates from other animal groups.
18. Classify human beings as to order, suborder, and family, and state our scientific name.
19. Discuss why each of the following traits is distinctive to man: (a) brain, (b) cranium, (c) bipedalism, (d) hair, (e) skin, and, (f) body form of males and females.
20. Tell how human beings are unique in respect to (a) cerebrum of the brain, (b) bipedalism, (c) reproductive anatomy, and, (d) cultural features.
21. Contrast man's social evolution with his biological evolution. Which now seems more important in shaping his future destiny?

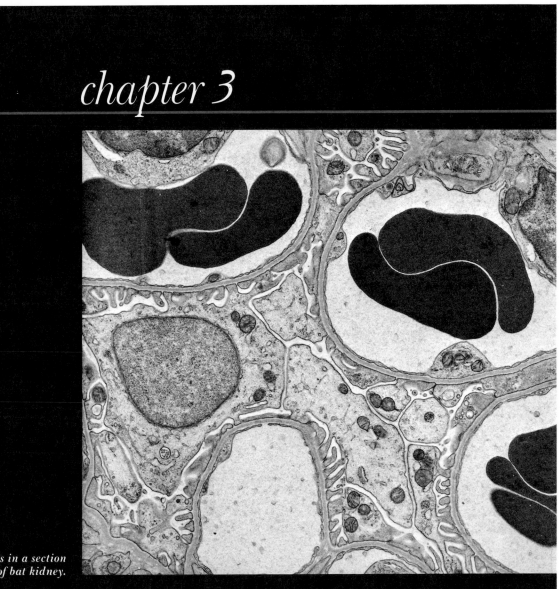

*Cells in a section
of bat kidney.*

CELLS AND
THEIR
FUNCTIONS

chapter 3

questions to think about

Why is the study of cells important?

How do you study microscopic-sized objects?

How are cells constructed?

What are the functions of the various cell parts?

What are the major events in a cell's life cycle?

What are the differences between prokaryotic and
   eukaryotic cells?

---

*—all organized bodies are composed of essentially
similar parts, namely, of cells; these cells are
formed and grow in accordance with essentially
similar laws—*

Theodor Schwann

The opening quote is a general expression of what we
now term the cell principle: *All life forms are composed of
cells and products of cells.* It was derived from the observations of Theodor Schwann, a German anatomist,
and other biologists over 100 years ago. Today we
know that virtually all functions essential to life start at
the cellular level: energy production, reactions to stimuli, growth, and reproduction. In fact, it is difficult to
discuss almost any biological system without involving
the cells that are a basic part of it. To take just one
example, we usually think of nutrition as involving
what an animal eats and whether this food will sustain
it adequately. This subject would seem to have little to
do with cells. In reality, however, nearly all of the
significant aspects of nutrition occur at the cellular
level. Digestive enzymes produced by cells in the gut
break down the food we eat into small molecules.
These molecules are then absorbed by intestinal cells.
All the other cells around the body eventually utilize

these molecules to sustain life. Therefore, a discussion
of nutritional concepts is really a consideration of what
cells do with small organic molecules.

The central role of cells in living processes is also
demonstrated dramatically (and tragically) by the occurrence of many diseases. One example is lung
cancer, a disease that begins with events triggered
within individual cells. The incidence of lung cancer
has increased sharply in recent years, especially among
middle-aged to older men, and it is now the principal
form of cancer in males. Abundant evidence links the
increased occurrence of the disease directly with the
use of tobacco, principally in the form of cigarette
smoking.

Most lung cancer actually occurs in the tissue that
lines the bronchial tubes leading into the lungs (Figure
3-1a); thus its medical name is *bronchogenic carcinoma* (a
carcinoma is a type of cancer). Its course of events at
the cellular level is generally as follows. The epithelium, the tissue that lines the bronchial tubes, is normally thin in a nonsmoker. It consists of two layers of
cells resting on an extremely slender *basement membrane* (Figure 3-1b). This membrane separates the
epithelium, which contains no blood vessels, from the
connective tissues beneath. The connective tissues
have an abundance of small blood vessels called capillaries.

Figure 3-1 *Human respiratory system showing some of its major parts* **(a)**, *and stages in development of lung cancer* **(b–e)**. **(b)** *Normal bronchial epithelium with cilia and a thin layer of epithelium cells;* **(c)** *smoking causes the basal layer of cells to form many layers and the cilia begin to disappear;* **(d)** *cilia disappear, basal cells continue to increase, and many cells develop the irregular-shaped nuclei found in cancer cells;* **(e)** *patches of cancer cells begin to break through the basement membrane to invade other tissues.*

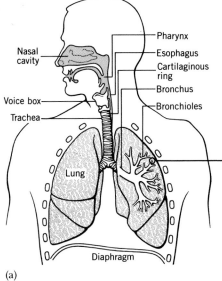

(a)

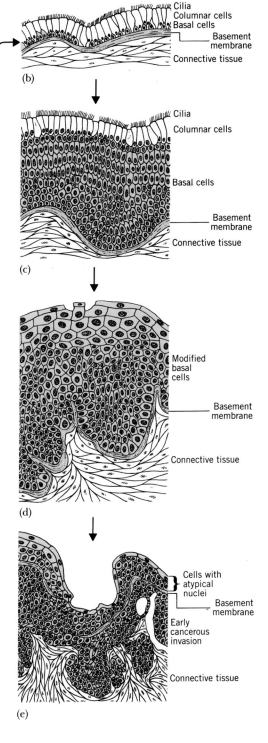

The outer cell layer has numerous tiny projections, or *cilia,* and it secretes onto its surface a sticky fluid (mucus). The motion of the cilia slowly but continually pushes the mucus up the bronchial tubes, into the windpipe, and eventually into the mouth. In this way, small particles in the inhaled air are caught in the mucus and moved out of the lungs. The picturesque term, *ciliary escalator,* is applied to this apparatus. It has a valuable cleansing and protective function for the lungs.

When an individual smokes cigarettes regularly, however, the bronchial epithelial cells undergo a series of changes. First, the cells of the deeper of the two layers, the basal layer, divide to form many layers (Figure 3-1*c*). This action produces a demonstrably thicker mass of epithelial tissue. Meanwhile, the tobacco smoke appears to inhibit the ciliary motion, so the mucus is no longer moved up the bronchial tubes. Substances in the tobacco smoke, including several cancer-causing chemicals, may now accumulate along the bronchial lining. Eventually the cilia as well as the cells containing them disappear altogether. As the basal layer cells continue to increase in number, some of the cells develop the large, irregular-shaped nuclei typical of cancer cells (Figure 3-1*d*). In a heavy smoker, many such patches of abnormal cells may be found in the bronchial lining. This is evidently a precancerous stage, which has a probability of becoming a broncho-genic carcinoma.

Evidence also indicates that when cigarette smoking is stopped, there is a slow disappearance of the abnormal basal-layer cells. The normal bronchial epithelium is at least partially rebuilt and regains its functions. Therefore, smokers can greatly enhance their chances of avoiding lung cancer simply by ceasing to smoke.

In the final stages of lung cancer development, the abnormal cell patches break through the basement membrane to invade surrounding bronchial tissues where they enter the bloodstream by passing through thin capillary walls (Figure 3-1e). From there they usually spread throughout the body. The chance of survival once this invasion has occurred is extremely poor.

Lung cancer is just one example of a serious disease that begins with changes produced in cells, in this case by the presence of tobacco smoke. No one yet knows exactly how or why the smoke triggers these changes. This will be discovered only through additional studies of cells, both normal and abnormal. But this discussion demonstrates that biological events have their roots in happenings in the cell. This is a major reason why a large amount of biological and medical research involves the structure and function of cells.

HOW TO STUDY EXTREMELY SMALL OBJECTS

How do you study objects that are so small that most of them cannot be seen with the naked eye? This is the problem that biologists have faced for centuries in relation to cells. Back in the 1600s, an Englishman named Robert Hooke looked at a thin slice of cork with a crude microscope and applied the term *cell* to the tiny empty compartments he saw (Figure 3-2). What he actually found were the walls of cells that had long since died. Thus, somewhat ironically, the term *cell*, so widely used today in biology, is actually a misnomer of sorts. What we now know as cells are tiny bits of living matter and not empty chambers. Nevertheless, the term *cell* is so firmly fixed in biological literature now that it would be futile to try to change it.

As time passed, microscopes were improved and the interest in studying cells grew. A new problem arose as biologists attempted to distinguish different parts

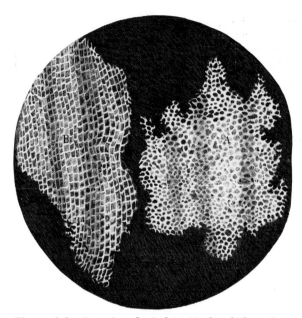

Figure 3-2 *Drawings by Robert Hooke of plant tissue such as cork (B) as he saw them with his primitive microscope.*

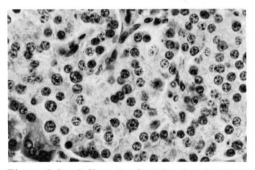

Figure 3-3 *Cells stained with a dye that shows up only the nuclear structures (dark round objects).*

within this extremely small unit. It was found that various dyes could be applied to cells and would stain specific parts therein. For example, there is a dye that imparts color only to the nucleus of a cell (Figure 3-3), another that stains granules in other parts of the cell, and so on. Gradually the cell's anatomy began to be explored.

Another important and necessary tool in cell study was the development of a special, razor-sharp knife that could make slices of a tissue that were less than one cell-layer in thickness (or thinness!). This highly precise tissue slicer, called a *microtome*, is invaluable to cell

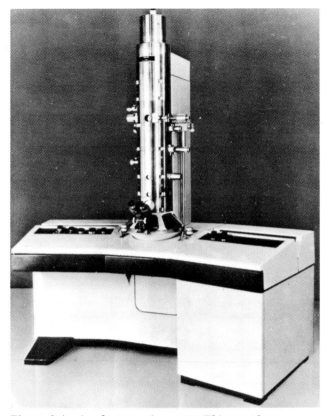

Figure 3-4   *An electron microscope. This complex, expensive instrument is used extensively in cell studies.*

biologists even today. In a typical procedure, a tissue is "fixed" (killed quickly) so that its cells do not disintegrate, embedded in paraffin or plastic to hold it rigid, and sliced with the microtome into a desired thickness. The slices are stained and mounted on glass slides for study under the microscope. All of these technical developments took many decades.

In 1931 the *electron microscope* was invented and eventually became a major tool in cell biology (Figure 3-4). Because this device uses beams of electrons rather than light waves, it has a vastly greater resolving power (ability to distinguish objects lying near one another) than does the optical microscope. In fact, most of our recent knowledge about the internal anatomy of cells came from observations made with the electron microscope.

Another basic and useful device in modern cell biology is the use of high-speed centrifuges. A centrifuge is a device that spins rapidly and separates particles of different densities in a mixture. If a biological material such as blood is placed in a test tube and centrifuged at the proper speed, the blood cells separate from the liquid part of the blood within a few minutes. These two portions of the blood can then be used for various purposes. An extension of this device is the *ultracentrifuge,* which can spin materials at speeds faster than 50,000 revolutions per minute. This force is sufficient to separate very small particles, such as the parts within cells, from one another (Figure 3-5). Thus

Figure 3-5   *The centrifugation of broken-open cells at successively greater speeds allows separation of cellular parts as indicated.*

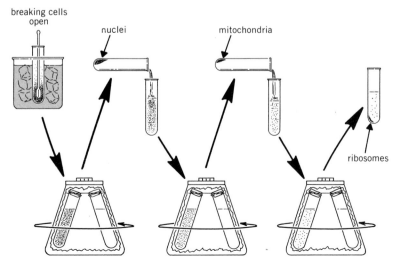

breaking cells open

nuclei

mitochondria

ribosomes

the cell biologist has a tool for obtaining relatively pure concentrations of specific cell parts (such as nuclei) for study. Because these parts are still "alive," their particular functions (chemical activities) can often be identified.

Cell biologists use a variety of sophisticated chemical techniques in studying cells today. An especially valuable one is the use of radioactive substances (materials that emit energy waves or particles). By using laboratory instruments that detect radioactivity, various chemicals that have been "tagged" with radioactive atoms can be traced or followed as they are incorporated into cells or cell parts. This procedure has immensely aided the understanding of many chemical reactions that take place within cells.

In recent years, after much experimentation, biologists have learned how to grow cells and tissues in laboratory glassware, a process called *cell* or *tissue culture* (Figure 3-6). This in turn allows biologists to investigate the action of living cells and to perform experiments on them that would otherwise be impossible. Much of our knowledge about nutritional requirements, cell growth and reproduction, drug effects, the structure and function of genes and chromosomes, and characteristics of cancer cells has been obtained from these living cell cultures.

Basic Concept

Cells are extremely small; knowledge about their structure and function has increased in proportion to the development of new scientific tools and techniques, particularly with reference to electron microscopy, ultracentrifugation, and use of radioisotopes.

THE ANATOMY OF A CELL

When a student looks at a cell through a microscope for the first time, the impression is often one of disappointment because there is so little to see. The main visible features are the general outline of the cell, some grayish, rather featureless material called *cytoplasm*, and a rounded, denser body within the cell, the *nucleus* (Figure 3-7a). This is not a true picture of a cell at

Figure 3-6 *Large containers for the growth of cells. Many conditions important for cell growth can be controlled with this apparatus.*

all—cells are literally crammed full of functional parts, but on a very minuscule scale. To see these parts, the viewer must turn to electron microscopy with its vast resolving powers. Instead of the 1500× magnification of the optical microscope, we need magnifications on the order of 100,000× or more to see the fine structure of a cell.

In an electron photomicrograph of a cell, as shown in Figure 3-7b, we get quite a different impression of cell anatomy. Note that the cytoplasmic region is full of circles, lines, and irregular blobs. All of these features are cell parts, and most are constructed of membranes

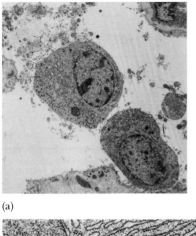

(a)

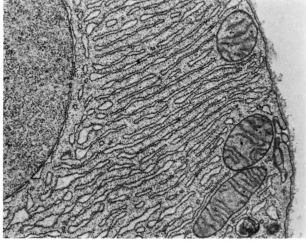

(b)

Figure 3-7 *Cells as viewed with* (a) *an optical microscope, and* (b) *with an electron microscope. The difference in amount of visible detail in* (b) *is due to the greater resolving power of the electron microscope.*

(the thin lines). Even the outer boundary of the cytoplasm is marked by a membrane. The largest, rounded object in the photograph is the nucleus, and it too is surrounded by a membranous envelope. The envelope is porous as denoted by the tiny openings in it here and there.

At this point, even without knowing the names of the parts, you can tell that the cell is separated internally by thin membranes into numerous compartments. This is an important concept to grasp because these compartments, called *organelles* ("little organs"), contain chemical systems that perform a variety of cell functions, such as harvesting energy, making more cell parts, and producing hormones. Compartmentalizing these systems prevents them from interfering with each other in their functions. The organelles are suspended in a slightly viscous material known as the ground substance of the cytoplasm. Recent studies show that it consists of exceedingly tiny filaments which help to support and move the organelles. Even the nucleus, the control center of the cell, is a large compartment, but its porous membrane allows it to communicate with the remaining cell organelles.

With this very general overview of cell anatomy, let us take a closer look at cell membranes and the different kinds of cell organelles.

Membranes form a highly adaptable building material for cell parts, and they also serve as the site of many biochemical reactions. The plasma membrane encloses the entire cell. Within this lie a vast number of subcompartments formed by a network of membranes called the *endoplasmic reticulum*. Within the subcompartments are found the organelles such as the *mitochondria, lysosomes,* the *Golgi complex* and, in plant cells, the *chloroplasts*. The membranous material making up all of these organelles is similar in structure and may be part of a single membrane system in the cytoplasm. Membranes are the basis of a cell's anatomy.

For many years, biologists have sought to learn more about the makeup of this thin but complex membrane. At present, the cell membrane (Figure 3-8) is visualized by many cell biologists as consisting of two adjacent layers of lipid molecules (a lipid bilayer) with patches of proteins on its inner surface and also penetrating through the lipid layers here and there. (Lipids are fatty substances found in cells; Chapter 5 describes them in more detail.) The proteins that protrude

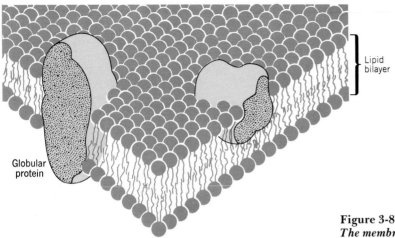

Globular protein

Lipid bilayer

Figure 3-8   *The lipid bilayer model of the cell membrane. The membrane is extremely thin but has numerous functions.*

through to the outer surface of the cell are often attached to complex sugar molecules. Such hybrid molecules (glycoproteins in this instance) serve to identify individual cells and are responsible for the phenomenon of immunity. For this reason, when cells are placed in another organism's body, they are quickly destroyed because the cells of the host organism recognize the new cells as invaders.

When growing on an artificial culture medium, cells multiply and spread out until they touch each other. This contact inhibits further growth of the cell mass— hence the term *contact inhibition.* In contrast, cancer cells in artificial media are not inhibited by contact with each other and thus grow into a jumbled, many-layered mass of cells, much as a cancer would inside a living organism. This difference in growth behavior between the normal and cancer cells is evidently due to some as yet unknown difference in their cell membranes. Thorough understanding of this difference might provide clues for more efficient cancer treatments.

Passage of Materials Through Cell Membranes

The passage of materials through the complex of molecules that comprise the cell membrane involves the phenomena of *diffusion, osmosis,* and *active transport.*

Diffusion refers to the movement of molecules from an area where they are abundant to areas where they are less abundant. This widespread physical phenome-

non occurs because all molecules are in constant and random motion. If packed together (concentrated), the molecules bounce off one another and gradually move outward until they are evenly distributed in their environment. If, for example, a lump of sugar is dropped into a cup of coffee and not stirred, two events take place. First, the sugar lump dissolves into separate sugar molecules in the coffee. Then, random motion of the molecules slowly diffuses them throughout the coffee. In a matter of days, your coffee would be evenly sweetened!

Small, uncharged particles such as water, carbon dioxide, and ammonia move easily through membranes. Larger molecules, such as sugars, pass through membranes slowly by diffusion. However, many membranes contain special protein molecules that act as carriers for some of these larger substances. An example is the transport of glucose into cells. The glucose molecules contact the appropriate carrier protein as they bump into the cell surface. The protein changes its shape and, in the process, transports a glucose molecule into the cell. After releasing the sugar inside the cell membrane, the carrier resumes its former configuration. This process, called *carrier-facilitated diffusion,* is limited because molecules are moved only from an area of high concentration to an area of low concentration. No energy is expended by the cell in this process.

Diffusion enables substances such as nutrients or gases to distribute themselves throughout the cytoplasm of a cell after they have been produced or have

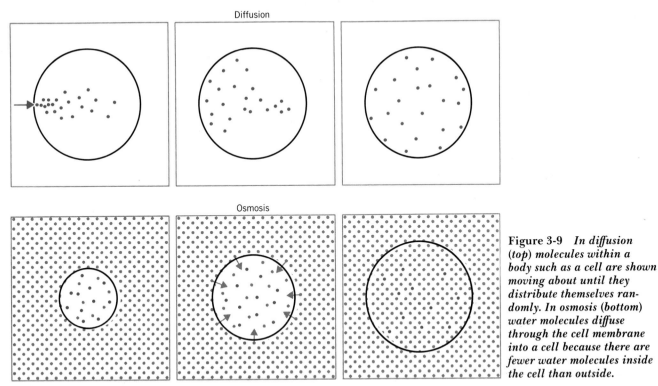

Diffusion

Osmosis

•Water molecules   • Dissolved materials

Figure 3-9 *In diffusion (top) molecules within a body such as a cell are shown moving about until they distribute themselves randomly. In osmosis (bottom) water molecules diffuse through the cell membrane into a cell because there are fewer water molecules inside the cell than outside.*

gotten into the cell (Figure 3-9). Diffusion also makes possible the phenomenon of *osmosis.* Osmosis is the diffusion, usually of water molecules, through a semipermeable membrane (a material containing tiny pores). It takes place when there are more water molecules on one side of the membrane than on the other side. The random movement of the closely packed molecules bouncing against each other results in their diffusion through the pores in the membrane into the area of the other side where there are fewer of them (Figure 3-9). Osmosis is a passive process requiring no input of external energy — the only requirement is an uneven distribution of water molecules in the two areas separated by the membrane.

The effect of osmosis on cells can be shown rather dramatically with human red blood cells. These cells are adapted to the slightly saline (salty) environment of the blood — normally about 0.85 percent salt. The same concentration of salt is found inside these cells. If the cells are placed in plain water, they swell rapidly and burst because they cannot control the rapid diffusion of water molecules through their membranes and

into their cytoplasm. In this case, water molecules were more abundant outside the cell than inside and thus diffused into the cytoplasm. On the other hand, if red cells are placed in a 1.5 percent salt solution, they shrivel up as water diffuses out of the cytoplasm. Can you explain why the water moved out of the cell in this instance? Perhaps these examples explain why it is important for materials to be in osmotic balance with the blood before they are injected into the blood stream of a patient in the hospital.

Cells are surrounded by membranes that are permeable to water molecules; hence osmosis moves water into or out of a cell as through a semipermeable membrane. The cell membrane, however, is *not* an inert semipermeable barrier. Instead, it is *selectively* permeable, in that it is capable of regulating the rate of passage of most ions and larger molecules.

Sometimes it is necessary for a cell to move molecules or ions from an area of lower concentration to one of higher concentration. Potassium ions, for example, are usually found in a greater concentration inside a cell than in the cell's immediate environment. The

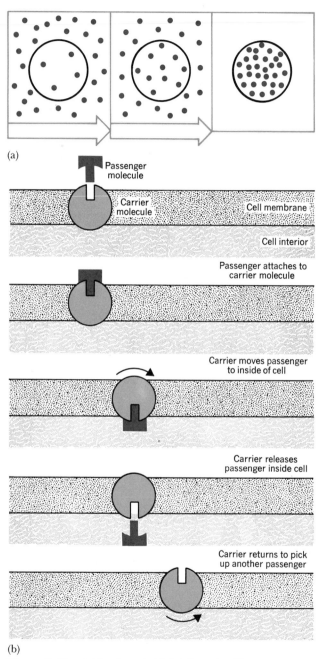

(a)

Passenger molecule

Carrier molecule

Cell membrane

Cell interior

Passenger attaches to carrier molecule

Carrier moves passenger to inside of cell

Carrier releases passenger inside cell

Carrier returns to pick up another passenger

(b)

Figure 3-10 *Active transport. (a) By expending energy a cell concentrates certain kinds of molecules within itself. (b) An example of how a carrier molecule might function.*

cell must expend energy in order to maintain this uneven relationship. The term *active transport* is applied to situations where a cell expends energy in transporting materials through its boundaries (Figure 3-10).

Many important substances such as amino acids, vitamins, glucose, sodium, potassium, and hydrogen ions may be moved through cell membranes by this mechanism. Exactly how the mechanism works is not clearly known except that it involves special carrier molecules and expends ATP energy.

Basic Concepts

All cells are bounded by a complex lipoprotein membrane, a highly adaptable structure with many specializations; most cellular organelles are composed of the same type of membrane.

A cell is divided internally into many membrane-bound compartments, hence allowing numerous diverse chemical systems to function simultaneously.

Materials move through cell membranes by diffusion, osmosis and active transport.

Organelles Made of Membranes in the Cell

Endoplasmic Reticulum. In most cells, the cytoplasm contains an extensive system of flattened, membrane-enclosed spaces called the *endoplasmic reticulum*, abbreviated *ER* (Figure 3-11). In electron microscope photographs, ER often appears as numerous parallel lines with small spaces between them; in reality these are probably membranous channels extending throughout the cytoplasm. Attached to the surfaces of the membranes are small granular bodies called *ribosomes* (Figure 3-11). Ribosomes are the sites where proteins are synthesized in the cell. After synthesis, these proteins are usually transported in ER channels to the Golgi bodies, for temporary storage and further chemical modification.

Golgi Bodies. The organelles called *Golgi bodies* appear under the microscope as a set of flattened sacs stacked closely together (Figure 3-11). The edges of the sacs extend into a network of tiny tubes that either end near the cell's surface or interconnect with other Golgi bodies.

Within the Golgi bodies various products produced in the cell are enclosed in small membrane-bounded sacs, which in turn may move to the surface of the cell,

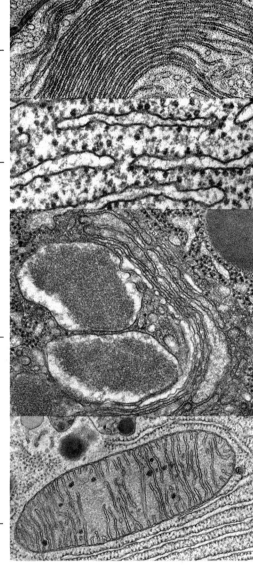

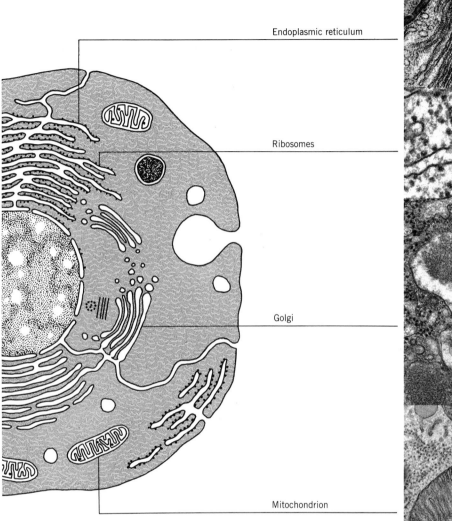

Figure 3-11   *Some major organelles found in cells.*

fuse with the cell membrane, and empty to the outside of the cell. In this manner a cell can concentrate and package its special secretions and then release them to the exterior for use elsewhere in the organism.

Lysosomes. These are tiny sacs distributed throughout the cytoplasm of cells. These small organelles consist of packets of enzymes that are capable of breaking down (digesting) a variety of organic materials: pro-

teins, carbohydrates, lipids, and nucleic acids. It is assumed that the membrane enclosing a lysosomal bag is resistant to these enzymes and thus prevents them from harming the cell itself.

Medical scientists are interested in lysosomes because they seem to be implicated in a variety of disease conditions. More than 30 hereditary diseases are known that are caused by ineffective or absent enzymes in an individual's lysosomes. Other diseases may be the result of lysosomal chemicals being released accidentally, such as in the inflammation of joints in arthritis. The drug cortisone evidently relieves the inflammation by stabilizing the lysosomal membranes in the joint area and preventing the excess release of enzymes. Two serious lung diseases, silicosis and asbestosis, may be caused by tiny fibers piercing the cell's lysosomes, thus releasing their destructive enzymes.

Mitochondria. These are large bodies, as compared with many organelles, and can be seen with an optical microscope. They are often in the shape of short rods but may also be spherical. Some cells contain only a few thousand mitochondria, whereas the number in a sea urchin egg has been estimated at 150,000. A mitochondrion is composed of two membranes and an internal matrix (Figure 3-11). The outer membrane is smooth and encloses the organelle; the inner one is folded to form numerous internal plates. The spaces between the plates, the *matrix,* contain enzymes, ribosomes, and other substances.

Mitochondria are vital to the cell's function because they are the sites of cellular respiration, or energy production. For this reason, mitochondria are sometimes termed the "powerhouses" of a cell. Even a minuscule sperm cell contains, in the neck region, an array of mitochondria lined up almost like a series of storage batteries (Figure 3-12). These organelles provide the energy supply for the rapid motions of the sperm's tail.

Mitochondria are sensitive indicators of injury to cells. If cell damage occurs, the mitochondria may either break down and disintegrate or swell into large bubblelike vacuoles. Abnormal-looking mitochondria are typical of scurvy and possibly of other diseases. Some cell biologists have suggested that it would be appropriate to speak of a cell being afflicted with "mitochondrial disease," a condition that would affect a cell's energy output.

Nuclear Membranes. The nucleus is enveloped by two membranes, probably derived from endoplasmic reticulum (Figure 3-13). This envelope is penetrated by pores which evidently can be opened and closed, thus controlling the passage of materials to and from the nucleus. The contents of the nucleus itself will be described shortly.

Figure 3-12 *The head and midpiece of a bat sperm. The tightly packed mitochondria in the midpiece provide energy for the motion of the sperm's tail.*

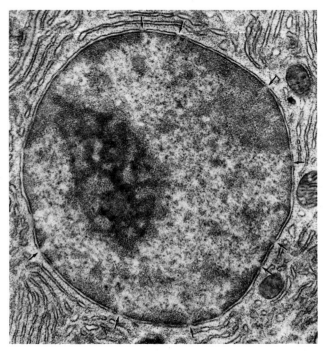

Figure 3-13 *An electron photomicrograph of the nucleus. The double membrane of the nuclear envelope and its pores (arrows) are evident. The nucleolus is the large dark mass at the left. Most of the remaining gray areas in the nucleus are chromatin.*

Basic Concept

Organelles composed of membranes include the cell membrane, endoplasmic reticulum, Golgi apparatus, lysosomes, mitochondria, and the nuclear envelope.

Nonmembranous Organelles

Ribosomes. Ribosomes are minute granular bodies visible only in highly magnified electron photomicrographs (Figure 3-11). They are usually attached to ER membranes, and their numbers reach the millions in many cells.

Ribosomes are the sites where amino acids are assembled into proteins, an event discussed in greater detail in Chapter 15. When engaged in this process, ribosomes sometimes associate in clusters called *polyribosomes*. This event is of special interest to medical researchers, because a malfunctioning of the ribosomal system affects the entire cell. The antibiotic streptomycin offers one example. It inhibits bacterial growth by interfering with synthesis of the proteins needed for growth.

Microtubules and Microfibers. Careful studies of cells by electron microscopy show the presence of numerous, extremely small, long, tubular bodies termed *microtubules*. They provide mechanical support within cells, especially in cells with unusual forms. In this case the microtubules form a sort of cell skeleton.

Microtubules are also the basic building units of cilia, flagella, and centrioles. *Cilia* and *flagella* are organelles of motion associated with certain kinds of cells. They may move the whole cell, as does the tail of a spermatozoan. They may create currents in fluids, as the cilia in bronchial epithelium do. They may also function as sensory receptors in some cases. *Centrioles* are associated with the formation of an organelle made of microtubules, the *mitotic spindle*, which appears during cell division.

In addition to microtubules, cells also contain numerous slender *microfilaments* thought to be composed of protein. The best known of these are the extremely fine filaments found in muscle cells. The ability of a muscle to contract is due to complex actions of these microfilaments within the cell.

Basic Concept
Nonmembranous organelles include ribosomes, microtubules and microfibers, cilia, flagella, and centrioles.

Cytoplasmic Structures: A Summary
The cytoplasm of a cell, with its many compartmentalized organelles, obviously has numerous functions, some of which have been described on the preceding pages. But it is important to realize that the cytoplasm is not just a collection of standardized parts; it is also the most plastic and adaptable part of a cell. It is the region that differentiates (specializes) for particular cellular functions, making up the long extensions of a nerve cell or the contractile elements of a muscle cell. Because of this adaptability of the cytoplasm, it has been possible for cells to specialize in various functions, such as transporting gases (red blood corpuscles), conducting impulses (nerve cells), and contracting (muscle cells), to mention only a few (Figures 3-17 through 3-19). These examples illustrate a very important concept, the relationship between *form* and *function*. In other words, the adaptations shown by the cytoplasm of a specific type of cell reflect the special functions performed by the cell. This form-function relationship is a basic one in biology, occurring at all levels of organization, from the molecular (as in the lipid bilayers of membranes) through cells, organs, and organisms. Think of your own body for a moment and reflect on how closely matched (adapted) are your various anatomic features to the functions that they perform.

The Nucleus and Its Structures
The nucleus in a cell appears as a relatively dense body, usually spherical or disclike in shape. As we have seen, it is enveloped by a double, selectively porous membrane that allows it to communicate with the remainder of the cell. The body of the nucleus consists mostly of a mass of intertwined threads called the *chromatin* network. Also there are one or more *nucleoli*, darker bodies made of tightly packed granules, and a fluid *nuclear sap* (Figure 3-13).

Chromatin Network. Chromatin in a nondividing cell forms an intertwined, tightly packed mass of threads consisting of DNA (hereditary material) and proteins. During cell division, however, each thread coils into a short rodlike body called a *chromosome*. This is evidently an efficient way to package the DNA and other chromatin material for distribution into daughter cells at the conclusion of the cell division process.

Nucleoli. These are tiny, dense-looking bodies representing ribosomal subunits that are being assembled and will eventually pass into the cytoplasm. The nucleoli disappear during cell division (perhaps by providing some of the chromosomal material) only to

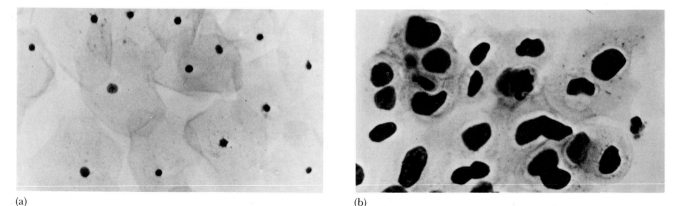

(a)   (b)

Figure 3-14 (a) *Normal cells from the cervix of the uterus;* (b) *cancer cells from the same organ showing large nuclei. This characteristic helps pathologist distinguish cancer cells from normal cells.*

reappear after division is complete, when the chromosomes uncoil back into chromatin.

The overall function of the nucleus is to act as the control center for many of the activities of the cell. The nuclear DNA is responsible for this control, because it regulates the production of special proteins called *enzymes,* which in turn regulate all the chemical reactions in the cell. DNA is also the hereditary reservoir of a cell, the "blueprint" of the organism, that is passed from one cell generation to another.

The appearance of a cell's nucleus tells a trained observer quite a bit about the health of the cell. This fact is particularly helpful to the pathologist in diagnosing cancer. In cancer cells the nucleus is usually large and irregularly shaped, and the nuclear envelope often has deep indentations (Figure 3-14). The nucleoli of cancer cells are also abnormal in shape and size. The chromosomes in cancer cells often are abnormal in number and in form, and the process of cell division itself may proceed in an abnormal fashion.

Basic Concept

The nucleus, made of a chromatin network, nucleoli, and nuclear sap, functions as the control center of the cell.

THE LIFE CYCLE OF A CELL

Cells, like organisms, have a life cycle. They are "born" from the division of a parent cell, undergo a period of growth, reach maturity, may reproduce, or eventually die. Four distinct phases can be distinguished in many cells and are customarily labeled G_1, S, G_2 and M (Figure 3-15).

G_1 (*Growth Phase One*) begins with the daughter cells produced by the division of a parent cell. At first these cells are small (half the size of their parent cell), and then increase in size (grow) by the addition of new cytoplasmic material. For many cells this is also a period in which they become highly modified (specialized) for performing specific functions in the organism—a nerve cell, for example. This is also called differentiation. The life cycle for these cells is essentially completed. They will perform their special tasks until they die. The length of G_1 varies greatly in different tissues and in different organisms. In human skin, for example, G_1 lasts about 11 hours, but it takes only 5 hours in some types of white blood cells. Specialized cells do not reenter the cell cycle.

The S (*synthesis*) *phase* is the period during which the chromatin material (DNA and proteins) in the nucleus is duplicated by chemical reactions. At the end of the synthesis stage, the nucleus contains precisely twice as much DNA and other chromatin materials as were present at the beginning of the stage. In terms of duration, the S phase varies considerably in different tissues. In human skin it takes about 5½ hours.

G_2 (*Growth Phase Two*) is essentially a preparation stage for entering the following M (mitosis) stage. Its duration is brief, seldom exceeding 2 hours. The end

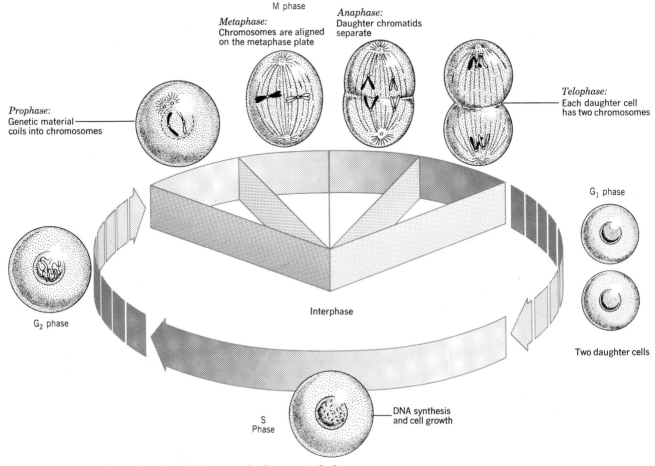

M phase

Metaphase:
Chromosomes are aligned
on the metaphase plate

Anaphase:
Daughter chromatids
separate

Prophase:
Genetic material
coils into chromosomes

Telophase:
Each daughter cell
has two chromosomes

G_1 phase

G_2 phase

Interphase

Two daughter cells

S
Phase

DNA synthesis
and cell growth

Figure 3-15 *The life cycle of a cell showing the four typical phases.*

of G_2 and the onset of stage M (mitosis) is signified by changes in the chromatin network. The stages of mitosis are described in Chapter 13.

The M stage is the shortest of the life-cycle stages, requiring less than 2 hours in many tissues. At the conclusion of the M stage, there are two identical daughter cells, each of which enters another G_1 stage. This is the "typical" life cycle of cells that are continuously reproducing, such as those in the growing tissue of an embryo. It is also typical of the deep layer of the skin, which continually produces new skin cells to replace dead ones, in the formation of new blood cells, and in the cells making up part of the lining of the gut. Interestingly, even some cancer cells go through the four stages and with about the same time intervals typical for normal cells. The abnormal feature of the

cancer cell life cycle is that the cycling never ceases. Hence a large mass of cells builds up that may eventually damage surrounding tissues.

A majority of the cells in an adult organism do not reproduce continually. Instead, they specialize in form and function for a particular task in the body. How do these cells fit into the four-stage growth cycle? Cell specialization usually begins during the G_1 phase and these cells do not enter an S phase. In a sense they will spend the remainder of their lives in a suspended G_1 stage. A nerve cell, for example, is committed to its fate permanently and cannot reenter the cell cycle at a later time. (Entry into the S stage normally signals that the cell intends to divide; in other words, it will go through G_2 and M stages to form two daughter cells.)

The ability to distinguish the four distinct stages is

extremely helpful in studying cells. It also raises many additional questions. What triggers the end of one phase and the start of the next? What determines whether a cell will differentiate or enter the S stage? Why do some cells continually recycle, whereas others do not? The answers to these questions are very incomplete at the present time.

Evidence is beginning to accumulate about growth-regulating substances in cells and tissues. One group of compounds, called *chalones,* has been shown to inhibit the S or M stages (or perhaps both), thus preventing cell division. A current hypothesis is that chalones are formed in tissues in concentrations that maintain cell division within limits normal for that particular tissue. If something happens to remove the chalones, such as an injury that causes loss of tissue fluid (and hence, of chalones), then the cells in the area are released from their "holding pattern." They again enter S, G_2 and M, and multiply, healing the injury. Some success has been achieved with tissue extracts containing chalones in temporarily reducing tumor growths in lab animals.

Basic Concepts

The life cycle of a cell usually consists of four phases: Growth Phase One, Synthesis, Growth Phase Two, and Mitosis.

Cell specialization usually takes place during Growth Phase One.

CELLS AND TISSUES

The ability of cells to assume different forms is well demonstrated by cells that cease to divide and, instead, specialize for a particular function in the body. These cells start out looking almost identical but end up extremely unlike in form and function. Examples are the cells shown in Figures 3-16 through 3-19. How these radical changes are brought about is only partially understood by biologists.

There are four general types of tissues in the body: (1) epithelium, (2) connective, (3) muscle, and (4) nerve. Each of these tissues contains a variety of specialized cells but only a few examples will be discussed. As you look at the illustrations for these cells, see if you can describe the form-function relationship that is demonstrated.

Epithelial Tissue. This tissue covers the surface of the body and its organs, lines the various cavities of the body, and forms glands. It is therefore a commonly-found tissue with a variety of functions including protection, secretion, and absorption. Its cells may be flat and tilelike as at the surface of the skin, column-shaped with a border of cilia in the trachea, or arranged in the form of complex glands as shown in Figure 3-16. Numerous additional cell types are known.

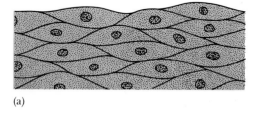

(a)

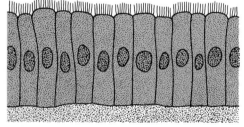

(b)

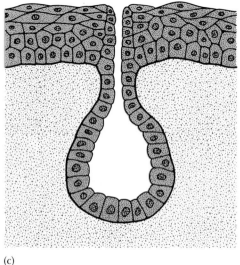

(c)

Figure 3-16 *Examples of epithelium cells. (a) Flat, tilelike covering cells; (b) ciliated cells as found in the respiratory system; (c) a gland made of epithelium cells.*

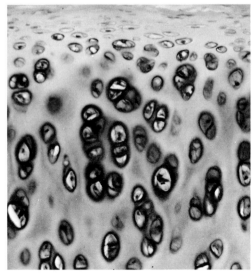

(a)   (b)

Figure 3-17 *Connective tissue cells. (a) Collagen fibers highly magnified (×20,000); (b) cartilage cells surrounded by matrix material.*

Connective Tissue. This tissue supports the body, because it includes the skeletal system, holds various organs together with fibers and ligaments, and forms blood. It is the most abundant tissue in the body. Included are the tough collagen fibers (Figure 3-17*a*) that hold may tissues together, fat cells, cartilage cells (Figure 3-17*b*), bone cells, and blood cells. Bone and cartilage cells are unusual in that they are embedded in a matrix of noncellular material.

Muscle Tissue. This tissue forms the *skeletal muscles* that move the bones of the body, the *smooth muscles* of the internal organs, and the unique *cardiac muscle* of the wall of the heart. All muscle tissue functions by contracting when stimulated, thereby bringing about motion. All three types of muscle tissue have a distinct microscopic appearance and some distinctive features. For example, skeletal muscle tissue, as in the biceps muscle, is made up of bundles of muscle fibers that have a banded appearance and are under our conscious nervous control. Internal organs such as the stomach consist of smooth muscle cells arranged in sheets or layers. Smooth muscle cells are spindle shaped, nonbanded, and are not under our conscious control. Cardiac muscle is banded like skeletal muscle but contains branched muscle fibers and unique disc-like bodies. Unlike skeletal muscle, cardiac muscle contracts rhythmically and automatically and never

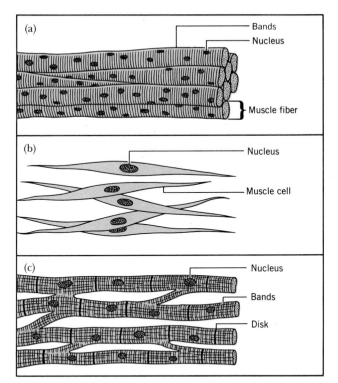

Figure 3-18 *Types of muscle cells. (a) Skeletal; (b) smooth; (c) cardiac.*

shows fatigue. Figure 3-18 shows these types of muscle tissue.

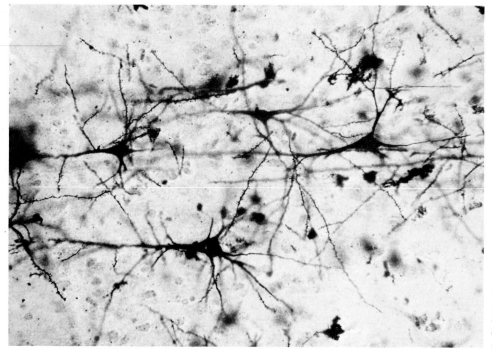

Figure 3-19 *A group of neurons from the brain. These long, slender cells are specialized for transmitting nerve impulses.*

Nerve Tissue. Nerve cells are uniquely adapted to transmitting impulses about the body in order to help coordinate the functions of other body systems. Its cells are highly specialized for impulse transmission (frequently being long and slender) but there are not as many cell types as in the other body tissues. Its basic cell type, the neuron, is shown in Figure 3-19. Neurons are frequently organized into cordlike bundles called nerves.

Basic Concepts

There are four types of tissues: epithelium, connective, muscle, and nerve.

Each tissue consists of a variety of specialized cells.

CELLS AS BASIC BIOLOGICAL UNITS

We have said more than once that cells are the basic building units of organisms. Perhaps, as some biologists contend, all the important problems concerning life processes can be solved by studying cells.

Until recent years, biologists considered all cells to be alike in their most basic features. Hence all cells were depicted as containing a membrane-enclosed nucleus, chromosomes, and a variety of membranous organelles, and they were all said to reproduce by mitosis. Indeed, this *is* the basic cell pattern for all multicellular plants and animals. Any such cell is termed *eukaryotic* (meaning "true nucleus").

However, eukaryotic cells are not the only type of cell found in the living world. Bacteria and blue-green algae are tiny, unicellular organisms whose internal organization is much simpler than that of eukaryotic cells. They do not have distinct, membrane-bounded nuclei, their DNA is not associated with a protein complex into chromatin threads or chromosomes, no membranous organelles are present, and cellular reproduction occurs by a rather simple process of binary fission—splitting—rather than by the complicated sequence of mitosis. These simple cells are termed *prokaryotic* ("prenucleus") (Figure 3-20). The two basic cell types may seem, at first, to have little kinship with each other. However, it has been determined that both share many chemical and metabolic features. For ex-

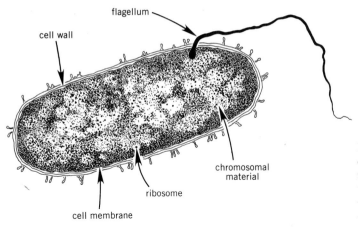

Figure 3-20 *A diagram of a bacterial cell as an example of a prokaryotic cell. Notice that there are few, if any, membranous organelles in the cell: no nuclear membrane, no mitochondria, no ER. Most have a cell wall outside the cell membrane. The flagellum is composed of protein but has no microtubules.*

ample, in both types of cells DNA controls the making of proteins and is the hereditary material. Both types of cells utilize surprisingly similar enzymes, carbohydrates, lipids, and amino acids. Prokaryotes are considered to be extremely primitive organisms, and their structure suggests that they were probably the evolutionary forerunners of eukaryotic cells.

Basic Concept

Bacteria and blue-green algae are prokaryotic cells; the remainder of the living world contains more complex eukaryotic cells.

KEY TERMS

active transport

carrier-facilitated
   diffusion

cell

cell life cycle

cell membrane

cell principle

centrifuge

chalones

chloroplast

cilia

ciliary escalator

chromatin

chromosomes

connective tissue

contact inhibition

cytoplasm

diffusion

electron microscope

endoplasmic reticulum

epithelium

eukaryotic cell

flagella

golgi body

growth phase one

growth phase two

lysosome

microfiber

microtome

microtubule

mitochondria

nucleoli

nucleus

organelle

osmosis

prokaryotic cell

ribosome

synthesis phase

tissue culture

SELF-TEST QUESTIONS

1. What contribution did the German anatomist, Theodor Schwann, make to our knowledge about cells?

2. State the cell principle.

3. In what respect is nutrition a *cellular* phenomenon?

4. (a) In what way might lung cancer be considered a disease of cells?
   (b) What is meant by the ciliary escalator?
   (c) Tell generally what happens to the cells along the bronchial tubes when they are exposed to tobacco smoke over a period of time.

5. What was Robert Hooke's contribution to the study of cells?

6. Why is the term *cell* actually a misnomer?

7. Why is it helpful to stain cells with various kinds of dyes?

8. In what way is the microtome a valuable device for studying cells?

9. What great advance was made in microscopy in 1931?

10. In what respect is the electron microscope superior to the optical microscope?

11. Explain how the ultracentrifuge has improved our knowledge of cell biology.

12. How are radioactive substances useful in studying cells?

13. Why is it important to be able to grow cells in laboratory glassware?

14. To what does the term *organelle* refer?

15. What is the importance of the concept that the inside of a cell is separated into many small compartments?

16. What is the "ground substance" of a cell?

17. (a) Describe the structure of the cell membrane.
    (b) What are some of its functions?
    (c) Describe diffusion, osmosis, and active transport.

18. Give a major function for each of the following organelles: (a) endoplasmic reticulum; (b) ribosomes; (c) Golgi bodies; (d) lysosomes; (e) mitochondria; (f) nuclear envelope; and, (g) microtubules.

19. How do cells demonstrate the form-function concept?

20. (a) Define chromatin network, nucleoli, and nuclear sap.
    (b) Tell the main functions of each.

21. Name two major functions of the nucleus.

22. List the stages in the life cycle of a cell and briefly tell what occurs in each.

23. Tell where cell specialization fits into the cell cycle described above.

24. What is thought to be the role of chalones in cell division?

25. (a) List the four general types of tissues in the body and their functions.
    (b) Describe at least one type of cell found in each of the above tissues.

26. Contrast the structure of eukaryotic and prokaryotic cells.

chapter 4

THE ROLE OF
NUTRIENTS IN
THE BODY

chapter 4

questions to think about

What are proteins used for in the body?

What is the consequence of protein deficiency in children?

What is the Green Revolution?

What is the basic function of vitamins?

Why are minerals necessary in the diet?

Tell me what you eat and I'll tell you what you are.

A. Brillat-Savarin

Nutrition is a topic of widespread interest today for a variety of reasons. Americans are markedly diet conscious partly because of vanity and partly for health reasons. Nutritionists tell us that we show a definite tendency toward obesity, and that this contributes significantly to diseases of the circulatory system. Thus tens of thousands jog and diet and try to attain a state of "cardiovascular fitness" and a slim silhouette. Yet, is it *quantity* or *kinds* of foods that is important?

Most of the rest of the world, ironically, has quite the opposite problem; namely, how to obtain enough food in order, simply, to survive. Obviously these people are not concerned with vanity or health. Yet, the *kinds* of foods that they obtain are critically important, in the opinion of many nutritionists, to their future health and welfare. Imagine for a moment that you are an administrator who has the power to help an impoverished country improve its nutritional standards. What kinds of food would you advise them to eat? Don't be disappointed if you don't come up with a ready answer — nutrition is an exceedingly complex

field and there are no standard answers. The Food and Agricultural Organization (FAO) of the United Nations struggles constantly with this problem and has no universal solutions.

Most of us feel that we know quite a bit about nutrition. After all, we eat two or three (or more!) meals a day, have been taught or guided by our parents how to eat properly, and have survived up to this point. If all of this is true, then why are we having problems with obesity and coronary heart disease? Why is our culture almost obsessed with dieting and exercising? Part of the answer to these questions is that our fundamental knowledge of sound nutritional concepts is lacking. These concepts are rarely taught at any academic level (does *your* institution offer any nutrition courses?) and thus students seldom have an opportunity to become involved in the knowledge derived from this modern, dynamic, scientific field.

As we noted in the previous chapter, nutrition begins basically at the cellular level. Organisms take food material into the digestive tract by various means but from then on it is a cellular event. Cells produce the chemicals that break down the food into smaller molecules, absorb the molecules, and eventually use them in some manner. In fact, nutrition is a study in chemistry (biological chemistry) and we must understand these concepts in order to understand basic nutritional principles.

NUTRIENTS AND THEIR FUNCTIONS

As a beginning, we should define the term *nutrient.* A nutrient is any material taken into the body that has a useful function. Broadly these include proteins, lipids, carbohydrates, vitamins, minerals, and water. These various substances have multiple functions in an organism and in essence constitute its nutritional requirements.

One of these requirements, a major one in fact, is that of providing sufficient energy to support an organism's activities and bodily functions. Most of this is met by the carbohydrates and lipids that constitute the bulk of the diet of many organisms, including human beings. Details about our energy requirements are described in the next chapter.

In this chapter we need to look at some of the additional functions served by nutrients, especially the growth and maintenance of various parts of the body. A young, growing organism obviously needs a variety of nutrients to increase the bulk of its tissues and organ systems. But even in an adult organism, no longer experiencing overall growth, certain tissues never stop producing new cells.

For example, a layer of cells beneath the skin produces new cells steadily. These cells are pushed toward the surface of the skin. They die before reaching the outermost layer, but the dead skin cells on the surface protect the living layers underneath. The continual growth of hair and fingernails is another example. All these continual replacement processes require a steady influx of nutrients. Even tissues that do not reproduce in the adult, such as muscle or nerve cells, undergo a constant turnover of their chemical components. Their atoms and molecules are slowly but continually replaced by "new" atoms and molecules. These substances are provided in an organism's diet. Even in the normal, everyday functioning of an organism's body, a large number of new substances, especially enzymes, are constantly being synthesized. These compounds are used for a while and then broken down and re-

placed by new ones. A certain number of these compounds are recycled in the process, but others are excreted from the body, and their raw materials must be replaced from the environment.

The necessity for many nutrients can be dramatically demonstrated by withholding them from an animal's diet. The result is almost invariably a *deficiency disease* with very specific symptoms. The occurrence of such a disease signifies that some part of the animal's biochemical maintenance process has been interrupted. Many such diseases are well known in human beings, especially in relation to mineral and vitamin deficiencies. Some of these conditions will be described later in the chapter.

Not all organisms require exactly the same nutrients from their environments. For example, unlike people, laboratory rats do not require vitamin C in their diets; they make their own. Hence rats show no deficiency symptoms when fed a vitamin C-free diet. On the other hand, many of the rat's other nutritional requirements are similar to those of human beings. Hence, rats are commonly used in nutritional experiments designed to answer questions about human nutrition. Despite minor variations in nutrient requirements, all higher organisms require water and certain amino acids, vitamins, and minerals. These are the major items we want to examine in this chapter.

Basic Concept

In addition to providing energy, nutrients are necessary for growth of the organism and maintenance of its parts and metabolic functions. The absence of a required nutrient produces a deficiency disease.

PROTEINS

Proteins are the most abundant organic substances in animals and are among the largest molecules known to biologists. They are used extensively as structural materials in the body, making up muscle tissue and organs, but they also have many additional functions.

Proteins are made of carbon, hydrogen, oxygen, nitrogen, and, frequently, sulfur and phosphorus atoms. These elements are organized into building units called *amino acids.* Twenty-two different *amino acids* are found in proteins, but all have in common at least one amino group (—NH$_2$) and one organic acid group (—COOH) attached to the same carbon atom in this manner:

$$\text{``R''}-\overset{\displaystyle H}{\underset{\displaystyle COOH}{\vert}}\overset{\vert}{\underset{\vert}{C}}-NH_2$$

The "R" is a symbol that stands for "replacement" for the side group that makes each amino acid distinct. For example, if the "R" is replaced with a hydrogen atom, then we have the simplest amino acid, *glycine:*

$$H-\overset{\displaystyle H}{\underset{\displaystyle COOH}{\vert}}\overset{\vert}{\underset{\vert}{C}}-NH_2$$

If the "R" is replaced with a sulfur-containing group, then the amino acid, *cysteine* is created:

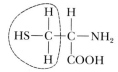

The amino acids in a protein are joined chemically into one or more chains called *polypeptides* (Figure 4-1*a*). Human insulin, for example, contains two polypeptide chains, one with 21 amino acids and the other with 30 amino acids. The two chains are held together by pairs of sulfur atoms (Figure 4-1*b*).

A polypeptide chain may contain any of the 22 different amino acids, in any sequence and number. Thus, the possible varieties of polypeptide chains, and of the proteins that contain them, are truly enormous. As a result, every person contains some unique proteins comparable to the distinctness of their fingerprints.

In addition, proteins also derive distinctness from the manner in which their amino acid chains are folded and coiled, sometimes in highly contorted forms (Figure 4-1*c*). This folding is accomplished by special sulfur bonds that form between adjacent portions of the polypeptide chains and by the attraction of positive and negative electrical charges that often exist on the chains.

On the basis of their overall shapes, some proteins take on a *globular* form because their polypeptide chains are tightly folded and coiled as described above. Still others consist of numerous polypeptide chains in a ribbonlike array somewhat like the fibers making up a rope. These are the *fibrous* proteins, as shown in Figure 3-17*a*.

Another common category of proteins consists of those forms that are combined with nonprotein substances. These *conjugated* proteins include the nucleoproteins (nucleic acid + protein) that help make up chromosomes; glycoproteins (carbohydrate + proteins) found in mucus; lipoproteins such as cholesterol; and others.

Proteins have numerous functional roles in organisms including their use as an occasional energy supply. The three most important kinds of proteins are structural proteins, blood proteins, and enzymes.

Structural Proteins

Structural proteins comprise the bulk of organic matter in animal cells. They form part of all cell membranes, connect various tissues and organs, make up the contractile material in muscle cells, and are a major part of all supporting protective tissues such as skin, hair, nails, claws, bone, and cartilage. Many of these proteins are fibrous in form and many exist as conjugated proteins.

Structural proteins play a lesser role in plant cells, but they are still a basic component in their membranes and organelles. Some plants, such as the legumes and alfalfa, contain as much protein as do many animals.

Blood Proteins

A variety of proteins with special functions are found in blood. Three of these are *albumin,* the *globulins,* and *fibrinogen.* They are important in maintaining the proper amount of water and the acid-base balance in the blood, and in transporting calcium ions. Fibrinogen plays a major role in blood clotting when a blood vessel is injured. Albumin transports a variety of items including amino acids and drugs; gamma globulin consists of numerous components that function as antibodies that help combat infections. Additional antibodies, also made of protein, are made by white blood cells in response to the presence of invaders such as bacteria.

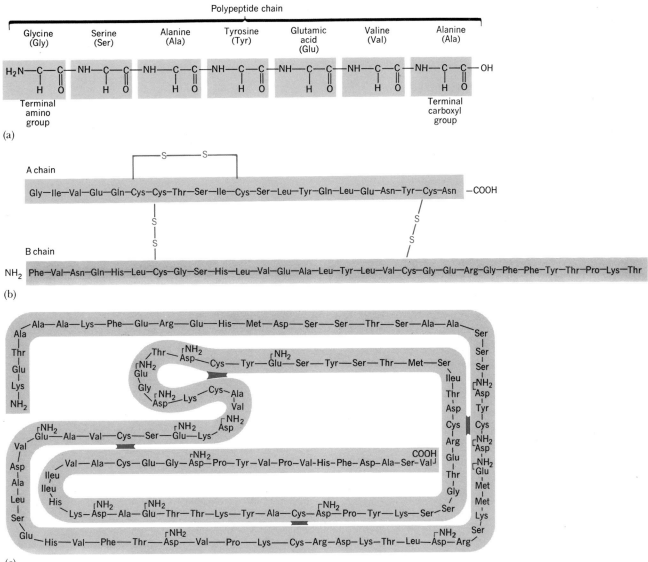

Figure 4-1 (a) *Short polypeptide chain showing how different amino acids link to one another. This linkage, a peptide bond, always occurs between the amino group of one amino acid and the carboxyl group of the next amino acid; (b) A small protein, insulin, consisting of two polypeptide chains linked by sulfur bonds; (c) Ribonuclease, a globular protein in which the polypeptide chain is folded and coiled.*

Enzymes

Enzymes are proteins with complex and varied surface shapes. Their function is to bring about virtually all of the chemical reactions in living cells. They perform this action by bringing together specific molecules on their surface (Figure 4-2). The surface of the mole-cules must match a portion of the surface of the enzyme. Because the surface of a given enzyme will fit with only a few different molecules, enzymes are specific in their action. This means that each of the many different chemical reactions in living matter requires a different (specific) enzyme. For example, the digestive

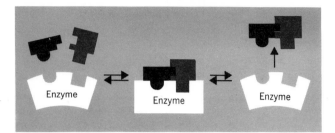

Figure 4-2 *Model of enzyme action. The surface shape of the enzyme determines the molecules with which it will react (center figure). An enzyme is capable of uniting two molecules (right) or may split them (left). In either reaction, the enzyme itself remains unchanged.*

enzyme, maltase, acts only on the sugar called maltose.

The action of an enzyme is reversible, that is, it may unite molecules or split them apart. In this remarkable action, the enzyme itself is not changed or used up. Substances that have this characteristic are called *catalysts*. Enzymes are *protein* catalysts. Because enzymes are proteins, they work best in rather narrow ranges of temperature, acidity, and various ion concentrations. If these conditions are not properly provided in a cell or an organism, many of its biochemical reactions are hindered. This knowledge is important in treating conditions that involve the tissue breakdown found in fever, blood loss, and surgery, for example.

Protein Requirements

All proteins are manufactured within cells from pools of amino acids that are normally available. These pools must continuously be replenished by the animal's dietary intake, that is, by eating protein in some form. The proteins are digested into amino acids, which in turn circulate about the body in the blood to be taken in by the cells. If necessary, a cell can also manufacture many of the amino acids it needs for a certain protein; however, there are a handful that the organism cannot make. These are called the *essential amino acids* because they must be obtained in its diet. Adult human beings require eight of these amino acids in the diet, listed in Table 4-1; laboratory rats have 10 essential amino acids. If these substances are not obtained in the diet, the synthesis of many cell structures and enzymes slows down and may stop. Eventually growth is inhibited, and the organisms's well-being is threatened.

How can an individual avoid such a deficiency?

TABLE 4-1. **Essential Amino Acids for Adults**

Lysine	Methionine	Tryptophan
Leucine	Phenylalanine	Valine
Isoleucine	Threonine	

TABLE 4-2. **Amounts of Protein in Different Kinds of Meat**

Meat	*Percent Protein*
Beef rump roast	24
Hamburger, lean, cooked	27
Sirloin steak, broiled	20–30
Chicken, fried	31
Eggs, fried (two)	14
Fish, ocean perch	19
Shrimp, cooked	20
Beef heart, cooked	31
Beef kidney, cooked	33
Beef liver, cooked	29

Source: From *Composition of Foods,* Agriculture Handbook No. 8, Rev. 1963, USDA.

First, by including an adequate amount of protein in the daily diet, and second, by eating proteins from a variety of sources in order to get all of the essential amino acids. The simplest way to do this is to eat meats from different kinds of animals—fish, poultry, pork, beef; and to include organ meats when possible in the diet—liver, heart, and kidneys, for example (Table 4-2). Unfortunately, many Americans have unreasonable prejudices against these nutritious organ meats and rarely use them in their diets. An attractive alternative or supplement is the use of high protein plant products such as beans and cereals. These are less expensive than meat, usually have fewer calories per gram, and provide a rich assortment of vitamins and minerals.

In terms of quantity, a relatively small amount of protein is needed in a daily diet if the quality (variety of amino acids) is high. The *Recommended Dietary Allowance* (RDA)* for an adult is approximately 60 grams per day—only about 2 ounces For a one-year-old child the RDA is about 25 grams. This is a small but critical

* Recommended Dietary Allowances (RDA) are recommendations of a study group of nutritionists and scientists, published by the Food and Nutrition Board of the National Academy of Sciences/National Research Council.

amount. In young animals, all organ systems are growing rapidly, especially the brain and nervous system. A lack of dietary protein not only slows general body growth but also greatly inhibits development of the brain.

Protein Deficiency

What happens to a human being who does not meet the RDA for protein intake? In an adult there is a slow wasting away of muscle tissue that can be reversed usually by adding more protein to the diet.

In infants and children the consequences are more severe. If a protein deficiency continues for months, the body wastes away until symptoms of a disease called *kwashiorkor* appear: apathetic behavior, discolored blotches and sores on the skin, puffy skin, and a swollen abdomen (the consequence of tissue fluid accumulating under the skin) (Figure 4-3*a*). If the child survives, it will be small for its age and will have diminished learning abilities.

As the map in Figure 4-3*b* shows, this tragic nutritional deficiency disease is extremely common in Latin America, Africa, and parts of Asia. The reason for its occurrence is similar in all areas: after a child is weaned from its mother's milk, it is put on protein-poor food. Gradually the symptoms of kwashiorkor appear. This could be avoided by the inclusion of relatively small amounts of meat in the diet, but this is considered a luxury in the diets of impoverished people.

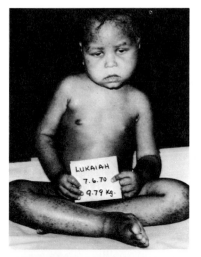

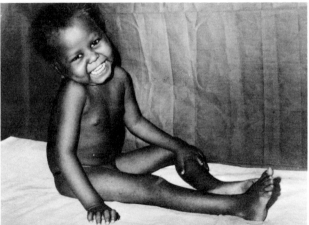

(a)

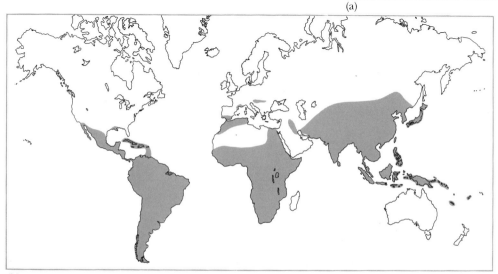

(b)

Figure 4-3   (a) *Child showing symptoms of kwashiorkor (top). Note the blotchy and puffy skin. With the proper diet, these symptoms disappear (bottom);* **(b)** *Map showing the distribution of kwashiorkor.*

The alternative solution is the use of less expensive protein-rich plants such as peas, beans, peanuts, and soybeans. Many of these plants are as high as, or higher than, meat in amount of proteins (Table 4-3). However, the variety of amino acids in them is somewhat poorer. Consequently these sources of protein must be supplemented with eggs or small amounts of meat from time to time. One hoped-for solution for underdeveloped countries is the production of more legume-type plants for food and the development of inexpensive, high-protein food supplements such as fish meal. Organizations like the Food and Agricultural Organization of the United Nations (FAO) are actively encouraging such programs.

People suffering severe protein deficiencies often also lack sufficient carbohydrates and fats for their energy needs. This caloric malnutrition that often accompanies protein deficiency creates the condition called *protein-calorie malnutrition* (PCM), a fancy term for the grim phenomenon of general starvation (Figure 4-4). PCM is tragically widespread in underdeveloped countries, and the U.S. Department of Agriculture, the Rockefeller Foundation, and the United Nations have made great efforts to alleviate it. The most promising strategy has been the development of new high-yield crops adapted to the particular climatic and soil conditions where crops are needed. One example was the development of a short-stemmed wheat adapted to the soils of the dry plateaus of Mexico. A new rice, designed for the Philippine Islands, produces spectacular yields there (Figure 4-5). For the tropical Americas, plant biologists developed a variety of corn with a high lysine content. Lysine, one of the essential amino acids is present in only small amounts in ordinary corn, which is often a major part of the diets of people in Central and South America. The substitution of high-lysine corn improves these people's diets.

The introduction of these spectacular new plant varieties into many tropical areas in recent years marked the beginning of what some optimists called the *Green Revolution*. An American plant scientist and Nobel Prize winner, Dr. Norman Borlaug, provided the leadership in developing some of these new plants.

TABLE 4-3.   Amounts of Protein in Several Types of Plant Products

Plant	Percent Protein
Beans, lima	8
Beans: red, pinto, calico	22
Peas, blackeyed	8
Peanut butter	26
Soybean flour	37–47

Source: From *Composition of Foods,* Agriculture Handbook No. 8, Rev. 1963, USDA.

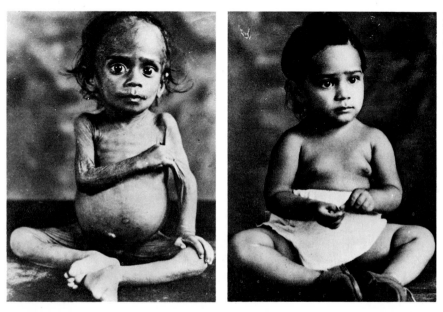

Figure 4-4   *A grim reminder of the consequence of PCM (left). With nutritional therapy, this infant was restored to normal health (right).*

He believes that such plant varieties, properly used, can feed most of the hungry peoples of the earth. Whether such an optimistic prediction will ever be a reality remains to be seen. Critics point out that these high-yield plants often require special fertilizers and may need to be irrigated. Farmers in many developing countries do not have these resources, and thus get poor results when they attempt to grow the new plant varieties.

Basic Concepts

Proteins are large molecular aggregates of amino acids. Three distinctive types include structural proteins, blood proteins, and enzymes.

The absence of essential amino acids in the diet causes the deficiency disease kwashiorkor in young children.

PCM, protein-calorie malnutrition, is widespread in underdeveloped countries. Plant scientists are attempting to alleviate PCM by developing special high-yield crops.

VITAMINS

Vitamins are small organic chemicals required by animals in extremely small amounts to maintain most of their essential physiological functions. With a few exceptions, animals cannot synthesize them and hence must obtain them from food. As a result, vitamins must pass through the process of digestion and be absorbed through the gut wall without alteration of their chemical structures.

While different vitamins are associated with specific functions in the body, these functions can be generalized by saying that vitamins are an essential part of most enzyme-regulated reactions. As such they are termed coenzymes. As explained earlier, enzymes are specialized proteins that function as catalysts for all chemical reactions in cells. Nearly all enzymes require the aid of a specific vitamin, or coenzyme, in order to function.

Table 4-4 is a reference list of the vitamins, their general functions, and some major sources of them in food. As noted in the table, we have followed a widely used scheme of grouping the vitamins as fat-soluble or water-soluble.

The *fat-soluble vitamins* (A, D, E, and K) dissolve in fats and are stored in fatty tissues in the body, especially in the liver. Vitamins A and D become toxic if greatly excessive amounts are ingested, as in overzealous use of vitamin pills. Vitamin A also illustrates the mistaken concept that each vitamin has one specific task, or affects only one particular tissue or organ. It is commonly said to be the "night vision" vitamin, which in fact it is, but this is only one of its many roles. Even a blind person still needs vitamin A for its other functions in the body. Among these are helping to maintain tissues such as the skin and the lining of the respiratory passageways. Vitamin A is of particular importance because it is frequently deficient in American diets. Its richest sources—yellow fruits and vegetables, green leafy vegetables, and liver—are too often neglected in our daily diets.

Figure 4-5 *Agricultural scientists explain the merits of a new short-stemmed variety of rice to two Indian farmers.*

TABLE 4-4. Vitamins

Vitamin	Chemical Name(s)	General Functions	Major Sources
Fat Soluble			
A	Retinol Carotene Cryptoxanthin	Vision in dim light; skin maintenance; essential for general growth	Yellow vegetables, green leafy vegetables; dairy products; egg yolk; liver
D	Calciferol Cholecalciferol	Calcium and phosphate absorption from intestine; prevents rickets[a]	Fish liver oils; irradiation of skin by sunlight; D-fortified dairy products
E	Tocopherol	Spares vitamins A and C and unsaturated fatty acids; prevents sterility in laboratory rats (but not in humans)	Numerous foods (salad oil and margarine are common sources)
K	Naphthoquinones	Aids in blood clotting	Intestinal bacteria; leafy vegetables; eggs and liver
Water Soluble			
B$_1$	Thiamine	Coenzyme in carbohydrate metabolism; maintenance of good appetite; normal functioning of nervous system (prevents beriberi)[b]	Pork; whole grains; legumes; enriched bread and flour
B$_2$	Riboflavin	Coenzyme in energy metabolism; skin maintenance; vision	Dairy products; meat; green vegetables
Niacin		Coenzyme in energy metabolism; synthesis of fatty acids; prevents pellagra[c]	Meats; cereals; vegetables
B$_6$	Pyridoxine	Coenzyme in a variety of metabolic reactions	In many foods
Pantothenic acid		Part of coenzyme A in metabolism of carbohydrates, fats, and proteins; used in many synthesis reactions	In many foods
Folacin	Folic acid	Coenzyme in many reactions; formation of nucleic acids	Leafy vegetables; organ meats
B$_{12}$	Cobalamine	Coenzyme in many reactions	Meats
Biotin		Carbohydrate metabolism; fatty acid metabolism; amino acid metabolism	Many foods
C	Ascorbic acid	Numerous general functions; prevents scurvy[d]; formation of collagen, an important connective tissue; synthesis of thyroxin; some functions debatable, such as preventing colds	Vegetables and fruits

[a] Rickets: a disease characterized by poor skeletal and muscular development.
[b] Beriberi: a disease accompanied by damage to the nervous and cardiovascular systems.
[c] Pellagra: a disease characterized by reddened, inflamed skin, diarrhea, irritability, and mental disturbances.
[d] Scurvy: a disease characterized by swollen, bleeding gums, swollen joints, anemia, and slow healing of wounds.

Vitamin D is another extremely important vitamin because it is essential for proper bone formation in infants and children, and for pregnant mothers. Commercially sold milk is now fortified with 400 units of vitamin D per quart—the RDA for all individuals (Figure 4-6).

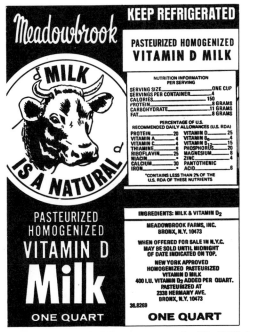

Figure 4-6 *Label from a quart of milk. Milk is fortified with 400 Units of vitamin D as indicated. According to the information shown on the label, is milk a good source of protein?*

Figure 4-7 *Label from a package of enriched flour. Most of the B vitamins removed by milling have been restored by the food processor.*

The *water-soluble vitamins* are not stored to an appreciable extent in the body, because any excesses are excreted in body fluids. Thus they must be replenished more frequently than the fat-soluble vitamins. As noted in Table 4-4, this is a large group of vitamins that affects virtually every metabolic activity in the body. Many of the B complex vitamins are removed from flour by the modern methods of milling. Enrichment —addition of approximately the same amounts removed—is practiced by most producers of flour and bread (Figure 4-7).

As you are undoubtedly aware, vitamin C has become the center of a controversy regarding its functions. In 1970 a world-famous chemist and Nobel Prize winner, Linus Pauling, proposed that massive doses of vitamin C would prevent the common cold. This claim provoked a number of experiments designed to test his proposal, most of which gave inconclusive results. Perhaps the most significant finding was that massive doses of vitamin C are toxic to some people. More recently, Dr. Pauling proposed that megadoses of this vitamin would cure and prevent cancer. This idea has not been verified.

Some recent experiments indicate that vitamin C depresses the cholesterol level in guinea pigs and prevents atherosclerosis in these animals. But a similar function for the vitamin in human beings has not yet been established. Because vitamin C does evidently participate in a variety of activities, it is not surprising to read of new claims for it from time to time. Unfortunately, the public often hears only the sensational side of such claims and wishes to share the supposed health benefits immediately. In many instances in the past, such benefits turned out to be illusions.

Of all the nutrients, vitamins are the least understood by most people. Health-food hucksters take advantage of this ignorance and make extravagant, sometimes fake, claims regarding the benefits of taking vitamin supplements. And as a result, millions of people consume vitamin pills that they usually do not need, and the hucksters gross millions of dollars. Fortunately for our health, excess intake of most vitamins damages only our pocketbooks.

Basic Concept

Vitamins are required by animals in small but essential amounts in their diets. Vitamins function basically as coenzymes.

MINERALS

By definition a mineral is an inorganic substance (element or compound) found in the earth and formed by nonliving processes; rocks, for example, are made of minerals. In nutrition, the term has a somewhat broader meaning. It includes the ionic forms of elements and compounds, and it also recognizes the fact that a number of inorganic substances such as iron and copper form complexes with organic materials like enzymes and hormones.

The mineral content of a food material or of a certain tissue is determined by burning the material to an ash. Burning does not affect minerals. The ash is then analyzed for the quantity and types of minerals present in it.

The overall quantity of minerals in a unit of food material or tissue is relatively small, usually less than 1 percent. Those found in amounts equivalent to one gram or more per kilogram of sample are termed *macronutrients*. Those present in smaller (microgram) amounts are called *micronutrients* or *trace elements*. The latter term was applied when chemists could detect only the presence or absence of an element but not its quantity in a sample.

This terminology does not mean that macronutrients are necessarily more important than micronutrients. In reality, the amounts relate more to dietary requirements than to relative importance. For example, an 18-year-old human male should take in about 1200 milligrams of calcium (a macronutrient) per day, but only 18 milligrams of iron (a micronutrient). Yet one element cannot be said to be more important than the other; the 18-year-old could not survive on either a calcium-free or an iron-free diet. This example also emphasizes another interesting aspect of mineral nutrition: the extreme divergence in the amounts of the different minerals required by organisms. For example, human beings require several grams per day of sodium, potassium, and chloride salts but only a tenth of a milligram or less of salts of iodine, selenium, and cobalt; in fact, larger amounts of these latter elements would be poisonous. Some of the most important functions of minerals are listed in Table 4-5. A few comments about some of these follow.

Calcium and Phosphorus

As indicated in Table 4-5, *calcium* is needed in considerable amounts in bones and teeth and for a variety of metabolic functions. Unfortunately, relatively few foods are high in calcium—for the most part, dairy products and some of the green leafy vegetables. People who do not use dairy products frequently show a calcium deficiency in their diets. In children, with their rapid growth, this deficiency may affect the formation of their bones and teeth. In adults, a deficiency may not be serious on a short-term basis because the body is able to draw calcium from bones for its metabolic needs. But a long-term deficiency may weaken the bones in older individuals. This condition is especially prevalent in females. Nutritionists recommend that adults include a pint of milk or its equivalent in other dairy products each day in their diets in order to meet the body's calcium requirements of 800 milligrams.

Calcium's companion, *phosphorus*, is needed in the same quantity as calcium, but it is found so widely in foods that deficiency is not likely.

Iron

The body's iron requirements are modest: 10 milligrams per day for adult males and 18 milligrams per day for adult females. Despite these low RDAs, the diets of many people tend to be deficient in this mineral because it occurs in such small amounts in most foods. The richest food sources of iron are meats, (especially organ meats such as liver and heart), clams and oysters, and whole-grain cereals. Most of the dangers of iron-deficient diets could be avoided by including these foods in one's regular diet.

Iron is an essential part of hemoglobin molecules in the blood, and a chronic iron deficiency in the diet may lead to a decreased capacity of the blood to transport oxygen and hence to a form of anemia. Infants are especially prone to iron deficiency because their high-milk diet is a poor source of iron. Pregnancy is also stressful, because the developing baby demands a considerable amount of iron from its mother's supply. Severe menstrual bleeding or hemorrhaging also drain the body's iron resources. But such depletions are better avoided by increasing iron-rich foods in the diet

TABLE 4-5. Some Important Mineral Nutrients for Human Beings

Macronutrients	Functions	Food Sources	Recommended Daily Intake for Adults
Calcium	Structure of bones and teeth; movement of ions across cell membranes; muscle contraction; blood clotting; part of many enzymes	Dairy products; green leafy vegetables	800 mg (milligrams)
Phosphorus	With calcium in bones and teeth; part of ATP; in cell membrane structure, DNA, and RNA	In many foods	800 mg
Magnesium	Part of many enzymes	In many foods	300–350 mg
Potassium	Transmission of nerve impulses; fluid balance in cells	Orange juice, dried fruits; bananas	1.9–5.6 g (grams)
Sodium	Major ion in fluids around cells; transmission of nerve impulses	Meats; table salt; milk	1.1–3.3 g
Chloride	Common in fluids around cells; forms hydrochloric acid in stomach	Many foods	1.7–5.1 g
Sulfur	Bonding in proteins; part of many B vitamins	In many foods, especially meats	not known
Micronutrients			
Iron	In hemoglobin and myoglobin, part of many enzymes	Organ meats, dried fruits; nuts and grains	10 mg: males 18 mg: females
Copper	Part of many enzymes	Cereals, nuts, legumes; shellfish	2.0–3.0 mg
Iodine	In thyroid hormone	Seafood; iodized salt	0.15 mg
Fluoride	Hardens teeth and bones; prevents tooth decay	Many foods; fluoridated water	1.5–4.0 mg
Cobalt	Part of vitamin B_{12}	Organ meats; eggs	not known
Manganese	Bone development; in several enzymes	Cereals, legumes	2.5–5.0 mg
Zinc	In several enzymes; component of insulin	Cereals; legumes; liver; eggs	15 mg

rather than by taking patent-medicine iron supplements, which are often ineffective.

Iodine

Iodine is unusual in that it is needed in amounts so small as to be measured in *micro*grams (ten-thousandths of a gram) and it functions only as a constituent of a single hormone, thyroxin. Yet a deficiency of this element causes a spectacular enlargement of the thyroid gland called a *goiter* (Figure 4-8). This deficiency disease is most often found in inland areas of the world where there is little iodine in the soil and where the commonly eaten foods are consequently lacking in iodine. The disease was once common in the Great Lakes area and the Midwest but it has greatly decreased as a result of the addition of small amounts of potassium iodine to table salt supplies. Such iodized salt is now sold everywhere. Although people living in coastal areas already have more than enough iodine in their diet, excess amounts are evidently not harmful.

Figure 4-8 *A group of people from a village in Paraguay showing a high incidence of goiters due to an iodine deficiency in their diets. How many people in this photo clearly have goiters?*

Nonessential Minerals

Trace amounts of many minerals with no known physiological functions can be found in animal tissues. These include lead, mercury, tin, gold, and cadmium. At one time, arsenic was also on this list, but recent studies indicate that it plays a role in growth and reproduction in some animals. More than trace amounts of arsenic are poisonous. It is thought that additional essential trace elements may likely be found in the future.

Basic Concept

Minerals are elements, compounds, and ions found in the earth and used for a variety of functions in the body. An absence of a required mineral in an animal's diet causes a deficiency disease.

WATER

It may sound strange to classify water as a "nutrient," but considering that it is the most abundant substance in living matter, organisms obviously need considerable quantities of it in their diets.

Water has several basic functions in cells. One of these is as a participant in chemical reactions such as the digestion of proteins, carbohydrates, and lipids into simpler materials, and, conversely, the synthesis of these molecules from their simpler building units. Water also serves as the major solvent for other chemicals within cells that helps them react with each other. When a substance dissolves in water, the number of its molecules or ions moving about in the water increases, improving the chance that they will come into contact.

Water is the basic transporting medium within all organisms. It surrounds and bathes all cells, helping to transport materials into and out of them. This liquid environment is so vital that the living cells of the body are constantly kept moist; if they were to dry out they would rapidly die. On a larger scale, water is the basic component of lymph fluid and blood, both of which have major transporting functions.

Obtaining sufficient water is not normally a problem in human diets. Most foods have a relatively high water content, and most people consume considerable quantities of various beverages.

Basic Concept

Water is the most abundant substance in living matter, functioning as a chemical reactant, a medium for other reactions, and as a transporting medium.

SOME DIETARY MISCONCEPTIONS ABOUT FOOD AND NUTRITION

People probably hold more misconceptions and erroneous beliefs about nutrition than about any other single aspect of their lives. From early childhood on we simply do not receive much correct nutritional information. This is unfortunate from a number of standpoints. Proper functioning of the body is tied to many aspects of nutrition that people should be aware of — the roles of vitamins and minerals, for example. Everyone should know how to select the most nutritious foods for their budgets. Cereal grains and legumes are far less expensive and contain more nutrients than do fatty meats such as hot dogs, for example. A nutritionally uninformed public is highly vulnerable to the lure of health-food stores, and to misleading advertising urging us to substitute "manufactured" foods for natural ones. Artificial eggs "low in cholesterol" (and nearly everything else), imitation milk with added vitamin D, and imitation orange juice "like the astronauts use" are common examples (Figure 4-9). All these products contain fewer nutrients than their natural counterparts, and all are *more* expensive than the products they are imitating. These and similar items support a very profitable food industry.

The following misconceptions are common ones found from various surveys related to food and nutrition. They are labeled "misconceptions" because they run counter to reasonably well-established (tested) nutritional concepts.

1. The many chemicals such as preservatives, emulsifiers, and coloring agents added to modern foods take away much of their nutritional value. This is a broad generalization that for the most part is not correct. For example, additives such as preservatives increase the shelf life of foods such as bread by preventing the growth of bacteria and molds, or by preventing undesirable chemical changes within the food product. Emulsifiers improve the texture of foods and thus make many foods more palatable. Although food dyes do not increase the nutritive value, they do improve the appearance of some foods considerably. A majority of these additives have been found not to pose a health hazard to consumers, at least according to the Food and Drug Administration's Generally Recognized as Safe (GRAS) list.

As a consequence of testing, a number of food coloring agents are no longer allowed in foods because they were found to cause cancers in laboratory animals. Sodium cyclamate, a sweetener, was removed from the GRAS list for this reason and can no longer

Figure 4-9 *An egg substitute. It contains fewer nutrients than do real eggs and costs more than the equivalent number of eggs.*

be used in foods in this country. Another widely used sweetner, sodium saccharin, has also been found to be carcinogenic when consumed in large amounts, and may be banned in the future.

On the other hand, additives other than preservatives do not actually increase the nutritive value of foods, and therefore are not really necessary. Moreover, these unnecessary additives do increase the cost of food. Critics of the F.D.A. point out that the organization has been slow to ban certain widely used additives that are known to be carcinogens. An example that has caused much controversy concerns the addition of nitrites to bacon, sausage, and other perishable meats to prevent the growth of the bacterium that causes botulism. No other known preservative provides this protection. Nitrites do cause cancer in laboratory animals and thus are a potential danger to human beings. However, this danger is judged less important than inhibiting growth of the very dangerous botulinum bacterium. Thus, small quantities of nitrites are allowed in foods.

People who wish to avoid foods that contain additives would have to forgo nearly all prepackaged food products and would find their selection of foods in modern grocery stores severely limited. A more prac-

tical approach might be to read food labels carefully and then avoid foods that contain controversial additives such as nitrites and saccharin. A useful reference in this regard is a paperback, *Eater's Digest: The Consumer's Factbook of Food Additives* by Michael F. Jacobson, published by Doubleday & Company, 1972.

2. Much of our food has been so processed and refined that it has lost much of its nutritive value. Again, this is a generalization that is not really valid. A few foodstuffs, such as grain products, are refined in such a way that they are less nutritious than in their original forms; however, many food processors now enrich and fortify their products to help remedy these losses. Some items, such as milk with vitamin D added, may be nutritionally superior to the original product. Other types of food such as meats, vegetables, and fruits are relatively unaffected by processing or refining.

Consumers who are concerned about the nutritional content of processed foods should carefully read the list of ingredients on the container. The label does not give the actual amounts, but it does list them in descending order of abundance (Figure 4-10). Thus, if a can of chicken soup lists chicken as the fifth or sixth ingredient, consumers know that they are buying

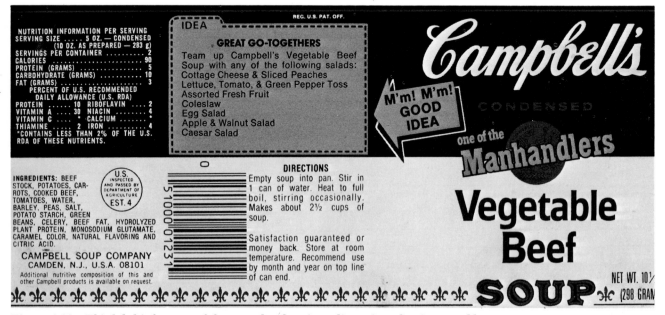

Figure 4-10 *This label informs us of the most abundant ingredients, in order, in vegetable beef soup. Would you term this a nutritious food product?*

TABLE 4-6. **Nutrients in 100 Grams of Milk and Eggs. (In most respects, eggs contain more nutrients than does milk.)**

	Calories	Protein	Carb.	Fat	Calcium	Phosphorus	Iron	Vita-min A	Thiamine	Ribo-flavin
Whole milk 3.5% fat	65	3.5	4.9	3.5	118	93	tr.	140	.03	.17
Fresh eggs	163	12.9	0.9	11.8	54	205	2.3	1180	.11	.30

water and chicken flavoring but very little chicken meat. Many cereal products list their ingredients in terms of their contribution to an individual's RDA, and this makes it easier for the consumer to compare various products.

3. Many foods lose a lot of their food value because they are shipped so far and stored so long. This may have been partly true some years ago, but modern refrigeration techniques and rapid transport by trucks and airlines invalidate this claim. Even for foods that might suffer long storage, their protein, carbohydrate, and lipid content would not change significantly.

4. Milk is nearly a perfect food and everyone should drink a large quantity of it every day. Milk is far from being a "perfect" food; in fact it rates below eggs in this respect (Table 4-6). Milk is much lower in proteins than are eggs, has very little iron, and is low in the B vitamins. In fact, eggs rather than milk are used as a reference food to which other foods are compared.

A number of individuals, especially adults, are unable to metabolize milk sugar (lactose). Large quantities of milk cause them acute gastric problems and diarrhea; hence milk is neither a perfect food nor one to be recommended for everyone's use.

5. A considerable proportion of common diseases and ailments can be attributed to poor diets, that is, poor nutrition. This statement is certainly what health-food store operators would like us to believe. In reality, however, most nutritional deficiency diseases take a considerable time to show up and are relatively rare, at least in our affluent society. Dietary factors may play a role in hardening of the arteries and coronary heart disease, but even this is debatable. Most of the common diseases in our culture are evidently related to factors other than poor diets.

6. Gelatin is a rich source of protein and its inclusion in the diet prevents fingernails from splitting, adds luster to

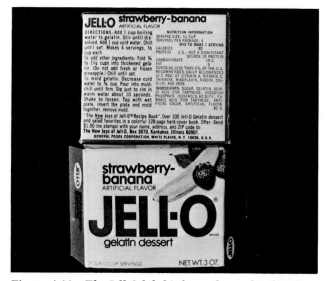

Figure 4-11 *The Jell-O label informs the reader that the most abundant ingredient is sugar and that it is a poor source of protein.*

hair, etc. An examination of a food composition table indicates that gelatin is a poor source of protein. In fact, it contains far more carbohydrate than protein and has a relatively high amount of sodium. In short, gelatin really has relatively little value as a nutrient (Figure 4-11).

7. Most people should take vitamin supplements because modern foods are so deficient in these nutrients. Again, this is the message of the health-food hucksters which, unfortunately, all too many people believe (Figure 4-12). The major sources of vitamins are fruits and vegetables, and they still contain as many vitamins as they ever did. Even the processed cereal products are so fortified and enriched that few individuals should ever need vitamin supplements. In terms of cost it is far less expensive to obtain vitamins through a regular diet than by purchasing them in pill form.

Figure 4-12 *Advertising on a typical health food store. The claims of B vitamins for energy and vitamin E for endurance are not supported by any reputable nutritional studies.*

8. *Regular use of vitamin pills gives a person more energy and vitality.* Although widely believed, this statement is erroneous in at least two respects. First, vitamins are not broken down and used for energy in the body. The only association between vitamins and energy is that some of the vitamins act as coenzymes in reactions in cellular respiration. Second, the probability that an individual would be deficient in the vitamins that function in cellular respiration (such as niacin, thiamin, and riboflavin) is remote because they are found in so many foods. The general opinion of many nutritionists is that vitamin supplements are necessary only when recommended by a physician.

9. *Organically grown foods are more nutritious than foods grown in chemically fertilized soils.* In recent years a certain mystique has enveloped the idea that organically grown food (plants grown with animal manures and no pesticides) is nutritionally superior to food from plants grown in the conventional manner. In fact, plants do not distinguish what source their minerals come from: calcium is calcium whether it comes from a

Drawing by Drucker © The New Yorker Magazine, Inc.

"Another thing. Let's lay off the health foods for a while."

cow patty or from a handful of chemical fertilizer. As any farmer knows from experience, animal manures are relatively poor fertilizers as compared with modern chemical fertilizers, and crop yields are always much higher with the latter. In terms of pesticides, fruits and vegetables grown without them are often wormy and unattractive.

As is often the case, hucksters have capitalized on the public's gullibility and general ignorance about nutritional matters to convince them to pay high prices for supposedly superior "organically grown" foods. In reality, the entire concept of health foods is a very misleading one, because no single food or group of foods can confer health on a person.

KEY TERMS

albumin	essential amino acid	green revolution	protein-calorie malnutrition (PCM)
amino acid	fat-soluble vitamin	kwashiorkor	polypeptide
blood protein	fibrinogen	macronutrient	recommended dietary allowance (RDA)
catalyst	fibrous protein	micronutrient	structural protein
conjugated protein	globular protein	mineral	vitamin
deficiency disease	globulin	nutrient	water soluble vitamin
enzyme	glycine	protein	

SELF-TEST QUESTIONS

1. Explain the text statement that nutrition begins at the cellular level.
2. (a) Define the term *nutrient.*
   (b) Name the six basic nutrients.
3. What is meant by a deficiency disease?
4. List the basic elements (atoms) found in proteins.
5. Define an amino acid.
6. What is a polypeptide chain?
7. What is the basis for the statement that a person's proteins are as unique as his fingerprints?
8. Define globular protein, fibrous protein, and conjugated protein.
9. List four places in the body where structural proteins are found.
10. Name the three important blood proteins and give a function for each.
11. (a) Define enzyme.
    (b) Describe how an enzyme functions.
12. (a) Define essential amino acid.
    (b) Describe three ways of avoiding an amino acid deficiency.
13. (a) What is an RDA?
    (b) How is an RDA value determined?
14. Describe the cause and symptoms of kwashiorkor.
15. What is the relation between PCM and kwashiorkor?
16. To what does the expression *Green Revolution* refer?
17. (a) What are vitamins?
    (b) What is their general function?
18. (a) List the fat-soluble vitamins.
    (b) Name three water-soluble vitamins.
19. What does the term *mineral* mean in reference to nutrition?

20. List three macronutrients and three micronutrients.
21. Give a major function for each of the following: (a) calcium; (b) sodium; (c) iron; and (d) iodine.
22. Name three important functions of water in living systems.
23. What are some advantages and disadvantages of food additives?
24. Why is milk not a "perfect" food?
25. What appears to be the role of dietary factors in common diseases?

chapter 5

Expending energy.

CALORIES,
FOOD, AND
YOU

chapter 5

questions to think about

What are the sources of energy for cells?

How is energy use measured in organisms?

What are the functions of carbohydrates in the diet?

What are the benefits and detriments of fats in the diet?

What is the cell's energy machine?

All people are made alike.
They are made of bones, flesh, and dinners.
Only the dinners are different.

Gertrude Louise Cheney

ENERGY AND LIFE

What do a scrambling quarterback, a wriggling earthworm, and a vibrating bacterium have in common? They all rapidly expend *energy*. Perhaps more remarkably, all use essentially the same energy sources and the same energy-producing reactions in their activities. This uniformity of energy resources and energy production in the living world is a fundamental concept in biology.

The basic energy sources for all organisms are *carbohydrates, fats,* and *proteins.* The diets of most organisms contain considerable amounts of carbohydrate. Some of this material is used directly for producing energy for the organism, and the excess is stored. Plants may store large amounts of carbohydrate in the

form of starch. Animals, however, store relatively little carbohydrate in their bodies; instead they convert excess carbohydrate into fat, which as we will see later, can be called out of storage for energy production if the carbohydrate supply in the diet ever becomes deficient. Proteins can also be used as an energy source if no carbohydrate is available. Carnivores, of course, are adapted to use proteins (meat) for most of their energy needs.

Green plants are unique in that they are the only form of life capable of using solar energy to synthesize carbohydrates from simple substances—carbon dioxide and water. By harnessing the energy in sunlight through the mechanism of photosynthesis, plants provide the basic carbohydrate supply for the living world. Without plant life and sunlight there could be no animal life, because all animals depend directly or indirectly on plants as an energy source. (The next time you pass a green plant, pay it the proper respect!)

Before proceeding any further we need to examine the general concept of energy, especially as it relates to living matter. Energy, as commonly defined, is the capability for performing work. In living matter, this

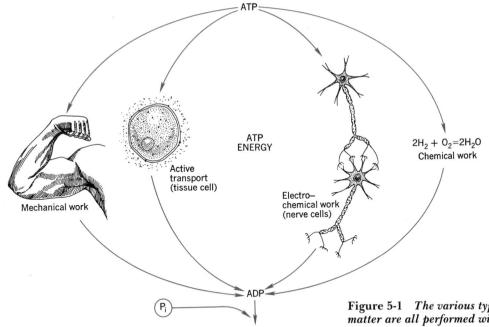

ATP

ATP
ENERGY

Active
transport
(tissue cell)

$2H_2 + O_2 = 2H_2O$
Chemical work

Electro—
chemical work
(nerve cells)

Mechanical work

P_i

ADP

ATP

Figure 5-1 *The various types of work performed in living matter are all performed with energy derived from ATP molecules.*

may be mechanical (muscle contractions), chemical (synthesis of various biological substances), electrochemical (transmission of nerve impulses), or the transport of materials through membranes (Figure 5-1). The immediate source of energy for performing all of these diverse biological jobs is chemical energy stored in the form of ATP molecules (adenosine triphosphate). Cells are equipped for converting the chemical energy of ATP into a host of other energy forms.

In theory, to determine the amount of energy used in performing work of a particular type, a biologist might measure the rate at which ATP molecules are used up. In fact, however, it is easier to measure other things that accompany the performance of work and to relate them to the energy expenditure. For one, all work is accompanied by a change in temperature. The reason is that every energy conversion (from chemical to mechanical, for example) is inefficient to some degree, and thereby produces some waste heat. This heat can be measured and correlated with work performance (energy expenditure). A second and more commonly used means of determining energy output involves *cellular respiration,* the chain of chemical reactions that replenishes energy as it is expended by a

cell. During cellular respiration, oxygen is consumed and carbon dioxide is given off. It is relatively simple to measure the rate at which an organism takes in oxygen and gives off carbon dioxide while performing work. From this rate it is easy to calculate the amount of work performed. Both techniques will be discussed later.

Basic Concepts

The basic energy sources for all organisms are carbohydrates, fats, and proteins.

Photosynthesis by plants provides the basic carbohydrate supply for the living world.

Energy is the capability for performing work. In living matter, energy is required for mechanical, chemical, and electrochemical work and for the transport of materials through membranes.

ENERGY AND THE ORGANISM

As stated in Chapter 1, we human beings have an internal metabolism, a chemical machinery, like that of other mammals of comparable size. We expend considerable amounts of energy in our daily activities and must replenish this by taking in food materials that can be converted into energy.

During its energy conversions, an organism's body produces heat as a consequence of three major physiological events. One is the normal functioning of all its internal organs when the body is at rest. This is called *basal metabolism*. The second is the digestion and utilization of food by the body, including cellular respiration. The third is the activity of the muscular system. Although the heat energy produced by these actions cannot be used except to warm an organism, it nevertheless provides an accurate indicator of the other energy events that are taking place.

It is customary to measure energy requirements in human beings and other organisms in terms of units called *kilocalories* (kcal). A kilocalorie, by formal definition, is the amount of heat required to change the temperature of a kilogram (1000 grams) of water one degree, from 14.5° to 15.5°C. This unit is sometimes also designated as the Calorie to distinguish it from the small calorie (always spelled with a small *c*), the amount of heat required to change the temperature of a single *gram* of water by one degree.

Devices that measure the amount of heat produced from any source are called *calorimeters*. An early type used in energy studies consisted of an insulated chamber large enough to hold a human being. Heat produced by the body was absorbed by water flowing through pipes embedded in the wall of the chamber. By measuring the amount of water flowing through the pipes and the change in temperature of the water, simple calculations told how much heat came from the experimental subject's body. Measurements could be made while the subject was resting, performing various physical activities, before and after eating different kinds of food, and so on. This device was cumbersome and expensive but provided valuable data on human energy expenditures and needs.

Simpler devices are now used for obtaining essentially the same information. These measure the rate at which oxygen is consumed and carbon dioxide is produced as a person breathes. From these measurements, the energy expenditure of an individual can be calculated for a specific period of time. Such devices are called *respirometers*. Some respirometers are small enough so that measurements can be made while a person performs various kinds of work (Figure 5-2). A

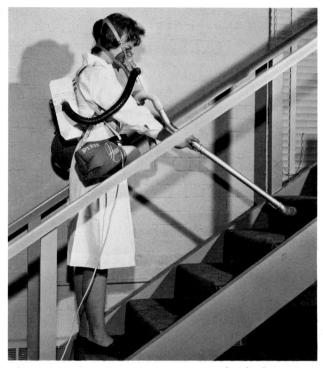

Figure 5-2 *A respirometer in use. An individual's energy expenditures can be measured while performing various tasks.*

sample of energy expenditures is given in Table 5-1.

The data in Table 5-1 are only approximate, because energy expenditures vary with a person's age and weight. In addition, some of the figures apply only when the activity is actually being performed. For example, the rate of 4.4 kcal per minute in bowling applies only to actually propelling the ball down the alley; sitting and recording the score do not count. An evening of bowling is not nearly as vigorous a form of exercise as it first appears. The same caution applies to many of the other activities: the kcal-per-hour rate in the table is often not a realistic one.

Several generalizations are evident from Table 5-1. Note the values under Children's Recreation. Even though most children are extremely active, their kcal rates are modest compared with those of adults. However, this is an effect of their much smaller body size. In reality, the BMR of children is higher than that of adults, and their daily energy expenditures are impressively high. As any parent knows, children never seem to "run down."

TABLE 5-1. Average Energy Expenditures for a Variety of Activities (Adult Males and Females Unless Otherwise Specified)

	kcal/min	kcal/hour
Light Indoor Recreation		
Lying down and relaxing	1.5	90
Sitting, relaxing	1.6	96
Sitting, listening to music	2.2	132
Sitting, playing drums	4.1	246
Standing at ease	1.8	108
Standing, playing a trumpet	2.1	126
Moderate Exercise		
Driving a car	2.8	168
Driving a motorcycle	3.4	204
Canoeing vigorously (4.0 mph)	7.0	420
Bicycling leisurely (5.5 mph)	4.5	270
Bicycling vigorously (13.1 mph)	11.1	666
Dancing to slow music	5.2	312
Dancing to rock music	What do you think?	
Weeding a garden	5.0	300
Golf	5.0	300
Bowling	4.4	264
Tennis	7.1	426
Walking normally (2.0 mph, 140 lb. person)	2.9	174
Walking rapidly (4.0 mph, 140 lb. person)	5.2	312
Vigorous Exercise		
Swimming, various strokes	5.0–11.0	300–666
Cross country running	10.6	636
Snow skiing, moderate speed	10.0	600
Walking on hard snow	12.0	720
Digging a trench in clay soil	8.5	510
Shoveling dirt	5.4–8.4	234–504
Pushing wheelbarrow with 220-lb. load	5.0	300
Climbing stairs	9.8–13.8	588–828
Working with an axe	6.9–24.1	414–1446
Felling a tree with a saw	10.7	642
Cutting logs into firewood	8.5	510
In the Home		
Sweeping floors	1.7	102
Scrubbing floors	3.4	204
Mopping	4.2	252
Picking up objects around home	5.9	354
Making beds	5.4	324
Cleaning windows	3.5	210
Washing dishes	1.2	72
Children's Recreation		
Sitting and playing games	1.0–1.4	60–84
Playing on school playground	1.6	96
Cycling	2.4–3.1	144–186

(Continued)

TABLE 5-1. *(Continued)*

	kcal/min	*kcal/hour*
Activities You Might Never Do!		
Walking across a ploughed field	7.6	456
Ploughing behind a horse	6.9	412
Carrying a 130-lb. load upstairs	30.7	1842
Drilling coal in a mine	7.2	432
Tending a furnace in a steel mill	10.2	612
Mowing wheat by hand	7.7	462
Milking a cow by hand	4.7	282
Trimming felled trees in a forest	10.0	600
Throwing hand grenades	3.8	228
Bayonet practice	4.0	240
Forced marching	9.7	582
Digging foxholes	4.6	276

Source: Modified from Passmore, R. and J. V. G. A. Durnin, "Human Energy Expenditure," *Physiological Reviews*, vol. 35, pp. 801–836, 1955.

Domestic chores often seem very tiring to those who must do them, yet in terms of caloric output only a few rate as even moderate exercise. Boredom may play as much a role in this fatigue as caloric expenditure.

The largest output of energy involves tasks that make use of numerous muscles (swimming, running, digging, or carrying a load), or that move the body against gravity, as in climbing stairs. But again, these figures are misleading in that some of the activities cannot be maintained for very long. Five kilocalories per minute is considered the maximum energy output for any sustained period of time, such as eight hours, and exercise of this intensity must be alternated with periods of rest. Few individuals could survive a regular work schedule even this vigorous, because their total energy outlay—including nonwork activities and basal metabolism—would amount to more than 4000 kcal per day.

Vigorous exertion causes individuals to breathe more rapidly and increases the rate of their heartbeat considerably. At one time, people were admonished not to "overexert" to avoid damaging their heart and blood vessels. But considerable evidence now indicates quite the opposite. Regular vigorous physical exertion for 15 to 30 minutes has many beneficial effects on the heart, blood vessels, and lungs of a normal, healthy individual. The American Heart Association now promotes exercise programs involving jogging, rope jumping, and bicycle riding. In their estimation, regular vigorous exercise would greatly decrease the high rate of heart attacks among Americans.

Many studies have been done on the total daily energy expenditures of people in various occupations. Thus an 18-year-old female college student averages around 2000 kcal per day, a clerk around 2400 kcal per day, and a construction worker on the Alaskan pipeline 4000 kcal per day. Tables have been constructed by the Food and Nutrition Board of the National Academy of Sciences—National Research Council showing the recommended caloric intake for healthy individuals at various ages, sizes, and sexes. A portion of such a table is shown in Table 5-2. The recommendations are somewhat on the generous side, to allow for a wide range of occupations. Obviously, extremely active people such as professional athletes have higher energy requirements, and extremely sedentary individuals have somewhat lower requirements than the table indicates. Even so, caloric intake represents perhaps the only area of nutrition in which an individual may personally make accurate estimates concerning his, or her own diet.

Basic Concepts

Energy expenditure (work) is measured in terms of the amount of heat produced, or the amount of oxygen consumed and carbon dioxide given off.

TABLE 5-2. Recommended Daily Dietary Allowances in Relation to Expenditure of Energy

	Years	*kcal*
Children	1–3	1300
	7–10	2400
Males	11–14	2800
	19–22	3000
	23–50	2700
	51+	2400
Females	11–14	2400
	19–22	2100
	23–50	2000
	51+	1800

Source: From Recommended Daily Dietary Allowances, Revised 1973. Food and Nutrition Board, National Academy of Sciences—National Research Council.

Energy expenditures and requirements in living organisms are customarily expressed in thermal units called kilocalories.

A calorimeter is used for measuring the amount of heat produced by an organism's body, and thus the amount of its energy expenditure.

A respirometer is used for measuring the amount of oxygen used and carbon dioxide given off by an organism's body, and thus the amount of its energy expenditure.

MEETING ENERGY NEEDS WITH FOOD SOURCES

As indicated several times in the chapter, an animal's energy needs are met by ingesting carbohydrates, fats, and proteins. The energy content of these nutrients, their various roles in the diet, and their relative abundance in various foodstuffs have been studied extensively by nutritionists and other scientists. By now much is known about specific foods and their importance in the diets of human beings and a number of other animals.

To determine the energy content of food, a useful research tool is the bomb calorimeter. It is a device in which a unit of food is burned, while the resultant release of energy is measured. From such measurements we learn that one gram of fat contains 9.45 kcal,

TABLE 5-3. Caloric Values for Some Foods

Apple, fresh, peeled, 2½″ diameter	66
Apple, baked with brown sugar	184
Bacon, 2 slices	98
Banana, one average	87
Beans, green, ½ cup boiled	16
Beans, lima, ½ cup	89
Steak, T-bone, 4 oz. broiled	539
Bread, 1 slice white	62
Butter, one tablespoon	100
Hershey chocolate bar, small	266
Cantaloupe, ½, 5″ diameter	58
Cheese, cheddar, 1 oz.	111
Chicken, broiled, 4 oz.	155
Corn on cob, 5″ ear, boiled	71
Egg, 1 large, boiled	81
Frankfurter, 1 average	151
Lettuce, 1 head	31
Milk, whole, 8 oz.	159
Orange juice, 8 oz.	117
Pork, roast, lean, 4 oz.	404
Tomato, 1 average	33

one gram of protein has 5.65 kcal, and one gram of carbohydrate releases 4.1 kcal. Recall, however, that living cells do not actually "burn" these substances, nor are the energy conversions in cells perfectly efficient. When corrections are made for these realities, the energy values actually obtained in the body are approximately 9 kcal for one gram of fat, and 4 kcal each for a gram of carbohydrate and a gram of protein.

With these values established, the caloric content of any other food material can be determined; indeed, calorie counts have been obtained for nearly all foods commonly eaten. Table 5-3 shows a portion of one such table. With such information you can easily calculate the amount of food necessary to sustain a human being in the various activities shown in Table 5-1.

An analysis of human eating habits indicates that the daily diet does not include equal amounts of carbohydrates, fats, and proteins. Some of the reasons are economic and some reflect personal preferences. For example, we find that carbohydrate (mostly in the form of starches) makes up from 43 percent to 73 percent of a person's diet. For most of the people of the world it averages around 65 percent. People at the bottom of the economic ladder, including many in our

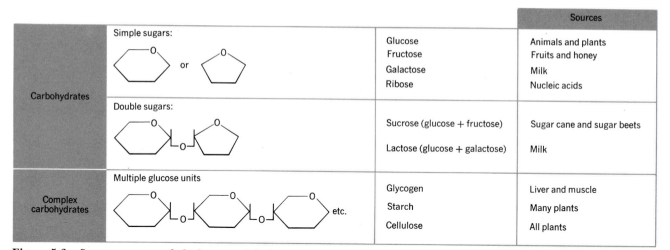

				Sources
Carbohydrates	Simple sugars:	Glucose	Animals and plants
		Fructose	Fruits and honey
		Galactose	Milk
		Ribose	Nucleic acids
	Double sugars:	Sucrose (glucose + fructose)	Sugar cane and sugar beets
		Lactose (glucose + galactose)	Milk
Complex carbohydrates	Multiple glucose units	Glycogen	Liver and muscle
		Starch	Many plants
		Cellulose	All plants

Figure 5-3   *Some common carbohydrates and their sources.*

own society, obtain most of their caloric intake from carbohydrates. Individuals in higher income groups eat markedly fewer carbohydrates and obtain more of their calories from fats. Protein intake also rises somewhat in the diets of these people.

Basic Concepts

The energy content of various foods may be determined by use of a bomb calorimeter.

The body obtains 9 kilocalories per gram of fat, 4 kcal/gram of carbohydrate and 4 kcal/gram of protein.

THE ENERGY FUELS

Carbohydrates

Carbohydrates are a class of chemical substances that includes sugars and larger compounds made up of linked sugar subunits: glycogen, starch, and cellulose (Figure 5-3). The smallest carbohydrates are *simple sugars* made of chains of from three to seven carbon atoms, attached to which are hydrogen atoms and hydroxyl (—OH) groups. In some of the sugars, such as glucose, the carbon chain forms a ring as shown in Figure 5-4.

Glucose is notable as the common sugar found in most animal bodies, where it is a major energy source. The energy in glucose (and other carbohydrates) lies in the chemical bonds that hold its atoms together, particularly those that link the hydrogen atoms to the carbon

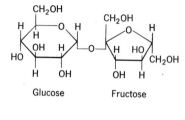

Figure 5-4   *Table sugar, sucrose, is composed of one molecule each of glucose and fructose. Cells can obtain and use the energy in the chemical bonds that link the hydrogen and carbon atoms.*

atoms (Figure 5-4). Cells can harvest and utilize this energy quite efficiently. Other simple sugars include fructose (from plants and honey), and galactose (from milk). These sugars are actually sweeter in taste than table sugar (sucrose), but they are not as easily obtainable in large amounts.

Sucrose is the most widely used sugar commercially because it is so abundant in sugar cane and in sugar beets. It is also the main sugar found in the juices of most other plants. Sucrose is called a *double sugar* because it consists of linked molecules of glucose and fructose (Figure 5-4). Other double sugars are lactose, found in milk, and maltose, found in many plants.

Very large and *complex carbohydrates*—glycogen, starch, and cellulose—are made of large numbers of simple sugar units linked to form long chains (Figure 5-5). Glycogen is an energy storage material in animals, mostly found in the liver and in muscle tissue; starches serve the same function in plants. Cellulose is a structural material used to strengthen plant cell walls. Though made up of glucose subunits, cellulose is not commonly used as an energy source because the glu-

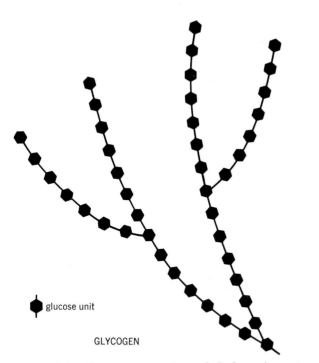

glucose unit

GLYCOGEN

Figure 5-5 *Glycogen, a complex carbohydrate, is composed of long chains of glucose molecules that branch at intervals. Why is glycogen considered a storage form of glucose?*

cose units are linked in a way that cannot be broken down by the digestive enzymes of most organisms. An interesting exception to this is the manner in which herbivores (plant eaters) utilize the cellulose in plant matter for food. In order to eat grass, which is largely cellulose, these animals have special adaptations, usually involving specialized bacteria and protozoans in their stomachs and intestines. These microorganisms degrade the cellulose into simpler forms of carbohydrates which the animal's own enzymes can then handle. Such animals as the cow have unusually large stomachs and long intestines to allow the bacteria to perform their helpful task.

Basic Concept

Carbohydrates consist of carbon chains to which are attached hydrogen atoms and hydroxyl groups. Cells efficiently utilize the energy found in these hydrogen bonds of carbohydrates.

Carbohydrates are the major energy source in a normal diet for several reasons. Being plant products,

they are the most abundant, most easily produced, and thus the least expensive of the major types of food. It is not surprising that high-carbohydrate crops such as wheat, corn, rice, sorghum, rye, barley, potatoes, sugar cane, sugar beets, and starchy roots form the bulk of the agricultural crops throughout the world.

Carbohydrate's predominance in the diet is related to several biological factors. First of all, the most commonly used and most available energy source in the body is the carbohydrate glucose. Its common designation as "blood sugar" indicates its ready availability to all cells of the body. A few parts of the body, including the brain and the red blood corpuscles, use glucose exclusively as an energy source because they cannot obtain or metabolize fats. Despite this dependency on glucose, however, animal bodies do not store glucose or other forms of carbohydrate to any appreciable degree. Approximately one day's energy supply is stored in the form of glycogen (animal starch) in the liver and in muscle tissue. Any dietary excess of carbohydrate is converted to fat and stored.

For these reasons, one might assume that a regular quantity of carbohydrate is necessary, which would in turn explain why it is such a large part of the diet. There are several drawbacks to this assumption, however. Many vertebrates are exclusively carnivorous. Laboratory animals and human beings alike have been maintained for long periods on diets that contained virtually no carbohydrate. Eskimos, in fact, normally live on such a diet. It is also known that a healthy human can survive starvation for about 80 days. In all of these instances, the body is obviously getting its energy from noncarbohydrate sources, namely, fat and protein. Fat cannot be converted into glucose. However, as we described earlier, fat can be taken from storage (or from the diet) and be digested into fatty acids and glycerol, and these components can enter the cellular respiratory cycle to provide energy efficiently.

In order to maintain the required glucose supply for the bloodstream and for the brain, substances such as amino acids, glycerol, and lactic acid can be converted into glucose by the liver. In a starved organism, amino acids are supplied by the breakdown of the animal's own proteins.

Thus, even though carbohydrate is the major energy source for people and for many other organisms, under normal dietary conditions alternative sources are available. For this reason, carbohydrate is not considered *essential* in the diet. But this does not decrease its *importance*. Carbohydrate is plentiful, cheap, and easily digested (usually). It "spares" protein, in that the body does not break down its own tissues as long as carbohydrate is present, and it provides bulk and fiber that aid in the movement of food through the intestines.

Basic Concept

Carbohydrate provides a valuable and major energy source in most diets because it is relatively cheap, abundant, easily metabolized, and aids in the movement of materials through the digestive tract.

Fats

Fats belong to a large, diverse group of substances, the *lipids,* that also includes cholesterol, phospholipids, and a variety of other substances that are important biologically. All cell membranes, for example, contain a phospholipid as a major structural element.

Fats are used for energy and as an energy storage material. Like carbohydrates, fats consist entirely of carbon, hydrogen, and oxygen. However, as you can

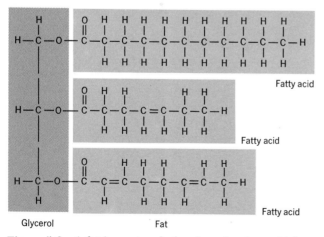

Glycerol   Fat

Figure 5-6 *A fat is composed of a glycerol unit to which are attached three fatty acid molecules. The fatty acid chains are much longer than shown here. The top fatty acid is saturated (no double bonds), the middle one is monounsaturated (one double bond), and the third one is polyunsaturated (two double bonds in this case).*

see in Figure 5-6 fats have a lot more hydrogen and much less oxygen, and consequently contain many more of the energy-rich carbon-hydrogen bonds. For this reason fats contain more energy per gram than do carbohydrates, over twice as much, as a matter of fact. Fats are a good energy storage material and secondary energy source when the normally abundant carbohydrates are not available.

Fats are made of two basic subunits, glycerol and fatty acids (Figure 5-6). *Glycerol* is a three-carbon relative of the sugars. A *fatty acid* consists of a chain of carbon atoms, usually 16 or 18, to which are attached hydrogen atoms and, at the end, a carboxyl group (—COOH). A glycerol molecule combines with three fatty acid molecules to comprise a fat (hence the more technical name, *triglyceride*). There are about 40 different known fatty acids, but only a handful are commonly found in most triglycerides. The most important aspect of fatty acids, in relation to energy, is that they provide a neat package, a string of hydrogens that the cell can use for energy whenever necessary.

Fatty acids are often categorized as *saturated* or *unsaturated,* depending on the number of hydrogen atoms that can be forced onto them chemically under laboratory conditions. Those that cannot take up any more hydrogen—mostly animal fats—are termed saturated. Those that can be hydrogenated—mostly from plants—are termed unsaturated or polyunsaturated (Figure 5-6). Animal fats with their long-chain fatty acids are usually solid (such as lard), whereas fats derived from plants (such as peanut oil) are often liquid at room temperature. Food manufacturers often hydrogenate plant oils in order to thicken them for use in products like peanut butter and margarine.

Basic Concept

Fats are composed of glycerol and fatty acids. They contain large amounts of energy by virtue of their high proportion of hydrogen atoms. Fats are often used as an energy source material.

We usually associate fats with meat products, and food composition tables tell us that certain cuts of beef, pork, and lamb are quite high in saturated triglyceride fats. Dairy products, especially butter, cheese, and ice cream are also rich in fats. Plant products are generally low, with the exception of nuts, chocolate, avocadoes, olives, potato chips (and other fried snack foods). Salad dressings, cooking oils and shortening are, of course, extremely high in fats.

Because the fats have such a high caloric value, it doesn't take much of these products to greatly increase a person's daily caloric intake. Snack food calories add up more quickly than we would like: 10 potato chips have 115 kcal, a handful of shelled peanuts is 400 kcal, a one-inch cube of cheddar cheese is 70 kcal, and two ounces of chocolate bar are 300 kcal. This adds up to 885 kilocalories—the equivalent of a full meal—and yet easily consumed between meals.

In addition to providing energy, fats have several other functions in the body. The fat-soluble vitamins, A, D, E, and K, are dissolved in fats, as their name implies. Good sources of these vitamins have high oil or fat content, and the vitamins are stored in the body's fatty tissues. In the diet, fats cause food to remain longer in the stomach, thus increasing the feeling of fullness for some time after a meal is eaten. Fats add variety, taste, and texture to foods, which accounts for the popularity of fried foods. Fatty deposits in the body have an insulating and protective value. The curves of the human female body are due mostly to strategically located fat deposits.

Whether a certain amount of fat in the diet is essential to human health is not definitely known. When rats are fed a fat-free diet, their growth eventually ceases, their skin becomes inflamed and scaly, and their reproductive systems are damaged. Two fatty acids, linoleic and arachnidonic acids, prevent these abnormalities and hence are called *essential fatty acids.* They also are required by a number of other animals, but their roles in human beings are debatable. Most nutritionists consider linoleic fatty acid an essential nutrient for humans.

In recent years evidence has accumulated that the polyunsaturated fatty acids function in protecting humans and some laboratory animals from diseases of the arteries and heart such as atherosclerosis. In this disease, small patches of fatty material, composed mostly of *cholesterol,* form on the inside lining of the arteries. As the deposits increase in thickness, they may cut down on the blood flow to the organs supplied by the arteries until the structures are severely damaged (Figure 5-7). If this occurs in a branch of the coronary artery supplying heart muscle, that portion of the muscle dies and the person experiences a painful and sometimes fatal heart attack. Another danger of *atherosclerosis* is that pieces of the fatty deposits may break free and travel in the bloodstream until they lodge in small vessels and block the flow of blood. This blockage may also cause heart damage, or, if it occurs in the brain, may damage brain cells and lead to a stroke.

Atherosclerosis is a major disease in our culture, because more than one half of the deaths each year in this country are related to diseases of the heart and blood vessels. Consequently, many studies have been conducted to determine the causal factors that are involved in it. One of the major factors appears to be the amount and source of triglycerides in the diet. Large amounts of animal fats, including cholesterol, cause a rise in blood cholesterol level. This in turn greatly increases the risk of coronary heart disease.

The role of cholesterol in atherosclerosis was controversial for a long time. Cholesterol is a common metabolic material made in the body, where it forms an important part of the bile salts that aid the breakdown of fats during digestion in the small intestine. Cholesterol is not an essential nutrient, because the body makes all that it needs. It seems contradictory that such a common and useful metabolite would influence the onset of a serious disease such as atherosclerosis. Nevertheless, two comprehensive studies involving numerous human subjects over a long time period (the Framingham Study and the Western Electric Study) provide strong evidence that such is the case.

The risk of *coronary heart disease* appears to be significantly less for individuals who decrease their dietary fat intake and use polyunsaturated fats when possible. This dietary pattern may be attained by decreasing the intake of whole dairy products and of meats such as pork, lamb, and beef while increasing the intake of poultry, fish, and plant oils.

Obesity. Nutritionists have often lamented the irony of our society's concern about eating too much food while so many people in the world worry about not having enough to eat. Even as the world's major exporter of food (chiefly in the form of grains), the United States still has enough left over to more than adequately feed its own people. Indeed, it has enough to permit many to become obese.

Obesity is defined rather simply and arbitrarily as a

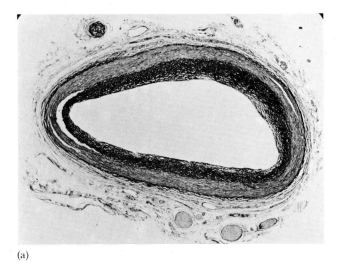

(a)

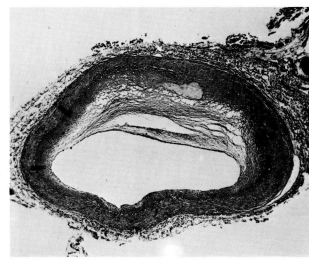

(b)

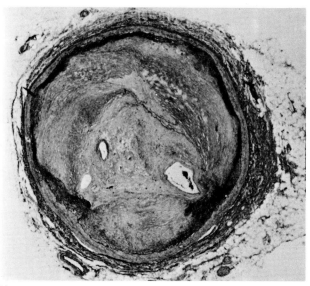

(c)

Figure 5-7 *Cross sections of arteries showing the development of atherosclerosis. Part* (a) *shows the appearance of a normal artery. In* (b) *fatty deposits of cholesterol are forming in the inner lining of the artery. In* (c) *the artery is almost totally filled with deposits.*

blame their problem on metabolic disorders, and a few are genetically inclined toward obesity, but most of us just fall into excessive eating and drinking patterns together with lessened physical activity, and consequently we experience a creeping weight gain. As one observer noted, most cases of overweight can be traced to sins of the fodder!

Despite the popularity of new kinds of diets that appear in magazines, and the increasing numbers of health spas and diet clubs, people seem to keep right on having weight problems. The ultimate (and only lasting) remedy is for an overweight person to find a daily style of eating that keeps the kcal intake within desired ranges.

Basic Concepts

Fats are high in calories, form an efficient energy storage material, provide a medium for the fat-soluble vitamins, add taste and variety to foods, help insulate and protect the body, and have a satiety value. Linoleic fatty acid is an essential nutrient for people.

Polyunsaturated fatty acids have a major role in protecting against diseases of the arteries and heart.

condition in which an individual is more that 20 percent heavier than is recommended for his or her weight, sex, and body build. The recommended values can be found in tables published by life insurance companies and medical groups. An example is Table 5-4. Even if you are only 10 or 15 percent over the recommended value ("overweight"), abundant statistics from health fields indicate that this condition is a threat to your health and longevity.

Like the definition, the basic cause of obesity is also simple: More kcals are taken in than the body is utilizing. Why individuals engage in this behavior, however, is not simple at all. A small percentage of fat people can

TABLE 5-4.   What the Longest-Lived People Weigh:

	Height	Sm	Med	Lrg
	5′2″	128–134	131–141	138–150
	5′3″	130–136	133–143	140–153
	5′4″	132–138	135–145	142–156
	5′5″	134–140	137–148	144–160
	5′6″	136–142	139–151	146–164
	5′7″	138–145	142–154	149–168
	5′8″	140–148	145–157	152–172
Men	5′9″	142–151	148–160	155–176
	5′10″	144–154	151–163	158–180
	5′11″	146–157	154–166	161–184
	6′0″	149–160	157–170	164–188
	6′1″	152–164	160–174	168–192
	6′2″	155–168	164–178	172–197
	6′3″	158–172	167–182	176–202
	6′4″	162–176	171–187	181–207
	(Weight at ages 25–59 in shoes and 5 pounds indoor clothing)			
	4′10″	102–111	109–121	118–131
	4′11″	103–113	111–123	120–134
	5′0″	104–115	113–126	122–137
	5′1″	106–118	115–129	125–140
	5′2″	108–121	118–132	128–143
	5′3″	111–124	121–135	131–147
	5′4″	114–127	124–138	134–151
	5′5″	117–130	127–141	137–155
Women	5′6″	120–133	130–144	140–159
	5′7″	123–136	133–147	143–163
	5′8″	126–139	136–150	146–167
	5′9″	129–142	139–153	149–170
	5′10″	132–145	142–156	152–173
	5′11″	135–148	145–159	155–176
	6′0″	138–151	148–162	158–179
	(Weight at ages 29–59 in shoes and 3 pounds indoor clothing)			

These are Metropolitan Life's 1983 weight tables by height and by small, medium or large frame.

THE ENERGY MACHINE

ATP—the Energy Currency of Life. The energy that a cell calls upon when performing its work is almost entirely one kind: the chemical energy of a phosphate group belonging to a larger compound, adenosine triphosphate (ATP). As its name implies, ATP contains three phosphate groups, but only the last one is customarily involved in energy transfers. When energy is needed, the bond of the terminal phosphate group is broken and a considerable amount of energy becomes available for use in the cell (Figure 5-8a). This energy transfer is usually brought about by attaching the released phosphate to another substance. If the substance happens to be glucose, the energy that bound the phosphate to the ATP molecule is now transferred to the glucose-phosphate complex. This energized, or *phosphorylated*, compound is then able to

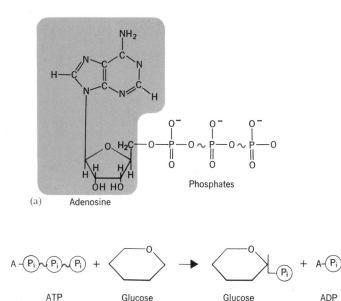

(a) Adenosine

Figure 5-8 *(a) The structure of adenosine triphosphate showing the high energy phosphate bonds (red squiggly lines). In (b) the terminal high energy phosphate is transferred to a glucose molecule, thereby energizing it to take part in additional chemical reactions.*

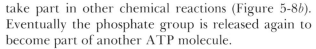

(b)

take part in other chemical reactions (Figure 5-8*b*). Eventually the phosphate group is released again to become part of another ATP molecule.

When ATP releases a phosphate, as described above, it becomes ADP (adenosine diphosphate). Obviously, cells must continually renew their ATP supply. They do so whenever energy becomes available to reconstitute the high energy bond between a free phosphate group and an ADP molecule. The course of energy for this reconstitution process is derived from the chemical events of cellular respiration.

Cellular Respiration

The series of chemical reactions known as cellular respiration begins in the cytoplasm of cells and is completed inside their mitochondria. The raw materials that can be used in the chemical "machinery" of cell respiration are simple sugars from the breakdown of carbohydrates, fatty acids and glycerol from fats, and amino acids from proteins.

Glucose ($C_6H_{12}O_6$) is the commonly used raw material shown in the conventional equation for respiration:

$$\text{glucose} + \text{oxygen} \rightarrow \text{ATP} + \text{carbon dioxide} + \text{water}$$

or

$$C_6H_{12}O_6 + 6O_2 \rightarrow \text{ATP} + 6CO_2 + 6H_2O$$

As a summary the equation is misleading because glucose is not actually combined with oxygen and there are numerous intermediate reactions before ATP, carbon dioxide, and water are produced. The correct interpretation is that the process starts with glucose and produces ATP energy. It also indicates that glucose is broken down to carbon dioxide and that oxygen is used as a hydrogen acceptor to produce water.

The chemical machinery in the above process is complex and still not fully worked out. The following very general description is intended to provide a glimpse of the awesome complexity of chemical events that occur mostly within the confines of tiny cellular organelles.

Prior to entering a mitochondrion, glucose is converted into small three-carbon molecules of *pyruvate.* Part of the energy from this "conversion reaction" is immediately used to reconstitute a few ATP molecules (Figure 5-9). These pyruvate molecules then enter a mitochondrion and undergo another change to become two-carbon acetate molecules. Then the fun really begins! The acetate now enters a series of reactions called the *Kreb's cycle* (a cycle means a circle; a set

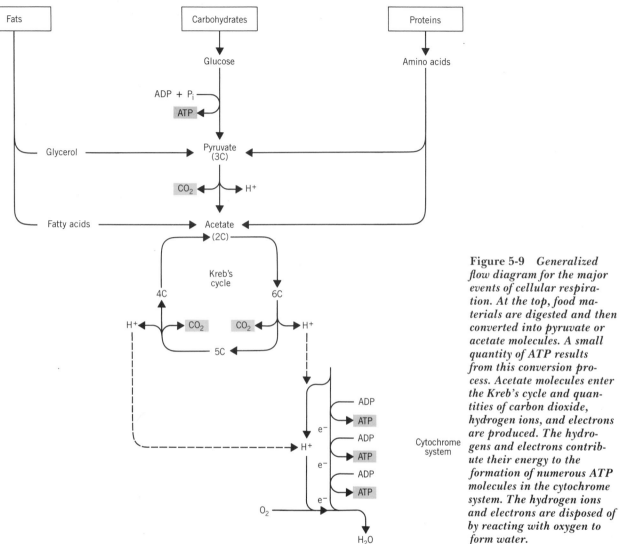

Figure 5-9 *Generalized flow diagram for the major events of cellular respiration. At the top, food materials are digested and then converted into pyruvate or acetate molecules. A small quantity of ATP results from this conversion process. Acetate molecules enter the Kreb's cycle and quantities of carbon dioxide, hydrogen ions, and electrons are produced. The hydrogens and electrons contribute their energy to the formation of numerous ATP molecules in the cytochrome system. The hydrogen ions and electrons are disposed of by reacting with oxygen to form water.*

of continuous events). In this cycle, acetate is combined with a four-carbon molecule to form the six-carbon citrate from which the cycle gets another of its names, the *citric acid cycle*. By the successive removal of carbon dioxide, the citrate is first changed to a five-carbon molecule and then to a four-carbon molecule. This, together with other reactions such as the addition of water and the transfer of hydrogen atoms to carriers, results in the reconstitution of the original four-carbon molecule. This reconstitution completes the cycle. The four-carbon molecule may now com-

bine with another acetate molecule, and the extensive series occurs again.

In many of the reactions to this point, carbon dioxide has been produced and hydrogen atoms have been transferred to special hydrogen-carrier molecules. The carbon dioxide is an end product of the reactions and diffuses from the cell as a waste product. The electrons from the hydrogen atoms, however, are transferred successively through a series of complex organic molecules (the cytochromes) to oxygen. This series of transfers constitutes the *cytochrome system*. At the same time that oxygen receives the electrons, it also picks up hydrogen ions from the cytoplasmic sap and water is formed.

It is important for the cytochrome system to release energy in small amounts rather than in one large amount. By doing this the cell is able to harvest a considerable amount of the available energy in the form of ATP. Students are sometimes surprised to learn that the primary function of oxygen in cells is as a hydrogen and electron acceptor — sort of a chemical trash can. Oxygen is never used to "burn" carbohydrates (or anything else) in the body as textbooks once commonly stated. Burning (combustion) is a form of oxidation, but it releases energy much too violently and in the wrong form (heat) to be used by cells.

As shown in Figure 5-9, lipids and proteins may also be used in the cell's energy machinery. Lipids are important in cellular respiration because they are high in energy content, easily stored, and easily adaptable to take part in the cycle. First they must be digested into their component fatty acids and glycerol. Most fatty acids are long chains of carbon atoms with some oxygen and a great number of hydrogen atoms attached. These chains are broken into two and three carbon fragments and fed into the respiratory cycle as acetate or pyruvate. Glycerol is a three-carbon compound which also enters the cycle after being modified slightly.

In order for proteins to participate in respiration, they must be digested into amino acids, which in turn are stripped of their amino groups and converted into fatty acids. These in turn enter the cycle as described above.

Respiration in Muscle Tissue

The events described above occur in most cells in the body, providing the energy required for all of their normal activities. At times, however, tissues such as active muscles use up all available oxygen in their surroundings. When this happens, additional methods of supply must be used. One solution to this problem (at least temporarily) is the use of *myoglobin*, a substance found in many muscles. Myoglobin is a protein closely related to hemoglobin, the oxygen-carrying chemical in red blood cells. Like hemoglobin, it contains loosely held oxygen. During peak activity, a muscle can obtain

extra oxygen for its respiratory reaction from myoglobin.

What happens when even this supply is depleted? Then a muscle may resort to still another adaptation, the use of a chemical "trashcan" other than oxygen. This is done by using the small pyruvate molecules, which in an emergency can become the muscle cell's temporary hydrogen and electron acceptors. As a result, the pyruvate is converted into another substance called lactic acid. In this way the muscle continues to obtain a supply of ATP energy and continues to function. Because oxygen is not required for these reactions, this sequence is termed *anaerobic* ("without air") *respiration*. Its drawbacks are that it is not nearly as efficient in producing ATPs as is normal respiration, and the end product, lactic acid, must be disposed of fairly rapidly in order to prevent muscle fatigue and soreness. As a sidelight on the efficiency of this whole operation, lactic acid can itself enter the respiratory mill of chemical reactions and make more ATPs.

One cannot help admiring the adaptability of muscle tissue in meeting its energy demands. Studies and experiments have shown that active, frequently used muscles are more efficient in their energy use and production than unexercised muscles. Part of this improved efficiency is related to improved blood circulation in active muscles, thus assuring a better oxygen supply for them and a more efficient disposal of lactic acid. This is one reason for encouraging people to participate regularly in vigorous physical fitness activities (Figure 5-10).

Basic Concepts

Adenosine triphosphate (ATP) provides the chemical energy that cells use in performing work. Cellular respiration refers to the series of chemical reactions that derives energy from short carbon chain nutrients for making ATP molecules.

SOME DIETARY MISCONCEPTIONS ABOUT FOOD AND ENERGY

In the previous chapter we noted a few of the misconceptions and erroneous beliefs concerning food and nutrition. You might recall that "erroneous" means that there is no verifiable evidence to support the

Figure 5-10 *The jogger and his spectators in the background provide a contrast in physical fitness.*

"A DAY? I THOUGHT YOU SAID 4,000 CALORIES A MEAL!"

belief. Some additional misconceptions related to the topics in this chapter are taken up here.

1. The best way to lose weight is to engage in a program of regular exercise. The error underlying this statement is related to the number of kcal involved in various physical activities. For example, an hour of vigorous tennis playing expends only 426 kcal—and this assumes that you are capable of a continuous hour of activity (see Table 4-1). This is a fairly insignificant amount compared with the 3500 kcal you must expend in order to lose one pound of body fat. Perhaps we should correct the statement to read as follows: "The best way to lose weight is to restrict the number of kcal each day to fewer than the body is using. A program of regular exercise aids this endeavor." In fact, experiments on laboratory animals and human volunteers indicate that moderate exercise also decreases the desire to eat—exercised organisms tend to eat less than inactive ones do. In addition, of course, there are a number of other health benefits in being physically fit.

2. Most people who reduce their carbohydrate intake will also reduce their caloric intake. As we saw earlier in the chapter, carbohydrates release 4 kcal per gram when broken down in the body for energy. This is the same rate as for proteins but far less than that for fats. The caloric values of some typical carbohydrate foods are not really very high. A medium-sized potato, for example, has only about 100 kcal, a slice of bread about 80, a serving of rice about 90, and so on. The kcal problem is not so much with carbohydrates themselves as with the various fats such as butter that we put on them, or the oils in which we fry them. For example, the 80-kcal slice of bread with a tablespoon of margarine becomes 180 kcal; a 100-kcal potato with sour cream or butter becomes twice as rich; a slice of cake with its butter, egg yolks, and concentrated carbohydrate in the form of sucrose adds up to 200 to 300 kcal. Therefore, it is not the carbohydrate per se but rather its method of preparation that increases the kilocalories. Recall also that carbohydrate is important in body metabolism for maintaining blood sugar, for energy production in the nervous system, and so on. Carbohydrates should not be completely avoided in an individual's diet; instead, one should be cautious about their preparation.

3. Table sugar is bad for your health because all of the nutrients have been refined out of it. You should use natural sugars such as honey instead. Table sugar is sucrose from sugar cane and sugar beets. No important nutrients have been refined out (except perhaps chromium); rather, it is merely a more concentrated nutrient than it was in the natural state. Honey is a mixture of glucose, fructose, and a smattering of minerals. It is actually higher in kcal, at 64 per tablespoon, than is sucrose, at 46 per tablespoon. Honey is thus more fattening and offers no real nutritional advantage over table sugar. All of this is not to deny that many Americans, especially children, use too much sugar in their daily diets. Sucrose is related to tooth decay and perhaps other diseases.

4. Candy bars have no nutritional value. In reality, quite the contrary is true — all candy is high in kilocalories. Depending on the kind of candy, it may also contain quite an array of nutrients. For example, two ounces of sweet chocolate such as a Hershey bar contain about 300 kcal, 2.5 grams of protein, 20 grams of fat, 33 grams of carbohydrate, 53 milligrams of calcium, 80 milligrams of phosphorus, 1.5 milligrams of iron, 19 milligrams of sodium, 152 milligrams of potassium, and small amounts of several vitamins. A chocolate bar containing peanuts has even higher values for these nutrients. This fact in no way makes candy a suitable substitute for vegetables and cereals, but neither is it absolutely nutritionally empty, as it is sometimes depicted. It becomes a highly useful and convenient food for backpackers, mountain climbers, and other individuals who need concentrated carbohydrate to balance their energy expenditures.

5. High-protein food contain no (or few) calories. As we have already seen, proteins have approximately the same caloric value as carbohydrates, and the body will use proteins for energy if fats and carbohydrates are not available. Unfortunately, many people believe that meats are low in calories (or have none at all), and hence they are led to try fad diets which claim that you can eat all the meat you desire if you just omit carbohydrates from your diet at the same time. Such a diet can lead to weight reduction *if* the individual restricts his or her meats to fish and poultry — but even then, only if the food is broiled or baked without sauces or gravies. Even on an exclusive diet of beef (1200 kcal per pound of hamburger), or lamb (1200 – 1600 kcal/pound of roast), or baked ham (1600 kcal/pound), a weight loss may occur eventually because of a lowered appetite and hence a reduction in caloric intake. However, this kind of diet could lead to a deficiency in several nutrients. As we learned in the last chapter, protein is a necessary dietary nutrient for several reasons, but not in large daily quantities. And because protein is expensive, it is foolish to use it for energy purposes in a diet when less expensive carbohydrates will do just as well.

6. Fats have no beneficial role in nutrition and should always be avoided in the diet. As explained, fats have a variety of important functions in the body. Also they are a part of so many foods that it would be impossible to keep them out of the diet anyway. Fatty foods do have two drawbacks: high caloric value, and a possible relation to coronary heart disease (at least for animal fats). Many Americans, especially males, would evidently improve their health by lowering their intake of fats, especially fried foods and fatty beef, lamb, and pork.

KEY TERMS

anaerobic respiration	cholesterol	fat	obesity
atherosclerosis	citric acid cycle	fatty acid	pyruvate
ATP	complex carbohydrate	glycerol	respirometer
basal metabolism	coronary heart disease	kilocalorie	saturated fatty acid
calorimeter	cytochrome system	Kreb's cycle	simple sugar
carbohydrate	double sugar	lipid	triglyceride
cellular respiration	energy	myoglobin	unsaturated fatty acid

SELF-TEST QUESTIONS

1. Name the three basic energy sources in foods.
2. (a) Define energy.
   (b) List four forms of it found in living matter.
3. What are three physiological events that produce heat energy in the body?
4. (a) Define kilocalorie.
   (b) Describe two ways of measuring the kilocalories produced by a human being.
5. What is a bomb calorimeter used for?
6. What are the kilocalorie energy values for fat, carbohydrate, and protein?
7. With reference to carbohydrates:
   (a) Describe their general chemical composition.
   (b) Name three sugars.
   (c) List three large, complex forms.
8. Why do carbohydrates provide such a valuable and important energy source in the diet?
9. Why is carbohydrate not classed as an essential nutrient?
10. Describe the basic chemical makeup of a fat.
11. Why do fats contain twice as much energy per gram as do carbohydrates?
12. Define the terms *saturated* and *unsaturated* fats.
13. Give several basic functions of fats in the body.
14. What appears to be the role of fatty acids and cholesterol in atherosclerosis?
15. Define *obesity* and give its major cause.
16. Why is ATP called the "energy currency of life"?
17. What is the relation of cell respiration to ATP?
18. Describe in your own words what happens to glucose in cell respiration.
19. What is the function of oxygen in cell respiration?
20. Tell two ways that muscle tissue obtains energy when its oxygen supplies are limited.
21. In what sense is exercise not the best way to lose weight?
22. What is erroneous about the idea of decreasing only the carbohydrates in a diet when an individual wishes to lose weight?
23. In what ways could a candy bar be considered a highly nutritious food?

Some examples of African herbivores.

FEEDING ADAPTATIONS AND DIGESTION

chapter 6

questions to think about

What is the digestive reaction?

Why are human beings considered to be omnivores?

What happens to food as it passes through the human digestive tract?

What is meant by filter feeding?

Why do herbivores need special digestive adaptations?

How does the structure of an animal reflect its feeding method?

An ingenious assembly of portable plumbing

Christopher Morley

If one looks at the great diversity of feeding mechanisms used by animals, and at the incredible variety of foods they ingest, one is almost forced to conclude that nature is determined not to let a single particle of organic matter go to waste. We commonly think of animals as being *carnivores* (flesh eaters), *herbivores* (plant eaters), or *omnivores* (generalized feeders). In actuality, there are many additional variations of these basic types. For example, many kinds of *fluid feeders* specialize in sucking juice from animals or from plants (female mosquitos and plant aphids are examples). Marine and fresh-water habitats abound with *filter feeders* adapted for obtaining stray bits of organic matter and microorganisms that drift into their surroundings; barnacles and bivalves such as oysters or clams are examples. Even the tiny morsels of organic material that sink into the sand and mud are strained out by *deposit feeders* such as earthworms and many marine worms. Should any parcel of food escape all of these organisms, it will become an energy supply for the microscopic decomposers, bacteria and fungi.

THE DIGESTIVE REACTION

Despite the great variety of foods that are ingested in one way or another, their fate inside the digestive tract is similar in all organisms. Proteins are broken down (digested) into short peptide chains and amino acids, carbohydrates into simple sugars, and fats into glycerol and fatty acids. Only in these forms can these nutrients pass through the gut wall and be distributed about the body. The term *digestion* refers to a specific type of chemical reaction called *hydrolysis* ("water-splitting") in which large nutrients are split into smaller units and water molecules are attached to the broken bonds. Specific enzymes, produced by the digestive system, are required for each different digestive reaction (Figure 6-1). Small molecules such as minerals, vitamins, and water pass into the gut wall without going through the digestive reaction.

Basic Concepts

Animals are adapted to utilize a wide variety of organic matter for food.

Digestion is a hydrolysis reaction and is essentially the same in all organisms.

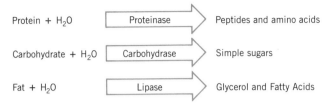

Figure 6-1 *A generalized summary of digestive reactions. Enzymes are denoted by the suffix -ase. Thus, proteinase designates an enzyme that digests proteins.*

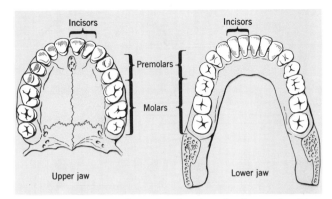

Figure 6-2 *Human dentition showing the sharp-edged incisors at the front of the jaws and the crushing-type premolar and molar teeth at the back. The form of these teeth is an adaptation for a generalized diet.*

THE HUMAN DIGESTIVE SYSTEM

Human beings are classed as omnivores because of their generalized diet. Out teeth also reflect this kind of diet, as indicated by the crushing molars and premolars in the back of the jaws, and the sharp-edged incisors at the front (Figure 6-2). Consequently, we can handle a varied diet of grains, fruits, plant leaves and stems, and meat. Indeed, our bodies demand the intake of a broadly based diet in order to function properly. This same sort of generalized pattern typifies the human digestive tract. The stomach is moderate in size, followed by a rather long small intestine and a relatively short large intestine. Carnivores have a similar gut, although it is somewhat shorter to accommodate the rather rapid digestion of proteins.

Mouth

The digestive process begins, at least in a minor way, in the *mouth,* as food is mixed with saliva from three pairs of salivary glands. Saliva contains an enzyme called *amylase* that begins the breakdown of starches to a slight extent. Like all digestive reactions, this is a hydrolysis reaction, in which water molecules are attached to the broken bonds of the material being digested.

Saliva also has a slight bactericidal action, but its major function is to moisten food so that it can more easily be chewed and swallowed. With the aid of the tongue, masses of food are pushed to the back of the mouth (*pharynx*) in preparation for being swallowed (Figure 6-3). Swallowing is a rather complex action, because it begins as a voluntary muscular activity, involves a reflexive closing off of the entrance into the voice box (*larynx*) by the *epiglottis,* and continues into

the *esophagus* partly by involuntary muscle action (Figure 6-4). This crossing of the food passageway and the air passageway is a potential disaster area, as you may have experienced by accidentally inhaling a piece of food. Each year a number of people die of suffocation after getting a large piece of food, usually meat, stuck in the larynx or upper part of the windpipe.

After reaching the esophagus, the food mass is pushed on down to the stomach by waves of muscle contraction called *peristalsis.* The 10-inch-long esophagus is normally collapsed and flattened except when material passes through it. No digestive events occur in it.

Basic Concept

The major digestive function of the mouth is to mix food with saliva and to render food into small portions.

Stomach

Even before food passes down into the *stomach* (Figure 6-3), digestive activities start to take place in the stomach wall. The anticipation of eating, or the smell or taste of food, starts the release of small amounts of *gastric juice.* Then food entering the stomach triggers the release of a hormone, *gastrin,* into the bloodstream from special cells in the stomach wall. *Gastrin,* in turn, stimulates several types of glands in the stomach wall to produce a copious flow of gastric juices. These juices include a slimy mucus material (*mucin*), *hydrochloric acid*

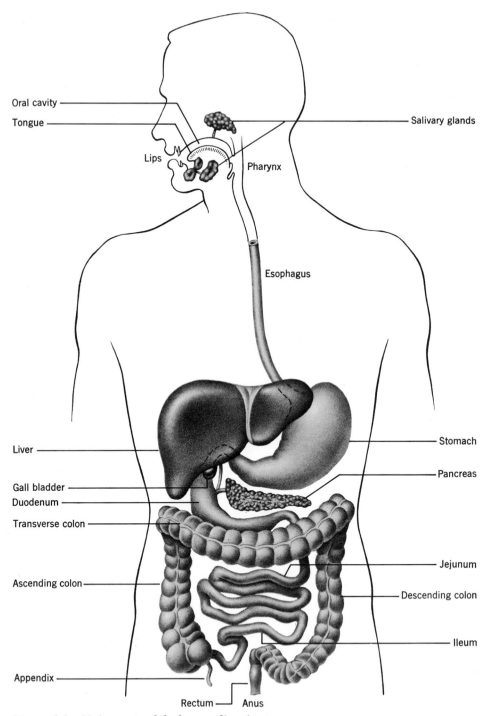

Figure 6-3   *Major parts of the human digestive tract.*

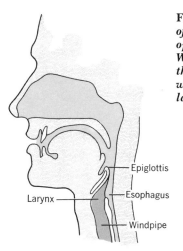

Figure 6-4 *Relationship of the esophagus and the opening into the larynx. When food is swallowed, the epiglottis moves downward and closes off the larynx.*

Epiglottis

Larynx — Esophagus

Windpipe

Figure 6-5 *A widely used antacid patent medicine. The most abundant ingredient, according to the label, is sodium bicarbonate.*

(HCl), and *pepsinogen.* The hydrochloric acid converts the pepsinogen into the enzyme *pepsin,* which then begins the digestive breakdown of large protein molecules into smaller fragments. The completion of protein digestion occurs later in the small intestine. Although we are never consciously aware of it, the muscular stomach wall thoroughly churns the mixture of food and digestive juices. In addition to starting protein digestion, a second major function of the stomach is to allow an organism to take in a sizable quantity of food during one feeding so that it does not have to eat continuously during its waking hours. Some of us, unfortunately, forget or ignore this nice anatomical adaption and persist in providing a continuous flow of food (called snacks) into the system!

The stomach contents become extremely acid during these digestive events, but mucin usually coats the walls in a protective manner. Because stomach acidity is a normal and necessary event, people should be cautious about ingesting various antacid patent medicines (such as Alka-Seltzer and Tums) for digestive upsets — TV commercials to the contrary. Recurrent indigestion may be symptomatic of a serious disorder such as heart disease, for which an antacid is the wrong treatment. For minor upsets due to overeating, drinking too much alcohol, or stress, ordinary baking soda (sodium bicarbonate) is as effective as the commercial antacids. In fact, sodium bicarbonate is the major ingredient in some of the widely promoted antacid remedies (Figure 6-5).

After being thoroughly mixed in the stomach for 3 to 5 hours, small amounts of the now semiliquid contents (*chyme*) are ready to pass by small squirts into the beginning of the small intestine, the *duodenum* (Figure 6-3). Except for water, sodium, and potassium, very few substances are absorbed through the wall of the stomach. A special case is alcohol, which passes quickly through the stomach walls unless the drinker has previously consumed some fatty materials such as milk, peanuts, or potato chips. Such substances delay the absorption of much of the alcohol until it reaches the small intestine, and the individual doesn't get intoxicated in such a rush.

Basic Concept

The stomach holds quantities of food, starts protein digestion, and creates a semiliquid chyme for passage to the duodenum.

Small Intestine

The squirting like passage of chyme into the duodenum, the first division of the small intestine, is partially regulated by the periodic relaxing of a sphincter muscle that separates it from the stomach. A sphincter is a circular mass of muscle that functions somewhat like a drawstring in closing off tubelike organs in the body.

Chyme is highly acidic when it enters the small intestine, but it is soon neutralized and then made alkaline as a consequence of juices added to it from the pancreas, gall bladder, and wall of the small intestine. An alkaline medium is required by the intestinal enzymes. As was true in the stomach, the secretion of these juices is controlled by several hormones produced in the wall of the intestine. The pancreatic secretions contain digestive enzymes that continue the breakdown of proteins (proteases) and carbohydrates (carbohydrases), and that split fats into fatty acids and glycerol (lipases). *Bile,* from the liver and gall bladder, aids the digestive process by separating fatty materials into extremely tiny particles that can be efficiently attacked by the lipases. These substances are secreted into the duodenum. The remaining portions of the small intestine (*jejunum and ileum,* in turn) also contain glandular cells in their walls that contribute proteases to complete the digestion of proteins into amino acids, and carbohydrases that convert carbohydrates into simple sugars.

After all of these digestive events are completed, the products are relatively small, absorbable molecules that can pass through the intestinal cell membranes. As an adaptation to increase the absorbing surface area of the small intestine, its inner surface contains numerous folds, and the folds themselves contain a mat of tiny projections called *villi* (Figure 6-6*a*). The villus is the basic absorbing unit of the intestinal wall. Each of the tens of thousands of villi contains a network of capillaries, tiny blood vessels, and a tiny lymph tubule called a *lacteal* (Figure 6-6*b*). Amino acids, simple sugars, short-chain fatty acids, water and water-soluble vitamins, and some minerals pass through the epithelial cells of the villi into the capillary beds. Diffusion and active transport are both involved in the absorption process. Glycerol, longer fatty acids (recombined as lipids in the villus), and the fat-soluble vitamins enter the lacteals.

The capillary network of each villus drains its contents into tiny veins that join and eventually form a major drainage system to the liver. This drainage is termed the *hepatic portal system.* The fact that blood passes directly from the intestines to the liver is very significant because it allows the liver to make various changes in some of the newly absorbed nutrients before they are distributed to the rest of the body. As a

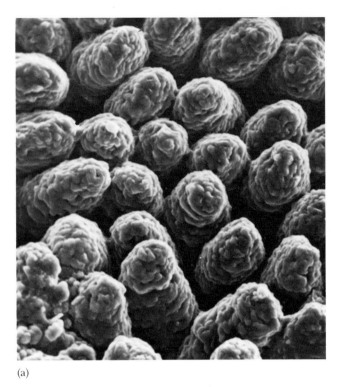

(a)

(b)   Intestinal glands

Figure 6-6 *Tiny projections, villi, increase the digestive and absorbing area of the small intestine. In (a) villi are shown in a scanning electron micrograph; (b) is a drawing of a section of a villus to show its epithelial cells, capillaries, and lymph tubule (lacteal).*

result, some of the glucose may immediately be converted into the storage molecule, glycogen; some of the amino acids may be converted into glucose; and the fat-soluble vitamins are stored in the liver for the time being. The lacteals drain into collecting vessels that in turn form larger ducts. Their contents finally empty into large veins near the heart.

In short, the small intestine is the major digestive and absorptive area of the digestive tract. It is about 12 feet long and, as mentioned previously, has a very large internal surface area. As its contents are being digested, they are mixed and churned thoroughly by muscular action. This action also results in materials being held in the intestine long enough for efficient and thorough absorption to take place. Consequently, material that is left in the small intestine after all of the digestive events have taken place consists mostly of plant fibers, other indigestible substances, water, and some minerals.

Basic Concept

The small intestine is the site of all major digestive reactions, using bile and enzymes. It is also the major absorption site for the breakdown products, for vitamins, and for most minerals.

Large Intestine

At the junction of the small intestine and the large intestine (*colon*) there is a sphincter, the ileocecal valve that periodically lets portions of the small intestine contents enter the colon (Figure 6-3). Here, bacterial action takes over and the contents are converted into fecal matter as they move through the colon. A considerable amount of water is absorbed along with some minerals, especially sodium. The vigorous bacterial growth in the colon results in the production of some B vitamins and vitamin K, which may also be absorbed. After traversing the 5-foot length of the large intestine, the feces enter the rectum and are held there by the anal sphincter muscles until defecation occurs.

It has been said that Americans are obsessed with bowel movements. Judging from the frequent TV commercials about "irregularity" and the more than 700 available laxative products, there must be considerable truth to the statement. Much of this excessive use of laxatives seems to be based on widespread but erroneous notions that a daily bowel movement is necessary for good health, or that one should purge the bowels at regular intervals. Neither of these ideas has any factual basis in human physiology. It is normal for some individuals to have daily bowel movements and for others to go several days between movements. Moreover, regular use of laxatives hinders normal bowel activity and often leads to even longer intervals between movements. A far better solution to the constipation problem, if one actually exists, is to increase the daily intake of high-fiber foods such as fresh fruits, vegetables, and cereals.

Many individuals experience the opposite problem —sporadic loose bowels, abdominal cramps, and excess gas. This condition, sometimes termed the "irritable colon syndrome," may be the result of emotional tension and stress, or it can result from an intolerance to specific foods such as milk. Many adults lack the digestive enzyme necessary for breaking down milk lactose and, therefore, have digestive problems after drinking milk. A close attention to the diet may help solve these problems.

A major danger of self-treatment with patent medicines for constipation or diarrhea is that these conditions are sometimes symptomatic of more serious problems, such as ulcers or cancer. A person should always consult a physician if a digestive problem persists for more than a week or two.

One of the most common and troublesome problems with the lower bowel and anal area is that of hemorrhoids, sometimes called piles. Hemorrhoids are clusters of enlarged veins in the region around the anus. These tissues become extremely painful, especially if blood clots form in some of the veins. Frequent warm baths may relieve the pain; surgical removal may be necessary in severe cases. Hemorrhoids are thought to be caused by chronic constipation and overuse of laxatives. As is often the case, the highly touted patent medicine remedies for this condition are not considered to be effective.

Basic Concept

The large intestine absorbs water, a few minerals, and the vitamins formed by the colon bacteria.

(a)

(b)

(c)

Figure 6-7   *Three kinds of animals showing feeding and protective adaptations. The lions in (a) show typical carnivore features such as large teeth and claws, and a powerful, muscular body. The zebras in (b) depend on camouflage and running ability to escape predators such as lions. The curious-looking armadillos in (c) dig for roots with their long claws and depend on their armor for protection.*

FEEDING ADAPTATIONS IN OTHER ANIMALS

The many types of feeding mechanisms found in animals require a large number of special body modifications (adaptations). In fact, nearly every major adaptation associated with an organism's body is related to the type of food it eats (Figure 6-7). A pleasant way to confirm this concept is to spend an afternoon in a zoo looking at feeding adaptations among the different animals. The lion's awesome teeth, huge claws, and muscular body clearly designate it as a carnivore. The zebra's horselike teeth, large belly, and relatively long, skinny legs (for escaping predators such as lions) indicate a herbivore. Even in the case of bizarre-looking animals, such as the armadillo, you might be puzzled by the functions of its long front claws, piglike snout, tiny eyes, and coat of armor until you observe the animal efficiently digging up the ground, and pushing its snout through the soil in search of tender roots and stems. The armadillo depends far more on its sense of smell than on its eyesight for locating food. If molested by a predator, the animal rolls up into a tight ball and is rarely harmed.

Filter Feeders

A vast number of freshwater and saltwater creatures utilize some type of *filter feeding* technique. These include sponges, bivalves, some gastropods (snails), worms, and barnacles. Water is circulated over or through the body by means of siphons, cilia, or body movements. Organic matter in the water is strained out or entrapped in sheets of mucus, which are then moved into the gut and digested. The most common type of filter feeding mechanism is seen in bivalves such as oysters or clams (Figure 6-8). In these forms, the beating of cilia causes water to flow over the animal's mucus-covered gill filaments. Food particles stick in the mucus and are moved to the mouth. The gills also extract oxygen from the water and thus serve two functions. A different mechanism is used by barnacles (Figure 6-9). They sweep their fringed appendages through the water like tiny nets to capture small organisms and organic matter.

Some of the organic matter in water consists of small living organisms, but the bulk of what is used for food by invertebrate filter feeders consists of tiny bits of

Figure 6-8 *A small filter feeding bivalve* (**Coquina donax**) *partially buried in the sand. Water is sucked in through one of the siphons, shown above, circulatd over the gills, and then pumped out through the other siphon. Bits of organic matter in the water stick to the mucus on the gills and are used for food.*

Figure 6-9 *Goose-neck barnacles filter feeding by sweeping their fringed appendages through the water to collect tiny bits of food.*

"dead" organic material (*detritus*) surrounded by bacteria feeding on it. These minuscule clumps of matter are evidently an extremely rich and abundant nutrient source, as is apparent from the vast populations of shellfish, barnacles, sponges, and other creatures that subsist on them.

Although most filter feeders are invertebrates, some fishes, mammals, and aquatic birds have also adapted to this type of feeding. Fishes such as mackerel and herring have slender, fingerlike gill rakers over the inner gill openings that act as strainers to hold small food particles in the mouth or pharynx. On a spectacular scale in terms of size are the whale shark and basking shark, both of which live entirely on small organisms strained through their gill rakers.

The huge whalebone whales, which are mammals, also have elaborate filtering devices made of horny plates with hairlike fringes. The plates are attached to the roof of the mouth. After scooping up a mouthful of small organisms such as the shrimplike krill, the mouth is closed, water drains out, and the krill are trapped by the fringed plates in the mammal's huge mouth. Blue whales, fin whales, and humpback whales feed in this manner.

In quite a few aquatic birds the bill is modified in various ways into strainer devices that enable the birds to utilize very small organisms for food. Examples include many ducks and the flamingos (Figure 6-10).

Herbivores

The most elaborate digestive strategies are probably those found in herbivores. The cellulose of plant cells is an extremely abundant, potential food supply, but animals lack the digestive enzymes necessary to break it down. Thus, groups that feed exclusively on plant products require a number of adaptations, including specialized teeth and enlarged chambers in the gut for storing plant matter while microorganisms break down the cellulose.

Many hooved mammals such as cows, deer, camels, and giraffes are *ruminants,* that is cud-chewing animals. Ruminants have the lower end of the esophagus and the stomach modified into compartments (Figure 6-11). The grass or other vegetation passes first into a chamber called the reticulum. From this chamber the animal regurgitates (ruminates) small compact masses of cud and chews them with seeming pleasure and contentment! During the cud-chewing process so much saliva is produced (60 liters per day in an average cow) that it literally drips from the mouth of the animal.

Figure 6-10 *The flamingo, a filter-feeding bird. Its odd-shaped bill is placed in the water upside down. Rows of tiny plates in the upper and lower bills form a fine-meshed filter for straining out diatoms and algae.*

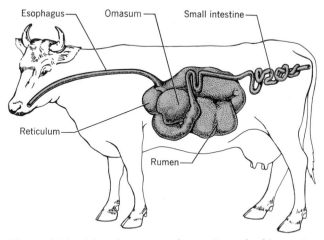

Figure 6-11 *Digestive system of a ruminant herbivore. In these animals, the front end of the digestive tract is highly modified for fermenting cellulose.*

The abundant saliva is needed in subsequent digestive processes.

The thoroughly chewed and moistened cud is swallowed again and passes into another chamber, the

The odd-toed hooved mammals such as horses and rhinoceroses have an entirely different type of gut for accomplishing the same ends. In these mammals, the stomach is relatively normal in size, but a portion of the large intestine is greatly enlarged forming the *caecum*. In the caecum, the fermentation bacteria bring about the cellulose digestion; hence, the designation of these mammals as *hindgut fermenters*. These animals do not ruminate their food. Rabbits and hares also have a large caecum (Figure 6-12) but with still another adaptation—instead of chewing a cud, they eat their own droppings and consequently pass the plant matter through the gut twice.

Although herbivore digestion seems far removed from what happens in our bodies, recent studies show that a similar fermentation process takes place in the human large intestine. Its function or significance is not fully known.

This review of a few of the diverse feeding and digestive strategies in the animal kingdom emphasizes the efficiency of nature in utilizing virtually all potential food sources. We will return to this basic theme later as we discuss food pyramids and energy flow in living communities in the environment.

Basic Concepts

The major structural features of an organism strongly reflect its means of obtaining food.

Filter feeding is used by a large variety of aquatic organisms.

Herbivores utilize microorganisms in converting cellulose into useful nutrients.

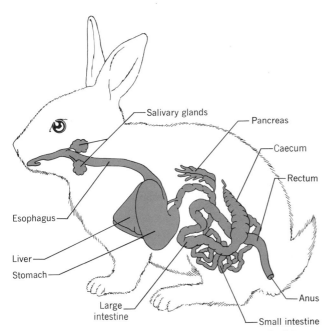

Figure 6-12 *Digestive system of a nonruminant herbivore. In these animals the back end of the digestive tract is modified for fermenting cellulose.*

Labels: Salivary glands, Pancreas, Caecum, Rectum, Esophagus, Liver, Stomach, Large intestine, Small intestine, Anus

rumen. The rumen harbors a large population of bacteria and protozoans. With the aid of the bacteria, a fermentation process breaks down the cellulose into a variety of small fatty acids that are absorbed into the bloodstream and used for energy. Following this process, the remainder of the plant matter passes through the other compartments of the stomach where the accompanying microorganisms are themselves digested to provide the ruminant with proteins. Animals such as these are termed *forestomach fermenters*.

KEY TERMS

amylase	deposit feeder	filter feeder	hindgut fermenter
bile	detritus	fluid feeder	hydrolysis
caecum	digestion	forestomach fermenter	ileum
carnivore	duodenum	gastrin	jejunum
chyme	epiglottis	hepatic portal system	lacteal
colon	esophagus	herbivore	larynx

| mucin | pepsin | peristalsis | ruminant |
| omnivore | pepsinogen | pharynx | villi |

SELF-TEST QUESTIONS

1. What are several methods used by organisms to utilize a wide variety of organic matter for food?

2. (a) What is the digestive reaction?
   (b) Why is it called a hydrolysis reaction?

3. How do our teeth reflect an omnivorous diet?

4. What are some digestive functions of the mouth?

5. Why could you probably not survive without an epiglottis?

6. List three functions of the stomach.

7. Why is it dangerous to counteract the acidity of the stomach by antacid remedies?

8. What major digestive events occur in the small intestine?

9. Define each of the following: chyme, bile, lipase, protease, carbohydrase, jejunum, ileum, and hepatic portal system.

10. Describe a villus and tell what happens there.

11. List the nutrients that can be absorbed through the villi.

12. What happens to nutrients in the large intestine?

13. What is the recommended procedure for dealing with constipation?

14. What are hemorrhoids?

15. Give some examples of how an animal's structure reflect its diet.

16. Tell how filter feeding is used by: (a) an invertebrate; (b) a fish; (c) a mammal; (d) a bird.

17. Describe (a) How ruminant animals digest cellulose and, (b) how this is accomplished in nonruminant animals.

chapter 7

*Dr. Robert Jarvik holding
an artificial heart as it
would be placed within the
thoracic cavity of a patient.*

CIRCULATION AND RESPIRATION

chapter 7

questions to think about

How can a multicellular organism supply its millions of cells with oxygen and nutrients?

Why do vertebrates require a closed-type circulatory system?

What kinds of respiratory structures are found in animals?

How is the human respiratory system like that of other mammals?

A man is as old as his arteries.

Thomas Sydenham

CIRCULATORY SYSTEMS

A basic problem faced by nearly every multicellular organism is how to supply its millions of cells with nutrients, oxygen, and other substances needed for normal functioning. Bound up with this is a second problem: how to transport away the waste products. To accomplish these rather formidable jobs requires an efficient distributing system and some sort of transporting medium—in brief, a circulatory system. We usually think of our own circulatory system, with its heart, blood vessels, and blood, as typical; but there are several totally different mechanisms in the animal world that achieve the same ends. It is worth looking briefly at some of these before taking up our own system in detail.

Simple Systems
If an organism is unicellular or so small that all of its cells are near the body surface, simple diffusion is able to move materials rapidly enough into and out of the cells to supply its needs. Protozoans, for example, do

not need a special transport system. Even simple multicellular animals, such as *Hydra* and small flatworms, solve the problem of combining the digestive and transport systems into a single gastrovascular cavity (Figure 7-1). Every cell in a *Hydra* or a flatworm lies near this central cavity, so that nutrients, oxygen, and wastes can be transferred to and from the cells by diffusion. The movement of materials within the cavity is limited to the sloshing of the contents. The transport efficiency is low, but this system is adequate for small and simple animals with their rather low rates of activity and metabolism.

Open-type Systems
Most of the remainder of the invertebrate world contain organisms of enough complexity to require some type of special transport system. The most common type is found in insects and most other arthropods. The grasshopper, for example, has one blood vessel, the *dorsal vessel,* extending the length of the abdomen and into the head region (Figure 7-2). A portion of the dorsal vessel is termed the *heart,* because it pulsates and drives the vascular fluid (called *hemolymph*) toward the head region. In the head region the hemolymph flows out of the vessel into spaces (*hemocoels*) around the organs. There materials are exchanged between cells

104

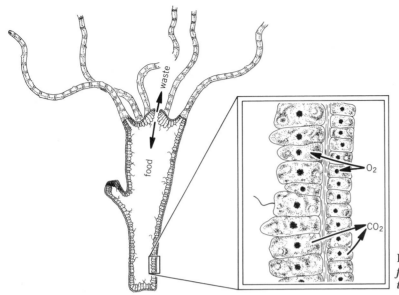

Figure 7-1 *The thin body wall of* **Hydra** *allows food, wastes, oxygen, and carbon dioxide to diffuse to all cells.*

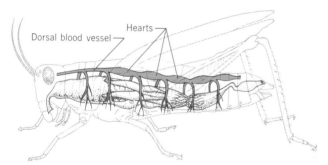

Figure 7-2 *An open-type blood system in a grasshopper. The hearts pump the blood through the dorsal vessel and out into spaces around the organs.*

and the "blood." The hemolymph gradually works its way back to the heart tube, which contains small openings with valves that let the hemolymph enter but close when the heart beats. Hemolymph does not transport gases to any great extent in most insects, because another system has that function. Although insect blood is usually colorless, it contains a variety of nutrients and various types of cells. Its functions include nutrient transport, coagulation, and wound healing, and it also acts as a lubricant around the internal organs.

Some arthropods (crayfish, for example) have somewhat more elaborate vascular systems, with a more distinct heart and several vessels extending from it to the hemocoels. The basic functioning of the system,

however, is the same as in insects. All of these systems are termed *open-type circulatory systems* because the blood flow is not confined to vessels.

Basic Concepts

Diffusion is an adequate transport device for very small organisms.

Open-type circulatory systems are found in many invertebrates. In this kind of system, blood flows into hemocoels rather than into capillary beds and is not returned to the heart by veins.

Closed-type Systems

The type of circulatory system found in all vertebrates is termed a *closed system* because the blood circulates through arteries, capillaries, and veins and does not directly bathe the cells and tissues of the body. These cells are separated from the direct flow of the blood not only by the walls of the blood vessels but also by fluid-filled spaces between the vessels and the cells. All exchanges between the cells and the blood must occur through this watery tissue fluid. The following description applies specifically to the human circulatory system, but it is typical of most other mammals.

Blood. Blood is the major transporting medium of the circulatory system. It is sometimes called a liquid tissue, because it consists not only of fluid but also of cells

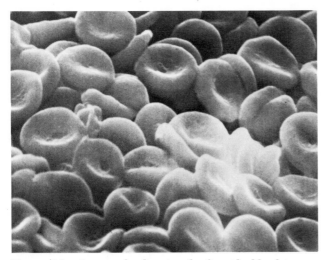

Figure 7-3   *A mass of red corpuscles from the bloodstream. These tiny, flattened sacs contain hemoglobin and enzymes to help them transport oxygen and carbon dioxide. Does their shape suggest how they might squeeze through small diameter capillaries?*

with specialized functions. The liquid portion, termed *plasma,* is mostly water and comprises approximately 55 percent of blood.

There are about 5 liters of blood in the body. Because blood is replaced rather rapidly by the body, a healthy donor can contribute a half liter and not experience any ill effects.

The most abundant cells in the blood are the *red corpuscles* (about five million per cubic millimeter of blood in an adult). When first formed in bone marrow, these cells have nuclei, but before being released into the circulatory system they lose their nuclei and become tiny doughnut-shaped sacs containing hemoglobin and several enzymes (Figure 7-3). Their lifetime is only three to four months, and hence large numbers of new red corpuscles must continually be produced to replace the old ones in the system.

Hemoglobin is a complex protein made of four polypeptide chains, each containing one heme ring with an atom of iron. When oxygen is plentiful, as in the lungs, it attaches readily, if loosely, with the iron atoms in the heme groups. When blood moves into body tissues where oxygen is deficient, the iron atoms easily give up their oxygen. Oxygenated hemoglobin is bright red, the color we associate with blood. Deoxygenated hemoglobin tends to be purplish in color as shown by venous blood. Hemoglobin also functions to help

transport carbon dioxide from tissues back to the lungs.

White blood cells (*leucocytes*) are far fewer in numbers (5000–9000 per cubic milliliter of blood) and are of several different types (Figure 7-4). They are produced in lymph tissue (lymph nodes, spleen, and thymus gland) and in bone marrow. Their functions are protective: some are capable of phagocytosis (engulfing foreign objects such as bacteria that get into the blood) and one type, the lymphocyte, usually produces antibodies whenever it encounters microorganisms or foreign proteins. Antibodies render these alien substances harmless.

A third category consists of extremely tiny, abundant fragile bodies known as *platelets.* The platelets contain several items necessary for blood clotting. When a blood vessel is injured or cut, platelets in the area of the injury disintegrate to release the chemicals that initiate the clotting reaction. This process seals the wound area and prevents excess loss of blood.

In addition to these elements, blood also contains proteins such as fibrinogen (a clotting agent), and albumin, globulins, and antibodies. These last three types of protein help protect the body against disease. Plasma also contains an assortment of ions, or *electrolytes.* These normally include sodium, potassium, calcium, magnesium, chlorine, bicarbonate, and phosphate ions, as well as certain proteins that have electrical charges and thereby function as ions. These electrolytes play many basic roles in body functioning, as noted in Chapter 4. Their importance is emphasized during prolonged vomiting or diarrhea, events that cause a rapid loss of electrolytes. This condition can eventually become life-threatening if the ions are not replaced.

Finally, blood contains various nutrients, such as amino acids and glucose, that are being transported about the body. It also carries hormones and cellular waste products such as CO_2, urea, and water.

Basic Concept

Blood is the transport medium for nutrients, waste products, electrolytes, gases, antibodies, phagocytic cells, and hormones.

Capillaries. Capillaries are microscopically tiny, thin-walled blood vessels that form networks called *capillary beds* throughout the body tissues. Only in capillary

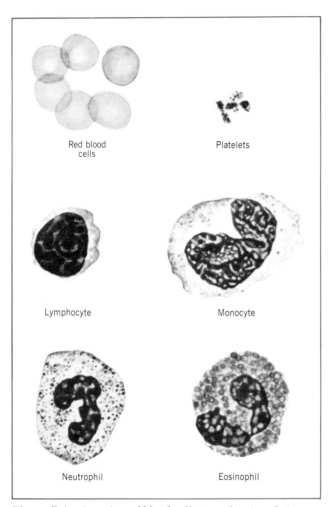

Figure 7-4 *A variety of blood cell types showing their forms and relative sizes. The two leucocytes at the bottom are distinguished by their large-lobed nuclei and granular cytoplasm.*

Red blood cells

Platelets

Lymphocyte

Monocyte

Neutrophil

Eosinophil

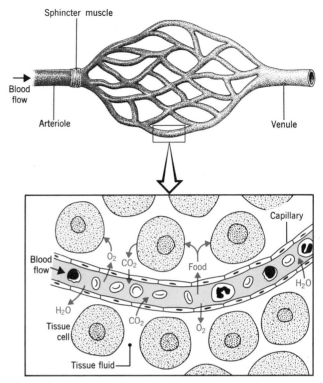

Sphincter muscle

Blood flow

Arteriole

Venule

Capillary

Blood flow

O_2 CO_2 Food H_2O

H_2O CO_2 O_2

Tissue cell

Tissue fluid

Figure 7-5 *A capillary bed (top). The rate of blood flow through it is regulated by constriction or relaxation of the arteriole. At bottom, tissue cells exchange materials with the blood in the capillary. What happens if water does not flow back into the capillary bed from the intercellular fluid?*

beds can materials enter and leave the bloodstream. Capillaries drain into tiny *venules* that eventually unite with other venules to become veins (Figure 7-5).

The tiny blood vessels that lead into the capillary beds (*arterioles*) can constrict to let less blood into the beds or dilate to allow more blood to enter. These reactions (called, respectively, *vasoconstriction* and *vasodilation*) are controlled by the nervous system and by hormones. On a hot day, nerve impulses cause vasodilation of the beds at the surface of the body, allowing heat to be dissipated from the blood more rapidly. On a cold day, vasoconstriction of the body surface capil-

laries helps conserve body heat. This same sort of control mechanism regulates the blood flow in muscle tissue, in the villi of the small intestine, and elsewhere throughout the body.

Most constituents of blood except the cells and proteins are forced by blood pressure through the thin-walled capillaries into the tissue fluid surrounding all cells. From the tissue fluid each body cell obtains the nutrients that it needs. The cells also receive oxygen released from the red corpuscles, even though the corpuscles themselves remain confined in the capillary vessels. Cellular waste products such as carbon dioxide and ammonia diffuse from the body cells into the tissue fluid.

As these products accumulate, osmotic pressure causes them to move along with excess water molecules into the capillary bed. This balance between substances leaving and entering capillaries is a delicate

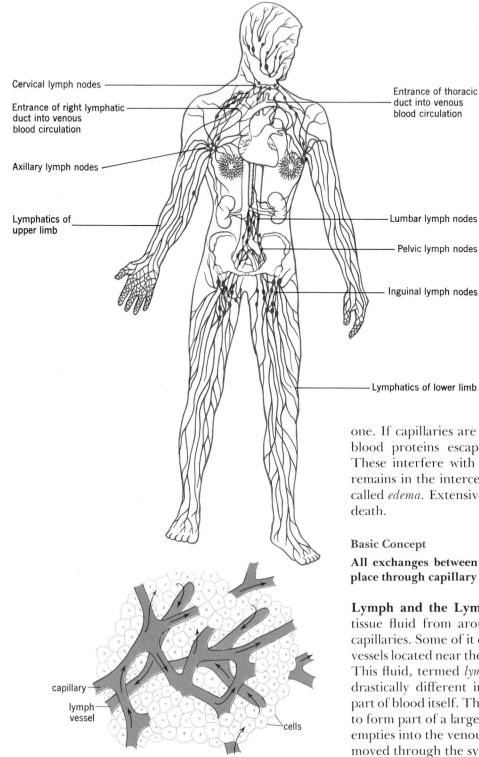

Cervical lymph nodes

Entrance of right lymphatic duct into venous blood circulation

Axillary lymph nodes

Lymphatics of upper limb

Entrance of thoracic duct into venous blood circulation

Lumbar lymph nodes

Pelvic lymph nodes

Inguinal lymph nodes

Lymphatics of lower limb

capillary

lymph vessel

cells

Figure 7-6 *The extensive lymphatic system (top) in the human body. The lymph eventually drains into large veins in the shoulder region. The close relationship between capillaries and lymph vessels is shown at the bottom.*

one. If capillaries are disturbed by injury or disease, blood proteins escape into the intercellular fluid. These interfere with osmotic balance so that water remains in the intercellular fluid to produce swelling called *edema.* Extensive edema in the body may cause death.

Basic Concept

All exchanges between blood and the body tissues take place through capillary walls.

Lymph and the Lymphatic System. Not all of the tissue fluid from around the cells drains back into capillaries. Some of it enters another network of small vessels located near the capillary beds, the *lymph vessels.* This fluid, termed *lymph,* is a thin, watery liquid not drastically different in composition from the liquid part of blood itself. The small lymph vessels join others to form part of a larger *lymphatic system* that ultimately empties into the venous system (Figure 7-6). Lymph is moved through the system by the combined action of

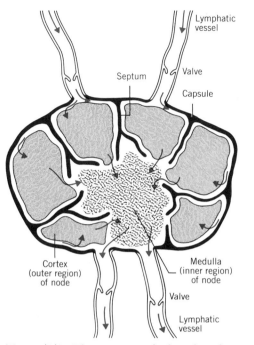

Figure 7-7 *The structure of a lymph node. Lymph enters node from lymph vessels, is "strained" through the cortex, enters the medulla, and then drains out the vessels at the bottom. Microorganisms or other foreign objects in the lymph are usually retained and destroyed within the node.*

body muscle contractions and the pressure of the accumulated tissue fluid itself. Valves in the lymph vessels keep the lymph flowing in one direction. The *lacteals* described in Chapter 6 in the intestinal villi form another important part of the lymph system for transporting lipids and fat-soluble vitamins.

Another important aspect of the lymph system is the presence of *lymph nodes* in the larger lymph vessels (Figure 7-7). All of the lymph must flow through these nodes before entering the venous system. The nodes act as filters to capture bacteria and even cancer cells at times. In addition, tissues in the nodes add lymphocytes and monocytes (types of white blood cells) to the lymph. These cells ultimately enter the vascular system of the body, where the monocytes function as phagocytes and the lymphocytes produce antibodies when necessary.

The lymphatic drainage system becomes a liability in some types of cancer. Cancer cells may enter the system and infect lymph nodes or else become distributed into other parts of the body to start new cancers.

Basic Concept

The lymphatic system aids the circulatory system in removing excess intercellular fluids, and in combating infections by phagocytosis and antibody formation.

Veins. *Veins,* by definition, are blood vessels that carry blood back toward the heart from capillary beds. They are composed mostly of connective tissue and a small amount of muscle. Veins contain valves that prevent the backflow of blood; they depend largely on the squeezing action of the body musculature to move the blood. An important function of veins is to serve as a reservoir for the body's blood supply. Approximately 70 percent of a person's blood is found in the veins at any given time.

Veins accompany the major arteries throughout much of the body, with a few exceptions such as the hepatic portal system. Also there are a number of superficial veins that lie at the surface of the body, as in the arms, that do not have a corresponding artery. These superficial veins provide convenient places to remove or add blood to the body for various medical reasons. It is relatively simple and almost painless to insert a hollow needle into a superficial vein; furthermore, the vein nicely seals itself after the needle is removed. Such a procedure cannot easily be performed on an artery.

Viewed from a general perspective, venous drainage can be divided into several main systems. The veins of the head region and the veins of the musculature of the upper limbs form one drainage unit, in the sense that all eventually join to make up the large *superior vena cava* in the upper chest region (Figure 7-8). The vena cava then empties into the upper right chamber (atrium) of the heart.

The lower-body drainage is more complex. It is composed of three units that are quite distinct in function as well as location. One of these units consists of the veins of the lower limbs and trunk musculature. They come together in the pelvic region to form the large *inferior vena cava*. This vein passes up toward the heart, receiving the veins from the abdominal area as it goes (Figure 7-8).

Another unit in this category is the *hepatic portal system*, which drains blood from the intestines and spleen to the liver. This blood, with its load of nutrients from the intestinal villi, enters capillary beds in

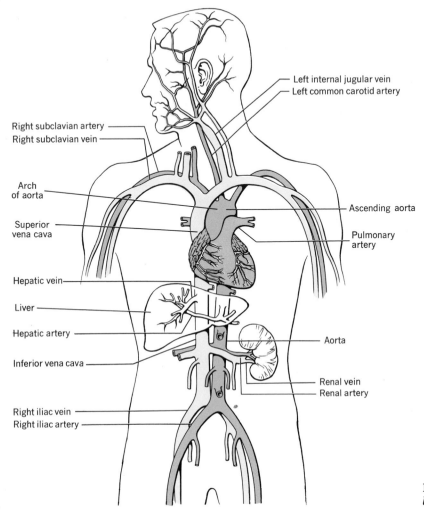

Right subclavian artery
Right subclavian vein

Arch
of aorta

Superior
vena cava

Hepatic vein

Liver

Hepatic artery

Inferior vena cava

Right iliac vein
Right iliac artery

Left internal jugular vein
Left common carotid artery

Ascending aorta

Pulmonary
artery

Aorta

Renal vein
Renal artery

Figure 7-8 *Major arteries and veins in the body.*

the liver. The liver, as mentioned before, alters the nutrients and may store some of them; the remainder drain into the hepatic veins and then into the inferior vena cava.

The third major unit of the lower body drainage involves the *kidneys.* Blood flows into the kidneys from the *renal arteries* and then enters millions of tiny, globular capillary beds within the kidney, termed *glomeruli.* Blood pressure forces the liquid portion of the blood through the glomerular walls into a labyrinth of tiny tubules. These tubules form a complex filtering system, retaining wastes and excess minerals and water, and returning the remainder of the blood to the capillaries that surround the tubules. These capillaries

drain into venules and ultimately into the *renal veins* that return the filtered blood to the inferior vena cava. This filtering action of the kidneys is vital to the life of the organism and any interference with it has serious consequences.

The final major category of venous drainage is that of the *pulmonary veins,* leading from the lungs to the left atrium of the heart. The *pulmonary veins* carry blood that has been oxygenated in the lungs, returning it to the heart for distribution to the rest of the body.

Basic Concepts

Veins are thin-walled blood vessels that transport blood toward the heart.

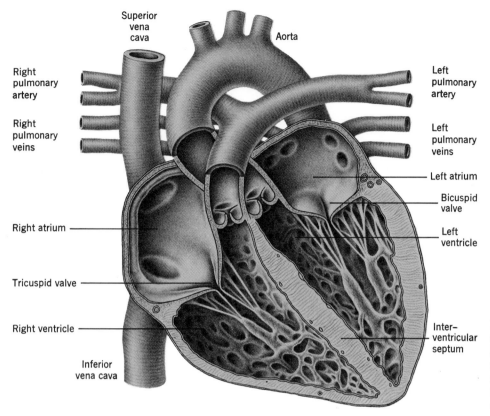

Superior
vena
cava

Aorta

Right
pulmonary
artery

Right
pulmonary
veins

Left
pulmonary
artery

Left
pulmonary
veins

Left atrium

Bicuspid
valve

Left
ventricle

Right atrium

Tricuspid valve

Right ventricle

Inferior
vena cava

Inter-
ventricular
septum

Figure 7-9 *A cutaway view of the heart showing its four chambers and other major parts. Note that the left ventricle has considerably thicker walls than the right ventricle. What function does this serve?*

Veins function as an important blood reservoir in the body.

Heart. The heart provides the pumping force that drives the blood through the circulatory system. It is made of a special type of muscle tissue unlike that found anywhere else in the body. Heart tissue never fatigues; indeed, it has an inherent, rhythmical beat. Even a single isolated piece of heart tissue pulses steadily.

Human beings, like other mammals, birds, and some reptiles, have a four-chambered heart (Figure 7-9). The advantage of this arrangement over simpler hearts is that it completely isolates the incoming venous blood, low in oxygen, from oxygenated blood destined to be sent out to the body.

Venous blood, depleted of oxygen, enters the *right atrium* from the superior and inferior vena cavae. The atrium fills with blood and passes it through the tricuspid ("three-flap") valve into the much larger *right ventricle*. Contraction of the right ventricle then sends the blood out through the pulmonary artery to the lungs. The tricuspid valve is forced closed by the pressure in the ventricle, preventing blood from flowing back into the atrium. The pulmonary artery has a similar set of valves that prevents blood from flowing back into the ventricle.

After giving up its load of carbon dioxide and being oxygenated in the lungs, the blood returns via the pulmonary veins to the *left atrium*. From here blood is pushed through a bicuspid ("two-flap") valve into the large, thick-walled *left ventricle*. Contraction of this powerful chamber forces blood out in a surge through the aorta to the body. The left ventricle pumps blood at a rate of about 4 quarts per minute. The aorta contains a set of valves similar to those in the pulmonary artery to help keep the blood flowing in one direction.

The heart, then, actually takes part in two distinct circulatory patterns, or loops. One of these is the

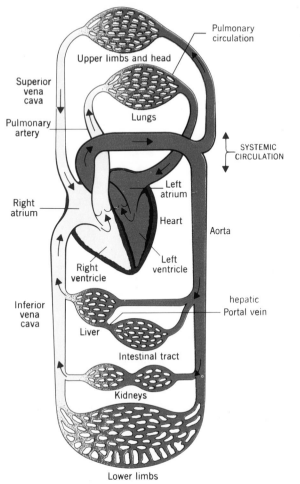

Figure 7-10 *The pulmonary and systemic loops in the circulatory system. Separate loops allow a higher pressure to be developed in the systemic loop, and to efficiently oxygenate the blood in the pulmonary loop. Do the pulmonary veins transport oxygenated or deoxygenated blood?*

pulmonary circulation, involving the right side of the heart, which serves to oxygenate the blood. The other is the *systemic circulation,* involving the left side of the heart, which serves to provide this oxygenated blood, as well as other substances, to all the rest of the body (Figure 7-10). This is evidently an efficient way to handle these tasks because all large, complex higher animals have it. One advantage is the relatively high blood pressure that can be generated for forcing blood into capillary beds throughout the body. The heart plays the initial role, forcing blood into the aorta. The

aorta expands with each ventricular contraction to accommodate the surge of blood, and then in turn constricts to help push the blood along. Blood pressure thus results from the combined action of the heart, the aorta, and the resistance of the blood vessels throughout the body to the blood being pumped through them.

Blood pressure can be determined accurately by inserting a needle in an artery and measuring the height to which it will push a column of mercury. Less accurate, but far simpler, is the use of the blood-pressure cuff and a stethoscope. This simple apparatus, with a complex name — sphygmomanometer, gives an approximate measure of the force of the ventricular contraction (*systole*) and a measure of the pressure present while the ventricles are relaxed (*diastole*). If the systolic measurement is 120 millimeters of mercury (mm Hg) and the diastolic pressure is 80 mm Hg, the blood pressure is noted as 120/80. Blood pressure changes with age, so that 120/80 is considered normal for a 20-year-old individual, but 150/90 is within normal range for an individual in the sixties. Many factors influence blood pressure on a day-to-day basis. However, a higher-than-average pressure for an individual over a period of time is considered an important danger signal of disease. Such a condition can damage the kidneys or the liver with their millions of tiny capillary beds, or it may rupture a blood vessel somewhere in the body. As artery walls lose their elasticity with age and gain fatty deposits in their walls, high blood pressure becomes a highly dangerous factor. *Stroke,* the rupture of vessels in the brain, is often attributed to this combination of factors.

In the wall of the heart are specialized masses of interconnected fibers called *nodal tissues* (Figure 7-11). These begin with the *sinoatrial node* (S–A) in the right atrial wall. It is called the *pacemaker,* because it sets the pace of the heartbeat. The S–A node connects with the *atrioventricular node* (A–V), located in the lower part of the right atrial wall. From this node, fibers extend into the walls of the ventricles. The S–A node initiates the basic heartbeat and causes the atria to contract. Impulses then travel from this node to the A–V node and in turn into the walls of the ventricles, causing them to contract just after the atria complete their contraction. In this way, the two sets of chambers

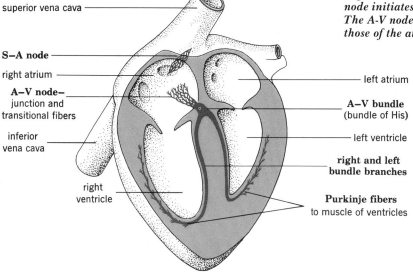

superior vena cava

S–A node

right atrium

A–V node—
junction and
transitional fibers

inferior
vena cava

right
ventricle

left atrium

A–V bundle
(bundle of His)

left ventricle

right and left
bundle branches

Purkinje fibers
to muscle of ventricles

Figure 7-11 *The pacemaker tissues in the heart. The S-A node initiates the heartbeat and causes the atria to contract. The A-V node coordinates the ventricular contractions with those of the atria.*

are coordinated to move blood into and out of the heart with maximum efficiency.

If the nodal tissue does not function properly, as happens in certain types of heart disease, an artificial pacemaker may be installed. This is a small electronic device that sends electrical signals into the ventricle walls to maintain a regular rate of contraction. Many thousands of people are able to lead normal, active lives with the aid of these.

Nearly one million people still die of heart disease in the United States yearly, despite the use of pacemakers and surgical procedures. An artificial heart, if available, would save a number of these victims. Dr. Robert Jarvik, a medical researcher, has designed a heart made of polyurethane plastic chambers and an aluminum base. In experiments it kept a calf alive and healthy for more than 3 months in 1981.*

The first attempt to replace part of a human heart with a Jarvik-type of artificial heart occurred in December 1982 at the University of Utah Medical Center. In this procedure, the badly diseased ventricles were removed from a 61-year-old patient's heart and replaced with plastic chambers operated by air-

driven diaphragms. The patient, Barney B. Clark, survived initial complications of the surgery and lived for about 3 months.

A rather large compressed air system and monitering unit is presently required to operate the artificial chambers. Such drive systems will probably be portable in the future.

Basic Concept

A four-chambered heart is an efficient device for separating venous blood from oxygenated blood, and for providing sufficient force to push blood through the circulatory system.

Arteries. *Arteries* are the vessels that transport blood away from the heart. Artery walls are much thicker than those of veins, due to the presence of elastic connective tissue in the larger arteries and considerable smooth muscle tissue in the smaller ones. Most arteries are buried deeply in the body's tissues for protection.

The arteries from the heart are usually divided into two categories—*pulmonary* and *systemic*. Pulmonary circulation to the lungs has already been described.

* "The Total Artificial Heart," by Robert K. Jarvik. *Scientific American,* January 1981.

The systemic circulation is the distribution to the remainder of the body. Its pathway is as follows.

Systemic circulation (Figure 7-8) begins with the *ascending aorta*, which exits from the left ventricle. Branching off from this portion of the aorta is the relatively small but exceedingly important *coronary artery*, which carries blood into the capillary beds in the muscular wall of the heart itself. As the first branch from the aorta, the coronary artery provides the heart with blood that is high in oxygen and rich in nutrients. This artery is the heart's only "lifeline," and any interference with the coronary circulation is quickly reflected by a malfunctioning of the heart—the all-too-familiar heart attack.

The short ascending aorta curves to the left as the *aortic arch* and then descends through the thorax and abdomen and into the lower limbs. Along this route, the aorta supplies blood to all major organs and muscles.

As arteries approach capillary beds in tissues and organs, they branch into tiny vessels (arterioles) containing smooth muscle tissue. These arterioles can constrict to divert blood away from certain capillary beds or relax to let more blood in, a regulatory function that is usually under nervous control.

Most principal areas of the body are supplied by a major artery, but they frequently have an additional supply derived from smaller arteries. Such additional blood supply is called *collateral circulation.* If the major artery should be cut or damaged, the collateral vessels can often take over and continue supplying blood to the region.

A major health problem in America and other affluent societies is that of atherosclerosis, the buildup of fat deposits (plaques) in arteries (Figure 5-7). The role of diet in preventing this condition was discussed in Chapter 5. An even more important factor seems to be that of exercise. Considerable evidence indicates that the incidence of many heart diseases, in addition to atherosclerosis, is considerably lower among individuals who engage in regular, vigorous physical activities. Such activities prevent the buildup of fat plaques in blood vessels, improve the collateral circulation in the wall of the heart, and enable the heart to pump more blood with less effort. Additional information concerning *cardiovascular fitness* is presented in Chapter 12.

RESPIRATORY SYSTEMS

Gas Exchange

Nearly all organisms require oxygen and use it for the same purpose, as a hydrogen acceptor in the cellular production of energy (see Chapter 5). In small animals, the oxygen may diffuse into the body through the surface, but more often it requires special respiratory structures such as gills or lungs. These structures also function to eliminate waste products such as carbon dioxide.

Requirements. All structures used for respiration must have certain basic features in order to allow for the exchange of gases. The structure must maintain a moist surface at all times in order for gases to diffuse through it. This poses no problems for water dwellers, because they are immersed in water, but is definitely a problem for land dwellers because air has a strong drying effect. Terrestrial animals must either live in moist environments (as many amphibians do) or else have their respiratory membranes enclosed in a chamber where they can be kept moist.

Because gases pass through respiratory membranes entirely by diffusion, the membranes must be quite thin, and they must also present as large a surface area as possible. These requirements explain why gills are arranged in thin filamentous plates and why lungs often are divided internally into many tiny chambers.

In addition to having a large surface area, the respiratory structure must be exposed to a continual flow of air or water in order to keep it in contact with oxygen molecules. This flow is attained by movement of the animal itself (as in a fish moving through the water), by waving the gills about in the water, or by pumping air into or out of the lungs.

The environment in which an organism lives plays a basic part in the availability of oxygen. Air contains a considerable amount of oxygen, about 210 ppt (parts per thousand, that is, parts oxygen per thousand parts of air). Fresh water, on the other hand, holds only 7 ppt even when saturated (at 15°C) and seawater only 5 ppt (Figure 7-12). This is a major reason why there are no warm-blooded water breathers: the metabolism of a warm-blooded animal demands a higher oxygen intake than is available from water.

The availability of oxygen in the environment raises an intriguing question. If enough oxygen were present

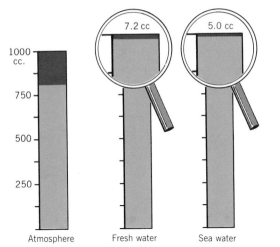

Figure 7-12 *A comparison of the amount of oxygen in several environments. What does this indicate about the efficiency of gills in obtaining oxygen from aquatic environments?*

in the water, could air breathers breathe water and survive? This question was answered by some experiments that began almost 20 years ago. In these experiments, mice were submerged in a pressurized chamber containing a weak saline solution similar to blood plasma. Pressure kept the solution charged with oxygen. The mice inhaled and exhaled the liquid and survived for a considerable period of time (Figure 7-13). After this original 1961 experiment, other lab animals such as rats, dogs, and cats have also successfully been made to breathe oxygenated fluids for short periods of time. However, the physical exertion of breathing fluid, which is far denser than air, and the disposal of exhaled carbon dioxide appear to be the major factors limiting how long an animal can survive. Recent experiments have made use of silicone oils and fluorocarbon liquids, because these substances hold far more oxygen and carbon dioxide than do water solutions and thus will support respiration at atmospheric pressure. However, a certain amount of lung damage occurs after a period of breathing these rather dense fluids.

If the various problems with fluid breathing could be solved, the process would have a number of useful applications. Deep-sea divers, breathing a liquid, could descend thousands of feet and rise rapidly to the surface without spending hours in a decompression

Figure 7-13 *A mouse in a pressurized chamber obtaining its oxygen by breathing a weak salt solution. Do you think that a human could survive under comparable conditions?*

chamber. The diver's liquid-filled lungs would not be crushed at great depths because liquids are incompressible. And, for the same reason, gases in the fluid-filled lungs would not be forced out into body tissues and into the blood as the diver descended. Hence he could safely rise to the surface as rapidly as he wished without danger of the "bends." A liquid breathing device would also be highly useful in escaping from disabled submarines.

Basic Concept

Respiratory structures must have the following features: (1) a moist surface, (2) a large surface area, (3) an extremely thin membrane, (4) exposure to a continual flow of air or water containing oxygen.

Respiratory Structures

Body Surface. Single-celled organisms such as protozoans and many small multicellular invertebrates exchange oxygen and carbon dioxide by diffusion through the surface of their bodies. In most of these cases, their cells are arranged in thin layers near the surface of the body. In an extension of this same principle, many amphibians exchange gases through their skin in addition to using gills and lungs. In fact, the skin is the sole respiratory organ in a few salamanders. It should be noted that amphibians also have a highly developed circulatory system to help in distributing the oxygen or disposing of the carbon dioxide.

Gills. Gills are respiratory organs found in a wide variety of animals that live in aquatic or very moist environments. Gills may be simple extensions of the body surface, such as the plumed *external* gills of some amphibians. They may be complex anatomical structures, such as the *internal* gills (Figure 7-14) found in molluscs and fishes. In either event, gills are always associated with an abundant blood supply.

Gills often serve more than a respiratory function. In mollusks and other feeders, gills are a basic part of the feeding mechanism, as discussed in the previous chapter. In fish, too, gills usually have accessory functions. In marine fishes they provide a site for the excretion of excess salts, and in freshwater fishes, a site for taking in salts.

In fishes, gills reach their most complex form, usually as a series of plates or fine filaments arranged in such a manner that a constant flow of water passes over them. Fish gills embody a very useful adaptive mechanism called *countercurrent exchange*. The blood flow and water flow in the gills occur in opposite directions. Blood flows into capillary beds in the gill from the rear edges, and toward the forward parts of the gill where the gills attach to the body. Water flows through the gills in the opposite direction, passing through the fish's mouth and out through the gill chamber (Figure 7-15). This means that gas exchange can occur all along the length of the vessel. The oxygen-poor blood entering the gill plates meets water that has already passed over the gills and has given up a lot of its oxygen. But because the blood itself has even less oxygen than the water, some of the oxygen that remains in the water can still flow into the blood at this

(a)

(b)

Figure 7-14 *Types of gills. In* **(a)** *a salamander amphibian is shown underwater using its external gills.* **(b)** *The internal gills of an Eastern brook trout.*

point. At the other end of the gill, well-oxygenated blood meets fresh, fully oxygenated water. Again, there is enough of a difference in oxygen content for some of the oxygen to move into the blood. Thus the blood steadily becomes more oxygenated as it flows through from the back edge to the front of the gills. Countercurrent exchange assures that the gas exchange between the water and the gills takes place with maximum efficiency.

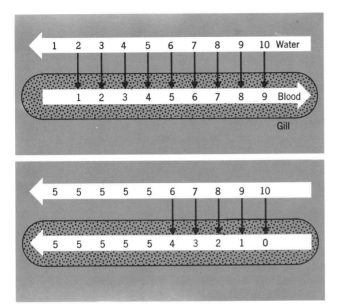

Figure 7-15 *Countercurrent exchange mechanism. As shown at the top, if the water flow and blood flow are in opposite directions, oxygen moves continuously from the water into the blood. If both flow in the same direction (bottom), the exchange is much less efficient because an equilibrium between the two is soon reached and no more oxygen is taken in by the blood.*

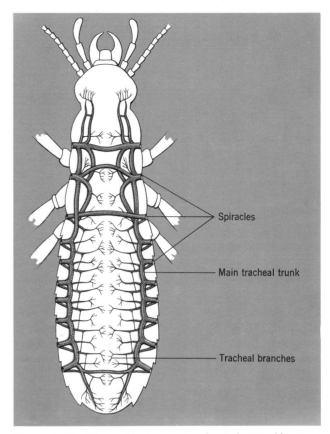

Figure 7-16 *The respiratory system in an insect. Air enters the spiracles located along the sides of the body. The air then moves passively through the tracheae that extend into the tissues of the insect's body. Why could you not drown an insect by holding its head under water?*

Tracheae. Insects and some of the other arthropods have a highly unusual solution to the respiratory problem. In these animals, a system of tiny air tubes from the surface penetrates all parts of their bodies. The tubes, called *tracheae,* open at regular intervals (Figure 7-16). The system is passive, depending on movements of the insect's body to squeeze air in and out. Oxygen diffuses from the tracheae into the tissues of the body where it is also distributed by the hemolymph (blood) to some extent. The blood of some arthropods contains respiratory pigments that are specialized to transport oxygen, and a few even have hemoglobin like that found in mammalian blood cells.

Lungs. Lungs are efficient devices for exchanging respiratory gases with the atmosphere. They are found in all of the major vertebrate groups, including many fishes. Some fish come to the surface and breathe air at regular intervals; others use their lungs when there is insufficient oxygen in the water to be obtained with gills.

In the lower vertebrates such as amphibians, lungs are simple, thin-walled bags. In mammals, however, the lungs are divided internally into numerous small, interconnected compartments which increase their surface area greatly. The metabolism of warm-blooded animals demands a large and rapid exchange of gases, which can be provided only by a huge surface area.

Basic Concept

Respiratory structures used by animals include the body surface, gills, tracheae, and lungs.

The Human Respiratory System

The Respiratory Tract. The human respiratory system is typical of that found in many mammals. It

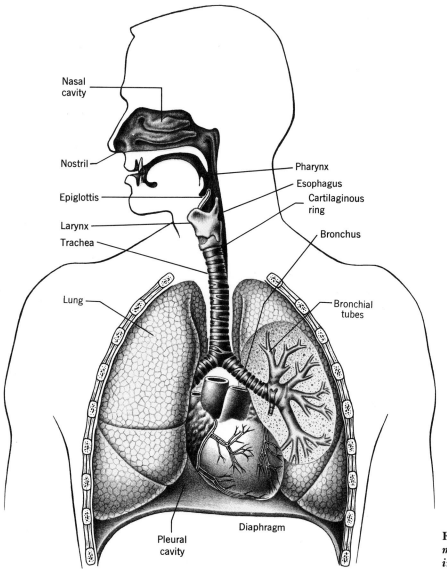

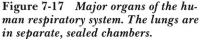

Figure 7-17 *Major organs of the human respiratory system. The lungs are in separate, sealed chambers.*

consists of an *upper respiratory tract* that delivers air to and from the lungs, the *lungs* themselves, and the muscles used in the breathing process (Figure 7-17). The circulatory system is also vitally important as the major distributing mechanism.

The upper respiratory tract includes the nose, mouth, pharynx, larynx (voicebox), and trachea (windpipe), and the bronchial tubes before they enter the lungs. At the back of the mouth (the pharynx), the air passes through the larynx and into the trachea. A

large flap, the *epiglottis,* helps seal off the larynx during swallowing.

The trachea, slightly over 4 inches long and three-fourths of an inch in diameter, is held permanently open for free passage of air by means of rings of stiff cartilage. The trachea branches into the right and left bronchial tubes. Each bronchial tube enters a lung, where it continues to divide into smaller and smaller branches called bronchioles, ending finally in extremely small air sacs called *alveoli.* There are about

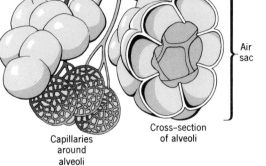

Figure 7-18 *Alveoli in the lung. Gas exchange takes place between these tiny air sacs and their surrounding capillary beds.*

150 million alveoli in each lung (Figure 7-18). It is through these thin-walled chambers that gases are exchanged with the capillary beds in the lungs.

Structures as tiny and thin-walled as the alveoli would normally tend to collapse, because the attractive forces between the molecules at the surface of a small, hollow, spherical object are high enough to pull the walls inward. To counter this tendency, special cells in the alveolar walls produce a chemical that has the ability to decrease greatly the surface tension at the alveolar surface. This chemical, called a surfactant, enables the alveoli to expand and contract without collapsing as air moves in and out of the lungs. The importance of this chemical is emphasized when infants are born lacking it. This results in a condition called hyaline membrane disease, which is frequently fatal to newborns.

Alveoli are closely enmeshed in capillary beds in the lungs so that the exchange of gases between the air in the alveoli and blood in the capillaries is rapid and efficient.

Basic Concepts

The respiratory tract consists of the nose, mouth, pharynx, larynx, trachea, bronchii, and lungs.

Alveoli are the sites of gas exchange with the blood.

Functions of the Respiratory Tract. The respiratory tract obviously provides a pathway for air to enter and to leave the lungs, but it has some additional important functions. One of these is to change the temperature of the inhaled air to approximately that of the body. Even very hot or cold air is adjusted to body temperature by the time it reaches the lungs. Moisture is also added to inhaled air. This air-conditioning and humidifying effect is brought about by the abundant blood supply in the mucus membranes of the nose, mouth, and pharynx. Another vital function of the upper respiratory tract is to filter particles from the air before they reach the tissues of the lungs themselves. The tract is coated with mucus, to which particles including microorganisms tend to stick as the air flows by. Tiny cilia continually push the mucus and its trapped particles upward into the pharynx and thus out of the respiratory system. The movement of air over the folds of tissue in the larynx (the vocal cords) provides a sound-producing mechanism of considerable versatility.

Basic Concept

Functions of the respiratory tract include providing an air passageway to the lungs; moderating the temperature of inhaled air; adding moisture to inhaled air; filtering pollutant particles from inhaled air and removing them from the tract; and producing sound.

Getting Air Into the Lungs. Relatively large volumes of air must be moved in and out of the lungs at frequent intervals. This movement is accomplished by the diaphragm and rib muscles. The diaphragm is a sheet of muscle in the form of a dome over the liver and stomach, which forms a partition between the thoracic region and the abdominal cavity. In the thorax, each lung lies sealed in a separate lung chamber. When an individual inhales, the diaphragm contracts to become flattened. This action pulls downward on the floor of the lung chambers and thus increases their size. At the same time the rib muscles usually rotate the ribs up and outward, thereby further increasing the volume of the chambers. Each lung is left, in effect, suspended in an expanding chamber in which the air pressure becomes less than that of the atmosphere around the organism. Therefore, atmospheric air rushes into the lungs, via the respiratory tract, and inflates them (Figure 7-19). The lungs themselves contain no muscle tissue and thus are passive during inhalation; atmospheric pressure alone inflates them.

Exchange of oxygen and carbon dioxide rapidly occurs. The diaphragm and rib muscles then relax, and air is exhaled with the aid of elastic tissue in the lungs. One of the dangers of a chest injury is that a lung chamber may be punctured, keeping the air pressure inside the same as the atmospheric pressure. A punctured lung cannot inflate and is said to be *collapsed*. An individual can breathe with only one lung but suffocates if both collapse.

Basic Concept

Atmospheric pressure inflates the lungs when the lung chambers are enlarged by the action of the diaphragm and the rib muscles.

Gas Exchange and Transport. The tiny alveoli are thin-walled, always moist, and closely surrounded by capillaries. When air enters the alveoli, oxygen molecules diffuse rapidly through the alveoli membranes and into the blood in adjacent capillaries. This movement occurs because the capillary blood is relatively poor in oxygen compared with the alveolar atmosphere. In like manner, carbon dioxide diffuses in the opposite direction out of the blood and into the alveoli.

Because gas exchange can occur only through alveoli, any damage or change in these tiny air sacs interfere with their function. Respiratory infections such as *pneumonia* cause the sacs to fill with fluid, thus decreasing their ability to exchange gases. Another serious condition is *emphysema*, in which the partitions between adjacent alveoli break down, resulting in large cavities with a decreased surface area for diffusion. In addition, the lungs lose their elasticity and breathing becomes very difficult. Emphysema is a degenerative disease that most often affects elderly individuals. However, constant exposure of lung tissue to cigarette smoke is known to aggravate and accelerate the breakdown of alveolar walls.

As oxygen enters the capillaries, most of it attaches to molecules of hemoglobin, an oxygen acceptor, within the red blood cells. Hemoglobin enables blood to transport far more oxygen—about 65 times more—than could be carried by a simple fluid. The oxy-

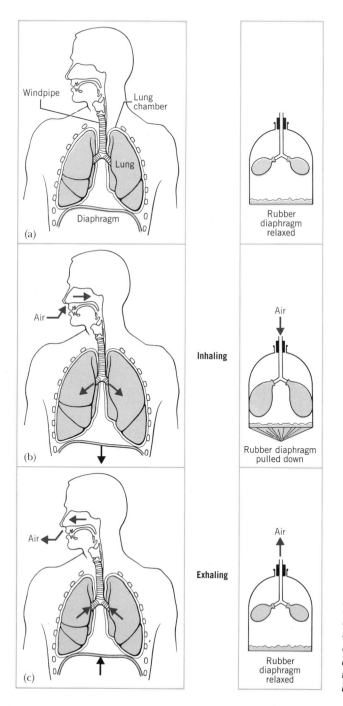

Figure 7-19 *How air enters the lungs. Diagram* (a) *shows the lungs in separated chambers and the diaphragm in its relaxed, convex-shaped position. In* (b) *the diaphragm flattens to help enlarge the lung chambers. Atmospheric pressure inflates the lungs. In* (c) *the diaphragm relaxes, the elastic lung tissue contracts, and air leaves the lungs. These actions can be simulated with a bell jar and balloons as shown at the right.*

gen-rich blood is returned to the heart and then is pumped out to capillary beds all over the body. Here the oxygen leaves the hemoglobin molecules, diffuses through the capillary walls, and becomes part of a cell's chemical machinery.

Carbon dioxide is much more readily soluble in blood plasma than oxygen is, and hence does not require a special carrier. Even so, from 10 to 27 percent of the waste CO_2 combines with hemoglobin and is transported by the red corpuscles. Most of the remaining CO_2 dissolved in the blood forms several simple compounds or ions such as carbonic acid (H_2CO_2) and bicarbonate (HCO_3-). A small amount, about 8 percent, is carried as dissolved CO_2.

Considering the billions of cells that constitute a human organism, and their absolute necessity to exchange oxygen and carbon dioxide continually, the efficiency of the systems that supply and distribute these gases is truly remarkable.

Basic Concept
Oxygen is transported in the blood by hemoglobin in the red blood cells. Carbon dioxide is transported by hemoglobin, in combined forms, and in the simple dissolved state.

Regulation of Respiration. A system as complex as respiration obviously needs to be closely regulated. The medulla of the brain and some of the major blood vessels contain CO_2 sensors, cells that are sensitive to carbon dioxide. When stimulated by the buildup of CO_2 in the blood, these sensors send impulses to the nervous system, which in turn increases the rate of breathing. In addition there are sensors for detecting increasing acidity in the blood, usually caused by the rise in CO_2, as well as sensors that detect a decrease in oxygen in the blood. In addition to all of these detectors, there are other cells that respond to changes in

blood pressure and cause the breathing rate to change accordingly.

The function of all these sensors is obviously to rid the body of excess CO_2 and to replenish its oxygen. However, many aspects of the functioning of this seemingly "obvious" system are not understood even after years of research. For example, when an individual exercises moderately, his or her breathing rate quickly increases. This is what one would predict, as the body's muscle cells use up oxygen and give off carbon dioxide. But the problem with this logical prediction is that blood samples taken from a moderately exercising individual show virtually no change in carbon dioxide or oxygen levels. The stimuli that cause the increase in breathing rate in this situation simply are not known.

Basic Concept
Breathing rate is controlled by chemical and pressure sensors located in the brain and in large blood vessels.

In closing this chapter, we might comment that the quote at the beginning of Chapter 6—about the digestive tract being an "ingenious assembly of portable plumbing"—would also be appropriate for the circulatory and respiratory systems. All three systems are amazingly well adapted to the particular functions they each serve. This observation reinforces again the principle that form follows function. The digestive and respiratory systems are so well integrated with the circulatory system that is hard to imagine any better way to design them or improve them. Even with the considerable abuse that human beings heap on the systems with air pollutants such as tobacco smoke, drugs such as alcohol, and diets that are minimal in essential nutrients, the body manages to function quite well for long periods of time.

KEY TERMS

aorta	atherosclerosis	capillary bed	collateral circulation
arteriole	atrioventricular node	cardiovascular fitness	coronary artery
artery	blood	closed-type circulatory system	counter-current exchange

diastole

edema

emphysema

epiglottis

heart chambers

hemoglobin

hepatic portal system

leucocytes

lymph

lymphatic system

lymph node

open-type circulatory system

pacemaker

plasma

platelets

pneumonia

pulmonary circulation

pulmonary veins

red corpuscle

renal arteries

simple circulatory system

sinoatrial node

stroke

systemic circulation

systole

tracheae

vasoconstriction

vasodilation

vein

vena cava veins

venule

SELF-TEST QUESTIONS

1. (a) Define what is meant by a "simple" circulatory system, and (b) give an example.
2. Describe an open-type circulatory system.
3. What is meant by a closed-typed circulatory system?
4. In what sense is blood a liquid tissue?
5. What is blood plasma?
6. Why is it not harmful for a healthy person to donate blood?
7. How does the form of a red blood corpuscle reflect its functions?
8. What is the function of iron in hemoglobin molecules?
9. What is the overall function of leucocytes?
10. How are blood platelets vital to the circulatory system?
11. Name some of the materials transported by blood.
12. (a) What is the major function of capillaries?
    (b) Discuss the functions of vasoconstriction and vasodilation.
13. Name the two forces that account for substances moving out of and into capillaries.
14. What produces edema?
15. Name some functions of the lymphatic system.
16. Under what condition is lymph drainage detrimental to the body?
17. List four major venous drainage systems and the area of the body that each one serves.

18. Trace the flow of blood through the heart beginning with the right atrium.
19. Explain the difference between pulmonary circulation and systemic circulation.
20. How would you explain what the figure 115/60 means to a person who knows nothing about blood pressure?
21. Briefly describe the nodal tissues of the heart and how they work.
22. Where is the coronary artery and what does it supply?
23. What is meant by collateral circulation?
24. Name the four basic features that all respiratory structures must have in order to function properly.
25. List four types of respiratory structures.
26. What is meant by countercurrent exchange?
27. Name the major parts of the human respiratory tract.
28. What are some functions of the human respiratory tract in addition to breathing?
29. Describe how atmospheric pressure is used to inflate the lungs.
30. How are oxygen and carbon dioxide transported in the blood?
31. Describe the system that controls the breathing rate in the human body.

HORMONES

chapter 8

questions to think about

What are the major roles of hormones in the body?

What happens when hormone production is abnormal?

Are hormones produced only from endocrine glands?

What is a feedback system?

How does a hormone affect only its target tissues?

Man may be the captain of his fate but he is also victim of his blood sugar.

Wilfred G. Oakley

What does the above quote have to do with hormones? The answer to this is not simple, but it dramatically emphasizes the basic role of hormones in our lives. Regulation of blood sugar is controlled by hormones. If this control mechanism fails, the consequences are serious and, not infrequently, fatal. In this sense we are truly victims of our blood sugar.

A hormone is a chemical that is formed at a particular site in the body, distributed by the blood, and produces a specific effect on cells (the target tissues) in some other part of the body. Hormones are immensely important biological materials that affect almost all aspects of an organism's life. Their influence extends to growth and development, many of the organism's metabolic activities, and even a lot of its behavioral activities. Yet despite their great importance, our knowledge about hormones has developed only slowly because they are so difficult to study. The gland or tissue that produces the hormone is often located far away from its target tissue. And the amount of hormone carried in the blood is so small that it cannot often be measured or even detected by laboratory tests.

Even today, most people have only vague and often distorted ideas about what hormones are, where they come from, and what they do in the body. One common misconception, for instance, is that obesity is often caused by hormones or glands that are not functioning properly. In reality, we know that obesity is almost always due to an excessive intake of calories, and hormonal disorders rarely contribute to it.

Other misconceptions abound. For centuries, sex glands (usually testicles) were reputed to restore male virility if eaten regularly. (Even today, "mountain oysters"—bull or ram testicles—are listed on the menus of many expensive restaurants.) Only a few years ago, a physician in western North Carolina became widely known for transplanting goat testicles into elderly men who suffered various sexual problems, usually impotency. This unusual treatment was eventually stopped by legal action, but not before the doctor became wealthy. Some of his patients strongly defended the transplant treatment, claiming that it restored their sexual activities. One suspects that the power of suggestion was more potent than the transplants, since there is no evidence that goat testicles survive very long in the human body.

More recently, hormones have frequently been in the news in connection with individuals who have had

(a)   (b)

Figure 8-1 *Results of a sex change operation. In (a) Dr. Richard Raskind is shown prior to the sex-change procedure. In (b) the same individual is shown after the sex-change transformation and a change of names to Dr. Renee Richards.*

sex change operations. With the skill of a plastic surgeon's scalpel and the proper administration of sex hormones, an individual can be changed into the opposite sex, at least superficially (Figure 8-1). To a biologist this is not especially surprising, since the great potency of male and female sex hormones was established long ago.

Basic Concept

Hormones are chemicals produced by special cells or tissues, transported in extremely tiny amounts by the bloodstream, which often affect tissues far from the cells or glands that produce them.

METHODS OF STUDYING HORMONES

Removal of Glands

The discovery and subsequent study of hormones began with investigations of the functions of various glands in animal bodies. Especially perplexing to early investigators were glands that had no drainage ducts, such as the thyroid. If there was no passageway to carry their secretions about the body, how did such glands function? The classic method of studying the ductless glands was to remove them from an animal and then observe what bodily functions were affected. Transplanting the same kind of gland into the body, or injecting an extract of the missing tissue should then restore the affected function. Although these techniques sound simple, in practice there are many complications. The operation itself may have side effects that obscure the purpose of the experiment, because structures in addition to the gland are usually damaged. And if a small piece of the gland is inadvertently left in the animal, the experiment will probably fail because the gland continues to produce the hormone. An additional complication is that some glands with ducts, such as the pancreas, produce hormones in addition to other products. Because of complications such as these, it took many years of experimentation to demonstrate conclusively the functions of the major ductless glands of the body.

One of the earliest investigations using the above approach was performed by an Austrian physician, A. A. Berthold, in 1849. His experimental subjects were six young male chickens (cockerels). Berthold removed both testes from two of the cockerels. Later he observed that their behavior became timid, their combs and wattles were pale and poorly developed, and their voices were unlike those of typical young roosters (Figure 8-2). In two other cockerels, Berthold cut the testes free, removing one from each bird but leaving the other one loose in the body cavity. These two birds subsequently developed as normal roosters in all respects. Finally, in the last two cockerels, both testes were removed and a testis from each was placed in the body cavity of the other bird. These two cockerels also developed normally. This simple but elegant experiment clearly demonstrated three things. First, complete removal of the glands resulted in abnormal sexual development and behavior. Second, the glands could function even when moved from their normal position in the body. Third, a transplanted gland could function as well as a normal gland. Berthold also concluded from his experiments, correctly, that the secretion from the testes must be carried by the blood.

The next step, identification of the hormone itself, took nearly another century, however. Not until 1935 was the hormone testosterone finally isolated and identified.

Clinical Studies

Another valuable source of information about the functions of endocrine glands and their hormones

(a)

(b)

Figure 8-2 *The results of castration in a young male chicken. In* (a) *a young cockerel (male) is shown with its typical comb, wattles, and tail feathers. After a young cockerel's testes are removed as in* (b), *the comb and wattles fail to develop and the bird resembles a young female.*

came from clinical studies on human beings. Long ago it was noted that degeneration or loss of the thyroid gland in children led to mental retardation and dwarfism, a condition known as *cretinism* (Figure 8-3). In adults such a loss results in a condition called myxedema, characterized by a puffiness of the skin caused by fluids accumulating there. As another example,

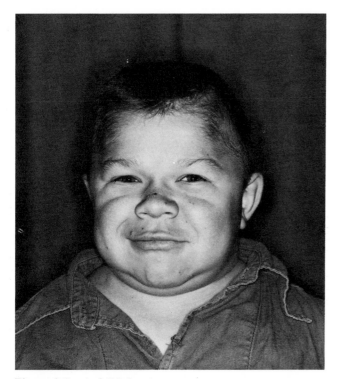

Figure 8-3 *A child showing cretinism, the tragic consequence of the loss of the thyroid gland. If detected in time, thyroid extract can be administered and the child will develop normally.*

diseases such as tuberculosis may destroy the adrenal cortex and produce a series of symptoms that is known as *Addison's disease.* As these and other glandular diseases became better known, it was found that some could be treated by administering extracts of the damaged glands. Today, purified hormonal extracts from the glands and synthetic hormones are available for such treatments. Table 8-1 lists the symptoms that result from a deficiency or an excess of certain hormones. Much of this information was derived from clinical studies.

Bio-assay Tests

If hormones are transported in the bloodstream, then why not simply analyze a blood sample for their presence or absence? The problem with this approach is how to detect the extremely small amounts of hormones that occur in body fluids. A technique that provides some method of doing this is the *bio-assay*— using living tissues or organisms that respond to small amounts of hormones.

One well-known bio-assay technique is the classical "rabbit test" for human pregnancy. The test is based on the fact that urine from a pregnant woman contains hormones associated with pregnancy. If a quantity of this urine is injected into a female rabbit, it causes the rabbit to ovulate within a short time. To see whether this reaction has taken place, the rabbit must be anesthetized and cut open in order for the lab technician to examine the rabbit's ovaries. This test has been replaced today with a relatively simple antigen-antibody test that can be performed in minutes, is more accurate than the older method, and is available in kit form.

The antigen–antibody reaction is a very sensitive bio-assay technique. Antibodies are used to test samples that may contain the hormone. If the hormone is present, it reacts with the antibodies to produce measurable results.

A modification of the antigen–antibody test can also be used to identify the cells that produce the hormone in a gland. To do this, some of the antibody is removed from the blood and "labeled" with a compound that fluoresces when viewed under ultraviolet light. When the fluorescent antibody is applied to sections of tissue containing the hormone, the precise location of the hormone in the tissue is revealed by the cells that show fluorescence. An application of this method confirms that insulin is produced only by special cells in pancreas tissue called beta cells.

Yet another method for locating the sites of hormone secretion is to use radioactive isotopes that are incorporated into the hormone. For example, if radioactive iodine is injected into an animal, its course can be traced through the body. The iodine concentrates in the thyroid gland and later shows up incorporated into the hormone *thyroxine.* This test helps verify the hypothesis that thyroxine is produced in the thyroid gland.

Radioactive isotopes can also be incorporated into synthetic hormones. When such a "tagged" hormone is injected into an organism, it can be traced to its site of action, that is, its target tissue. If radioactively tagged estrogen is injected into rats, it is later found in the anterior pituitary gland, the uterus, vagina, and

TABLE 8-1.   Major Sources of Human Hormones, and Their Functions

Source	Hormone	Functions	Deficiency	Excess
Thyroid gland	Thyroxine	Stimulate metabolism: regulate general growth and development	Cretinism	Graves' disease
	Calcitonin	Lowers blood calcium		
Parathyroid	Parathormone	Increases blood calcium; decreases blood phosphate	Muscle spasms	Calcium deposits
Pancreas	Insulin	Lowers blood glucose	Diabetes	Hypoglycemia
	Glucagon	Increases blood glucose	Hypoglycemia	
Adrenal medulla	Epinephrine (Adrenalin)	Increases metabolism in emergencies		
	Norepinephrine (Noradrenalin)	As above		
cortex	Glucocorticoids and related hormones	Control carbohydrate, protein, mineral, salt, and water metabolism	Addison's disease	Cushing's syndrome
Pituitary anterior	Thyroid stimulating hormone	Stimulates thyroid gland function		
	Adrenocorticotropic hormone (ACTH)	Stimulates adrenal cortex	Hypoglycemia	Cushing's syndrome
	Growth hormone	Increases body growth	Dwarfism	Gigantism, acromegaly
	Gonadrotropic hormones	Stimulates gonads		
	Prolactin	Milk secretion		
posterior	Vasopressin (ADH)	Water retention by kidneys		
	Oxytocin	Milk production		
Testis	Testosterone (Androgens)	Secondary sex characteristics, sperm production	Sterility	
Ovary	Estrogens	Secondary sex characteristics		
	Progesterone	Prepares uterus for pregnancy		
Hypothalamus	Hypothalamic releasing and inhibiting hormones	Release of hormones from anterior pituitary gland		
Kidney	Renin	Vasoconstriction		Increases blood pressure
	Erythropoietin	Production of red blood cells in bone marrow		
Gut wall	Digestive hormones	Digestion of food		
Thymus gland	Thymosin	Maturation of lymphocyte white blood cells		

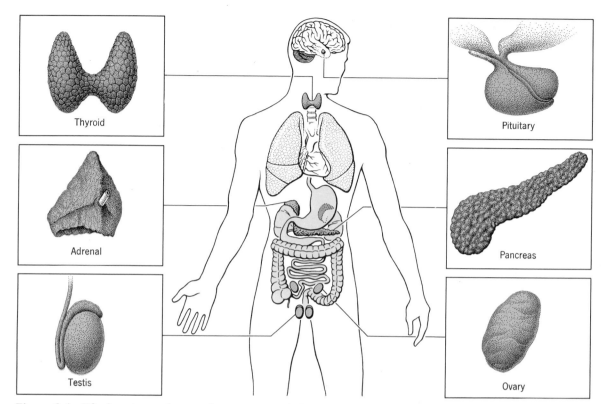

Figure 8-4 *The location and general appearance of glands that secrete some important hormones in human beings. Does each of these glands seem to be located in the part of the body where it brings about its action?*

certain areas of the brain. All of these organs are known to have a role in the function of estrogen.

Basic Concept

Methods of studying hormones include:

Removal of glands from laboratory animals and noting the subsequent effects on the animal

Observing the symptoms produced by diseased glands in human beings

Using living tissues that respond to small amounts of a hormone (bio-assay tests), and

Using radioactive isotopes.

WHERE HORMONES ARE PRODUCED

We usually think of the source of hormones as being specialized glandular tissue: the *endocrine,* or ductless,

glands. A list of these organs includes the thyroid, parathyroid, adrenal, and pituitary glands. However, a number of hormones are secreted by other kinds of tissues, including the wall of the digestive tract, the placenta, the seminal vesicles, kidneys, testes, ovaries, and the pancreas. Still other hormones are produced by neural tissues in the brain, such as the hypothalamus, and from nerve junctions throughout the body. Figure 8-4 shows the location of many of these glands in the body.

To complicate the picture even more, glands are known whose functions have been difficult to establish. The thymus gland, located in the chest region, is relatively large in infants but shrivels up in adults and becomes nonfunctional. In infants, the thymus produces a special type of white blood cell that functions in the immune system (discussed in Chapter 12). In addition, a hormone called thymosin occurs in thymus tissue, related to the production of white cells.

Until recently the function of the pineal gland in mammals was unknown, although the gland was known to produce a substance called melatonin. Evidence now indicates that melatonin inhibits the function of ovaries and testes in laboratory rats. In addition, its production is rhythmical, with more being produced in daylight hours than at night. (Such a day–night cycle is termed a *circadian,* "around a day," rhythm.) Whether human beings experience a circadian pineal rhythm is not known at present.

The following remarks are general comments about the locations and functions of some of the hormone-producing structures in humans.

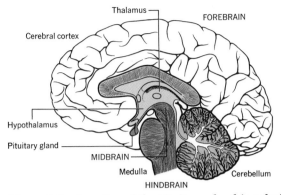

Figure 8-5 *Location of the pituitary gland in relation to other structures in the brain.*

Pituitary Gland

The pituitary gland, located at the base of the brain (Figure 8-5), is sometimes termed the master gland because its hormones have such widespread regulatory actions in the body (Figure 8-6). Its many hormones (see Table 8-1) affect the thyroid and adrenal glands as well as the ovaries, testes, and kidneys. Any damage or disease that affects the functioning of this small but vital gland may bring about drastic changes in the body.

The pituitary growth hormone, *somatotropin,* profoundly affects nearly every aspect of the growth of an individual. Its power is shown dramatically when the pituitary produces an excessive amount of somatotropin. In children, excessive growth hormone greatly accelerates the body growth and results in an abnormally large individual, a condition known as *gigantism* (Figure 8-7). In adults, the same condition causes rather grotesque changes in the form of various body parts, because by this time the ends of the long bones have ossified and can no longer increase in length. Thus, in adults, the hands and feet enlarge, and lower jaw protrudes excessively, the facial features become coarse, and body hair increases considerably. This unfortunate condition is termed *acromegaly* (Figure 8-8).

An insufficient amount of somatotropin inhibits growth and produces a type of dwarfism in which body proportions are normal in relation to the size of the individual. Growth is a complex event that is also influenced by several additional hormones, by heredity, and by nutrition. Hence dwarfism can be caused by a variety of factors in addition to a lack of somatotropin.

In African pigmies, for example, there seems to be an adequate level of growth hormone in their blood plasma, but for some reason the hormone does not stimulate their tissues to grow into what we think of as "normal" human size or height.

Thyroid Gland

The thyroid gland consists of two masses of glandular tissue, one on each side of the windpipe just below the voice box. When stimulated by TSH (thyroid-stimulating hormone) from the pituitary, the thyroid gland produces several hormones, of which thyroxine is the most abundant and best known. The thyroid hormones function primarily to regulate general growth and development of the body, and to control the rate of general metabolism of many body tissues. A list of additional secondary effects is found in Table 8-1.

An unusual aspect of the thyroid gland is that it is dependent on a sufficient dietary intake of iodine for synthesizing its hormones, which consist of iodine-containing amino acids. The thyroid's iodine demand is met easily in most diets. However, in the Great Lakes region of the United States and in Central Europe, where the soil is iodine-poor, crops are deficient in iodine and the residents may suffer iodine deficiencies (see Chapter 4). If the deficiency persists, the thyroid gland is unable to produce adequate amounts of thyroxine. The anterior pituitary gland responds by producing TSH (thyroid-stimulating hormone), but to no

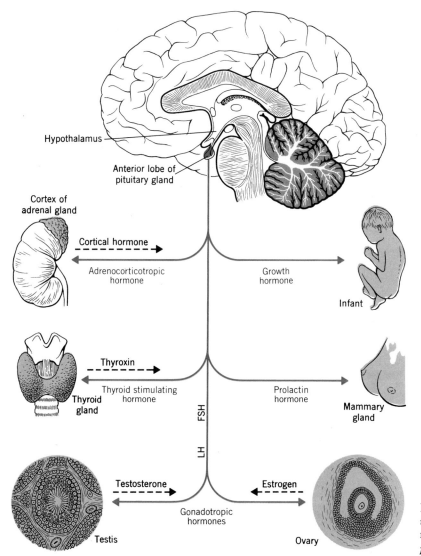

Hypothalamus

Anterior lobe of
pituitary gland

Cortex of
adrenal gland

Cortical hormone

Adrenocorticotropic
hormone

Thyroxin

Thyroid stimulating
hormone

**Thyroid
gland**

Testosterone

Gonadotropic
hormones

Testis

Growth
hormone

Infant

Prolactin
hormone

Mammary
gland

FSH

LH

Estrogen

Ovary

Figure 8-6 *Hormones produced by the pituitary gland. How does this support the term* **master gland** *that is sometimes applied to the pituitary?*

avail. Eventually the thyroid gland begins to enlarge, due to the excessive TSH stimulation. This swelling is called an iodine-deficiency *goiter* (Figure 8-9*a*). Unless properly treated, such goiters become huge lumps in the neck region. This symptom was so widespread 40 or 50 years ago in iodine-deficient areas that these regions were called "goiter belts." Today the deficiency is easily avoided if people use iodized salt (salt to which iodine has been added) or include marine fish in their diets. Because frozen seafood and iodized salt are readily available, goiters are no longer common.

A variety of diseases may affect the thyroid gland, causing it to enlarge and produce excess amounts of thyroid hormone (hyperthyroidism). A typical condition is *Graves' disease,* in which the symptoms include a goiter, protruding eyeballs (exophthalmos), nervousness, weight loss, and a trembling of the fingers and hands (Figure 8-9*b*). Hyperthyroidism may be treated by removing the thyroid gland (or a major part of it) or by administering antithyroid drugs that hinder the production of the thyroid hormone. Interestingly, cabbage, turnip, and rutabaga contain antithyroid

Figure 8-7   *Gigantism, the consequence of excessive growth hormone. The other members of the family shown here are of average height.*

Figure 8-8   *The development of acromegaly over a period of years.*

chemicals, and goiters have been recorded in individuals who ate large amounts of these vegetables over a period of time.

Adrenal Glands

An adrenal gland sits on the upper end of each kidney (see Figure 8-5). Each gland, however, is in reality two endocrine structures arranged so that one, the *adrenal cortex,* completely surrounds the other one, the *adrenal medulla.* The cortex and medulla produce different hormones and have no particular functional relationship to each other.

The adrenal medulla produces epinephrine (also called adrenalin) and a similar hormone, norepinephrine (noradrenalin). These hormones are not essential to life but evidently help the body during emergencies or in stressful situations. In general, they increase heart beat and blood pressure, stimulate the central nervous system for increased alertness, cause a rise in blood glucose, and increase the overall metabolic rate in the body. In performing these actions, the adrenal medulla mimics the action of the autonomic nervous system, because the nervous system also produces the chemicals epinephrine and norepinephrine at the nerve endings. The two systems, adrenal cortex and autonomic nerves, work together to help prepare an animal to fight off a predator or flee for its life. This action has been termed the "fight or flight syndrome." In actuality, the autonomic nervous system is more influential in this survival behavior than is the adrenal cortex. This is one of many examples of the overlap of regulatory functions between the nervous and endocrine systems.

The adrenal cortex is controlled (turned "on" and "off") by adrenocorticotropic hormone (ACTH), produced by the pituitary gland. Some of the adrenal cortex hormones are essential to life, having important and widespread metabolic effects throughout the body.

One group of cortical hormones, the glucocorticoids, of which cortisone is an example, have widespread effects on the metabolism of carbohydrates, fats, proteins, and water. An additional function of these hormones, not well understood, is their relation to stress in the body. For example, when an organism encounters a stressful (threatening or frustrating) situation, the pituitary gland increases its output of ACTH. This in turn causes a rise in the output of glucocorticoids. These hormones help the animal tolerate the stressful situation. So important is this function that experimental animals whose adrenal or pitui-

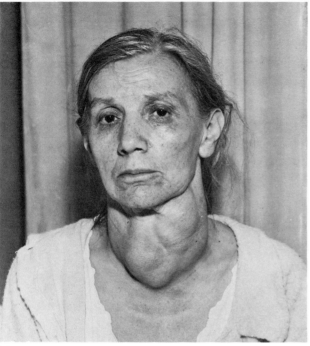
(a)

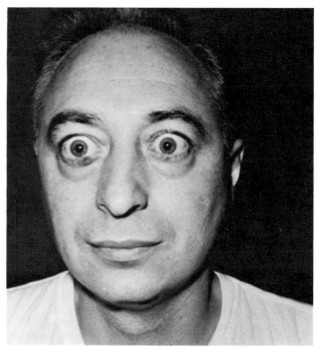
(b)

Figure 8-9 *Some symptoms of hyperthyroidism in two individuals. The person in* (a) *shows a goiter caused by growth of the thyroid gland. The individual in* (b) *shows greatly protruding eyeballs, another symptom of excess amounts of thyroid hormone.*

tary glands have been removed die when faced with stresses.

A second group, including aldosterone, is important in the regulation of the movement of sodium and potassium through various tissues in the body. The third group of cortical hormones includes small amounts of male and female sex hormones.

As you might imagine, an excess of glucocorticoids can have a drastic effect involving a variety of symptoms. The torso becomes obese, but the arms and legs remain thin. The face is rounded and full, but the hair is thin and scraggly. Wounds heal poorly and the skin bruises easily. Diabetes is a typical side effect, as is high blood pressure. These and additional characteristics taken together are termed *Cushing's syndrome* (Figure 8-10). One method of relief is to remove the adrenal glands altogether and then inject the hormones at the proper levels.

Addison's disease is the result of insufficient adrenal

hormones, sometimes caused by damage to the adrenal glands from cancer or tuberculosis. In this condition, the skin becomes diffusely tanned and develops spotty areas of dark pigmentation, the heart decreases in its size and capacity for work, and death results if the disease is not treated.

Basic Concepts

Hormones are produced from many sources including specialized glands, nervous tissues, and a variety of other tissues in the body.

The pituitary gland is termed the master gland because its hormones control the functioning of many other glands and organs.

Glandular disorders often produce abnormal conditions including gigantism and acromegaly (pituitary gland), goiters (thyroid gland), and Cushing's syndrome and Addison's disease (adrenal glands).

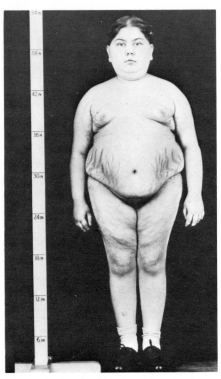

Figure 8-10 *Cushing's syndrome, the consequence of excess adrenal cortex hormones. The torso becomes obese, the face is rounded, and diabetes and high blood pressure are typical. One method of relief is to remove the adrenal glands.*

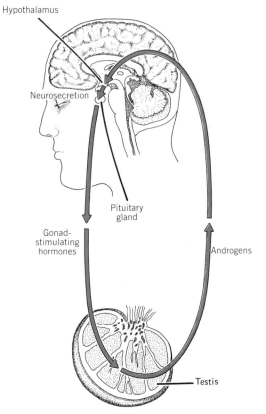

Figure 8-11 *The interrelationships of hormones produced by the hypothalamus, anterior pituitary gland, and testis in controlling reproduction in the male.*

HORMONES AND REPRODUCTIVE CYCLES

The development of secondary sex characteristics and the functioning of the mature male and female reproductive systems are brought about by sets of interacting hormones. This control is a complex one, involving the hypothalamus of the brain and perhaps other parts of the nervous system, the pituitary gland, and the ovaries and testes. This complex hormonal–nervous system interaction is evidently common to all vertebrates. The system functions generally as follows.

The hypothalamus, when stimulated, produces a hormone that is transported in the blood to the pituitary gland. This secretion stimulates cells in the anterior pituitary gland, causing them to release gonad-stimulating hormones. These hormones are in turn transported, via the blood, to the gonads (ovaries or testes).

In the male, the testes respond by forming sperm cells and by secreting male sex hormones called *androgens*. Androgens bring about development of typical male structures, body forms, and coloration as well as male behavior patterns typical for the species. As androgens build up in concentration in the blood, they eventually interact with cells in the hypothalamus to *inhibit* their production of neural hormones. This in turn slows down the anterior pituitary's formation of gonad-stimulating hormones. And as these hormones decrease, so does the production of androgens from the testes. With a drop in androgen concentration, the hypothalamus again stimulates the pituitary gland and so the cycle continues (see Figure 8-11). Such a mechanism, in which the controlling element is itself controlled by the element it controls, is called a *feedback mechanism.* In this case the androgens constitute the feedback element and the hypothalamus is the controlling element. Through feedback, hormonal levels are maintained at the level necessary for efficient functioning of the reproductive system.

A somewhat similar series of events occurs in females. The gonad-stimulating hormones stimulate egg formation in the ovaries and also help prepare the reproductive tract for the passage and care of the egg. Ovarian hormones (*estrogens*) parallel male androgens in bringing about the formation of typical female secondary sexual characteristics. The ovarian hormones also function in a feedback relationship with the hypothalamus. In most vertebrates — with the exception of man and some of our domesticated animals — the reproductive cycle is adjusted to function only during certain seasons of the year. These are the times when the young have the best chance for survival. The trigger for reproduction is usually an environmental stimulus such as increasing day length in the spring. In some manner not completely known, changes in the environment bring about the hormonal events that make the female receptive to mating. This is her period of *heat* or *estrous.* The same environmental stimuli usually also induce the male's typical reproductive behavior. This mechanism coordinates reproductive activity between the sexes and establishes a breeding season for the species.

Basic Concept

The reproductive cycle in vertebrates involves the brain, anterior pituitary gland, and gonads, all functioning together in a feedback system.

HORMONES AND TARGET TISSUES

Although hormones circulate throughout the body through the bloodstream, their actions are usually quite specific in that they affect only certain "target" cells or tissues. For example, insulin and glucagon affect only cells in the liver, fatty tissues, and muscle.

The mechanism for this specificity was a biological mystery for many years. Only recently have some of these mechanisms of hormone specificity become known. For example, in the case of the target cells for insulin and glucagon, there are special receptor molecules embedded in the cell surface membranes in the target tissues that bind specifically to the hormones. These particular receptor molecules have been identified as glycoproteins in the cell membrane. This knowledge is especially relevant to our understanding of diabetes. In one type of diabetes, which often occurs in older, obese people, there is a marked reduction in the number of insulin receptors in their target tissues, even though these individuals often have a higher than normal amount of insulin in the bloodstream. When such people lower their caloric intake, their insulin receptors *increase* in number, and their diabetic condition is greatly alleviated. Thus, in at least one type of diabetes, there is a significant relationship between obesity and insulin-binding sites.

Cell membrane receptors are known for a number of additional polypeptide hormones, that is, hormones made of amino acid chains. These include the hormones from the anterior pituitary gland as well as insulin and glucagon from the pancreas.

Second Messenger Concept

The presence of special receptor molecules or binding sites tells us why some cells and not others respond to a hormone, but it does not tell us how a hormone molecule, once bound to a cell, causes the cell to respond. Experimental evidence accumulated by Earl W. Sutherland and his coworkers at Vanderbilt University indicates what probably happens within the cell. When a hormone attaches to a receptor molecule, it activates an enzyme, *adenyl cyclase,* found in the membrane of the cell. This substance converts ATP in the cell into another chemical, cyclic adenosine monophosphate, AMP. The increased *cyclic AMP* level in turn affects certain cellular processes, usually the production of enzymes.

As an example, consider the case of the hormone glucagon, which functions in controlling blood sugar. The chain of events would occur somewhat as follows. In response to low amounts of blood sugar, glucagon is produced by the pancreas. When glucagon reaches the liver, via the bloodstream, molecules of the hormone attach to special receptor sites on cells in the liver. These cells, in turn, produce the enzyme adenyl cyclase in their cytoplasm. The enzyme stimulates the formation of cyclic AMP, which triggers the synthesis

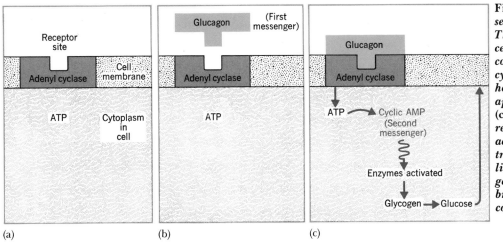

(a)   (b)   (c)

Figure 8-12 *Example of the second messenger concept.* **(a)** *The cell membrane with a receptor site for glucagon and containing the enzyme adenyl cyclase. In* **(b)** *glucagon hormone (the first messenger), approaches the receptor site.* **(c)** *Glucagon attaches to the receptor site, thereby releasing adenyl cyclase, which in turn triggers the formation of cyclic AMP (the second messenger). Cyclic AMP causes the breakdown of glycogen into glucose for use by the cell.*

of enzymes that can convert stored glycogen into blood sugar (glucose).

Other hormones would have different binding sites in different tissues. But in all cases, the hormone binding would produce cyclic AMP, and the cyclic AMP would then stimulate the tissue into the specific activity it was specialized to perform (Figure 8-12). Sutherland viewed the hormone in each case as the "first messenger" and the cyclic AMP as the "second messenger"; hence he called his theory the *second messenger concept.* For this work Sutherland received a Nobel prize in 1971.

Steroid Hormones

The steroid hormones (sex hormones and adrenal cortex hormones) have a different mode of action from that of the peptide hormones described above. Because of their small size and solubility in lipids, the steroids can diffuse into cells instead of attaching to their surface membranes. Inside the target cells, steroids encounter special protein receptor molecules in the cytoplasm and bind to them. The receptor–hormone complex then enters the nucleus, where it attaches to the DNA. In this location the hormones influence the DNA as it directs the activities of the cell (Figure 8-13). Thus, the steroid hormones bring about their regulatory activities in a way that is very different from that of insulin, glucagon, and other peptide hormones.

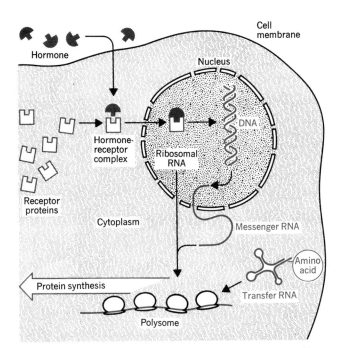

Figure 8-13 *Mode of action of steroid hormones. Small hormone molecules such as steroids pass through the cell membrane, into the nucleus, and influence the functioning of the DNA as it directs the making of proteins.*

Basic Concepts

Hormones of large molecular size attach to receptor molecules found in the membranes of their target cells. Cyclic AMP functions as the hormone's second messenger inside the cell.

Small hormones enter cells and attach to receptor molecules in the cytoplasm. The receptor–hormone complex enters the nucleus and affects the functions of the DNA found there.

HORMONAL FUNCTION: AN OVERVIEW

A major function of hormones is to regulate various aspects of the internal environment of the body. Hormones help control the amounts or levels of glucose, calcium, fluids, and other substances in the bloodstream; they influence the secretion of other hormones, cause the production of digestive enzymes in the gut, and stimulate smooth muscles to contract. They control the metabolism of carbohydrates, proteins, minerals, and salts, and they regulate numerous aspects of cellular respiration.

Certain other hormones play major roles in the development and maturation of the male and female sex organs, production of secondary sex characteristics, and regulation of the reproductive cycle, including the complex events of pregnancy and birth. Still other hormones regulate general growth, development, and specialization of tissue throughout the body. And finally, hormones play various roles in influencing the behavior of organisms. This role is especially evident in the changes in behavior that occur following sexual maturation, and in behaviors associated with courtship, mating, and care of offspring.

In short, hormones strongly affect almost all aspects of an organism's life. This is as true of human beings as of other animals. It is worth noting that even the life cycles of plants are considerably controlled by their own specific kinds of hormones. Events such as germination, growth, flowering, and reproduction are all chemically controlled within the plant. It appears that the use of chemical controls is a universal type of regulating device in all living things.

KEY TERMS

acromegaly	cretinism	gigantism	second messenger concept
Addison's disease	Cushing's syndrome	goiter	somatotropin
adenyl cyclase	cyclic AMP	Graves' disease	target cells
adrenal glands	endocrine gland	hormone	thyroid gland
bio-assay test	estrogens	pituitary gland	
circadian cycle	feedback mechanism		

SELF-TEST QUESTIONS

1. Define the term *hormone*.
2. Why have hormones been difficult to study in the past?
3. What was the early classic way of studying the hormone-producing glands?
4. What were the notable features of Berthold's experiments?
5. What is meant by a clinical study? Give an example.
6. Describe what is meant by a bio-assay test. Can you describe one?
7. How might a radioisotope be used to locate the site of hormone secretion?
8. Define what is meant by an endocrine gland.
9. Name some sites where hormones are produced, other than the endocrine glands.
10. List several hormones produced by the pituitary gland.
11. What is the link between diet and the production of thyroid hormone?

12. Are the adrenal glands very important to normal health? Defend your answer.

13. Although the pituitary gland is termed the master gland, what has control over the pituitary gland?

14. Describe how a feedback mechanism exists in the male and female reproductive cycles.

15. How do hormones interact only with their "target" cells?

16. Explain the second messenger concept.

17. How do the steroid hormones affect their target cells?

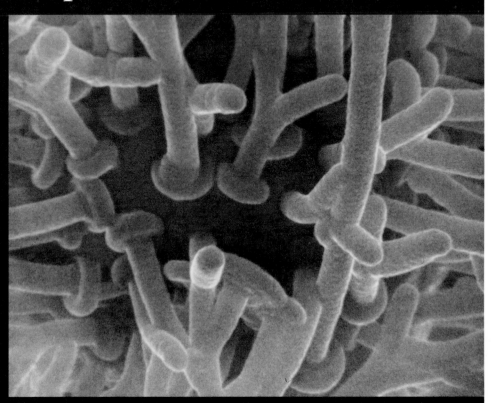

Synaptic knobs on the axons of a sea slug, Aplysia.

THE NERVOUS SYSTEM

chapter 9

questions to think about

What vital functions are served by the nervous system?

How is the nervous system organized?

What is a nerve impulse and how is it transmitted?

How does an impulse cross a synapse?

What is a reflex arc?

---

*Man is a machine into which we put what we call
food and which produces what we call thought.*

Robert G. Ingersoll

FUNCTIONS OF THE
NERVOUS SYSTEM

Nearly everyone knows that human beings and most
other animals have a nervous system, and that this
system has something to do with responding to stimuli
and initiating behavior. Indeed, one of the major characteristics defining "life" is *irritability,* the motor activity shown by animals when reacting to stimuli. Few of
us, however, are aware of the great extent of the vital
functions served by the nervous system. First of all, the
body (via the brain) is kept constantly informed of
changes in the environment by a variety of sensory
devices. These include the eyes, ears, nose, and temperature detectors, among others. Environmental
changes (stimuli) are converted into nerve impulses by
these organs and transmitted to the brain very rapidly

by a network of extremely specialized cells called
neurons.

Secondly, different parts of the nervous system
usually receive these stimuli and then make judgments
as to whether any further actions (responses) are necessary. If a response is required, the brain sends impulses
to the appropriate effectors (usually a set of muscles) to
act. Such actions require a considerable amount of
coordination, which is also accomplished by special
parts of the brain. Suppose at this moment you become
thirsty, rise from your chair, walk to the refrigerator,
remove and open a can, pour its contents into a glass,
return to your chair, sit down, and drink the beverage.
In doing these things you have performed a very complex series of neural and muscular actions involving
sensory inputs, balance, coordination of numerous
muscle groups, reflexes, memory, and probably some
decisions. (Hmm, should I drink a beer or a Coke?)

In addition, the brain has the incredible ability to act
as the repository for the memory of past events, a
function whose biological basis is not at all clear. Related to this capability is the equally outstanding array
of qualities we associate with our intellect: abstract

reasoning and problem solving, thinking, creativity of many kinds, esthetic appreciation, and a sense of morality, to name a few. Concentrated within the nervous system are clusters of nerve cells that are the centers of our emotions and desires, and other centers that control sleeping and dreaming, awareness, breathing, heartbeat, swallowing, and many other reflex activities. It is clear that the brain is the most complex and least understood organ in the body, and by far the most challenging research frontier in all of biology.

In addition to all of these functions, and often operating below our level of awareness, the nervous system is concerned with integrating the actions of the internal organs. Many of these organs, including the digestive tract, heart and arteries, and the bladder, are made of muscle tissue that must be coordinated in its actions. Glandular tissue is also connected with nerve fibers. All of this innervation (nerve supply) arises from a specialized part of the nervous system called the autonomic division.

In short, the nervous system represents the major controller and integrator of virtually all bodily actions. Its responses are usually extremely rapid and precise, and they can involve many parts of the body simultaneously. The nervous system is also closely tied in with the functioning of the hormonal system and, in fact, uses hormones extensively in its own activities.

MAJOR PARTS OF THE NERVOUS SYSTEM

The major organs of the nervous system are the *brain* and its associated *cranial nerves;* the *spinal cord* and *spinal nerves.* Some of these nerves control internal activities of the body, and are termed the *autonomic nervous system.* It is customary to refer to the spinal cord and brain together as the *central nervous system.* The cranial and spinal nerves are said to make up the *peripheral nervous system,* since they extend out into various parts of the body from the central nervous system (Table 9-1).

Spinal Cord

In human beings the spinal cord extends from the brain stem down through the vertebral canal (inside

TABLE 9-1   Major Parts of the Nervous System

I. Central nervous system
   A. Brain
   B. Spinal cord
II. Peripheral nervous system
    (cranial and spinal nerves)
   A. Nonautonomic portion
   B. Autonomic portion
      1. Sympathetic nerves
      2. Parasympathetic nerves

the bony backbone) for 16 to 18 inches (to about the level of the small of the back). The cord ends there, with only a small filament extending to the end of the backbone (Figure 9-1). The spinal cord is surrounded by protective membranes and bathed in a clear, watery spinal fluid. In a common medical procedure, a needle is inserted into the vertebral canal below the level where the spinal cord ends, and a small amount of this spinal fluid is withdrawn for diagnostic tests. Injecting anesthetics into the spinal fluid temporarily deadens sensations in the lower part of the body but does not render the patient unconscious. This procedure (called a saddleblock) is useful in aiding childbirth or during surgery on the lower part of the body.

A cross-section of the spinal cord (Figure 9-2) shows a central H-shaped portion, the *gray matter.* This material consists of neurons, including association nerve cells and the cell bodies of the spinal motor nerves. It appears gray because these cells lack the white myelin coating found around the axons of many nerve cells. The gray matter is the site of reflex actions in the spinal cord, a function described later in the chapter.

The remainder of the spinal cord consists of bundles of motor or sensory nerve fibers (nerve tracts) extending to and from the brain to various levels in the spinal cord (Figure 9-2). Because the neurons in these tracts are myelinated, this part of the cord is termed the *white matter.* The location and specific function of each tract became known as a consequence of experiments involving staining.

When neurons are damaged and begin to die, they stain differently from normal neurons when treated with certain biological dyes. In one procedure, a particular bundle of neurons in the spinal cord of a rat is

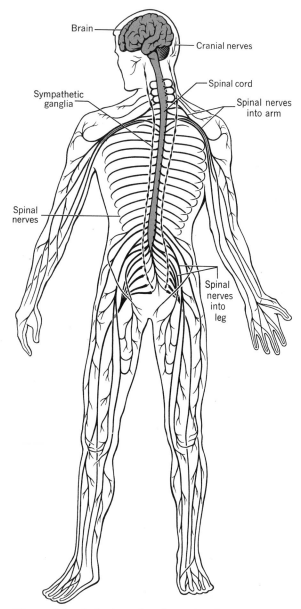

Figure 9-1 *Spinal cord and associated structures.*

cut. Later the animal is sacrificed and extremely thin slices of its spinal cord and brain are prepared. When stained appropriately, the cut nerve tract can be followed through different areas of the spinal cord and brain. A similar technique has been used with human patients who have suffered damage to their spinal cords and subsequently died. By means of such experi-mental and clinical experiences, the pathways of major nerve groups in the central nervous system have been traced. This enables neurologists to locate accurately the site of damage to the spinal cord from accidents or tumor growths.

If the spinal cord is completely severed, all body sensations and muscular activity are lost below the level of the cut. Reflex (involuntary) muscular activity eventually returns within a few days in some animals, and in several months in a human patient. However, sensation and voluntary motor activity are perma-nently lost. If the spinal cord in the neck region is severely damaged, the patient usually experiences pa-ralysis of both arms and legs and is termed a quadriple-gic. Injury to the cord at a lower level may result in paralysis of the lower extremities only, a paraplegic state.

Basic Concepts

The spinal cord consists of bundles of nerve fiber tracts (the white matter), and association neurons and nerve cell bodies (the gray matter).

The spinal cord is the site of reflex actions and the trans-mission of nerve impulses to and from the brain.

Damage to neurons in the spinal cord is usually perma-nent.

Peripheral Nerves

Nervous tissue located outside the central nervous system is designated the peripheral nervous system. Twelve pairs of *cranial peripheral nerves* are connected with various parts of the underside of the brain. Most of them innervate structures in the head region. One, the vagus nerve, is a major nerve supply for many of the body's internal organs.

Joining the spinal cord are 31 pairs of *spinal nerves* beginning in the neck region and continuing to the end of the cord (Figure 9-1). Spinal nerves are mixed, containing both motor and sensory neurons. They innervate the skeletal muscles, skin, and internal organs. Just before entering the spinal cord, each spi-nal nerve divides into two parts: a *dorsal root*, which contains a swelling, and a *ventral root* (Figure 9-2). The dorsal root contains the sensory (incoming fibers) and the ventral root contains the motor axons going to muscles. The swelling on the dorsal root, the *dorsal root*

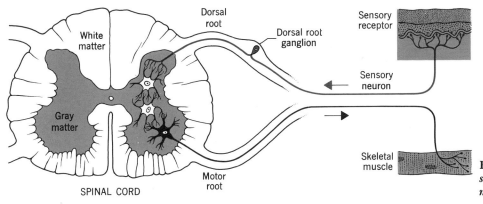

Figure 9-2 *Cross section of the spinal cord showing some of its major features.*

ganglion, contains the cell bodies of the sensory fibers.

If a peripheral nerve is cut, the structures that were connected to the nerve, such as skin and muscles, lose their sensation and ability to move. In vertebrates this damage is often permanent. The portion of the nerve on the far side of the cut slowly disintegrates because its fibers are no longer in contact with their respective nerve cell bodies. In the case of the motor fibers, however, the stumps of the axons attached to the cell bodies may slowly grow and attempt to reestablish contact with the muscles they formerly served. This regeneration can succeed if enough remnants of the old nerve remain to more or less guide the new axons as they grow. In this fashion, the useless muscles in a hand (resulting from an injured nerve in the lower arm) may slowly regain their ability to contract. Some degree of sensation may also be regained if sensory nerves in adjacent areas grow into the skin formerly served by the cut nerve. Today it is not rare for a surgeon to sew a severed finger or hand back on, provided that the appendage is not crushed or mangled. If the repair is performed promptly after the accident, there is a reasonable chance that the individual will regain use of the appendage.

Basic Concepts

Peripheral nerves lie outside the central nervous system, and consist of cranial nerves from the brain, and spinal nerves attached to the spinal cord.

Spinal nerves innervate the skin, muscles, and viscera of the torso of the body. Each spinal nerve divides into a dorsal (sensory) root and a ventral (motor) root before entering the spinal cord.

The autonomic nerves innervate the smooth muscles found in the internal organs and all glandular tissues. The number of organs served by autonomic nerves is impressive: the salivary glands, heart and blood vessels, lungs, the entire digestive tract, the kidneys, bladder, and external genitalia, to name a few (Figure 9-3).

In general, autonomic nerves enter and leave the spinal cord within the spinal nerves. In addition, there are two chains of autonomic ganglia (enlargements), one on each side of the backbone in the chest and abdomen (see Figure 9-1). These ganglia contain motor neurons of the autonomic nervous system. Additional autonomic ganglia can also be found out near the organs they serve. One such clump of ganglia lies in the abdomen near the diaphragm and is known as the celiac or *solar plexus.* You may have had the unpleasant experience of being struck in the abdomen and being unable to breathe for a brief period of time (which probably seemed like hours). Such a blow affects the nerves in the solar plexus that supply the diaphragm so that this large breathing muscle is, in effect, temporarily paralyzed.

From the functional standpoint, the autonomic nervous system is divided into two distinct parts. One is made up of fibers from the brain and the sacral region and is known as the *parasympathetic* division of the system (Figure 9-3). This unusual combination of "brain-tail" nerves functions as a unit because the ends of its motor nerves produce the transmitter agent *acetylcholine.* The functions of transmitter agents are discussed later in the chapter.

The other part of the system consists of autonomic

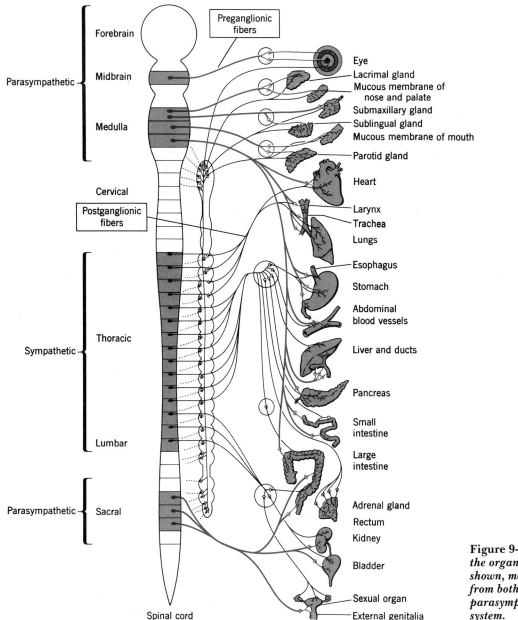

Preganglionic fibers

Forebrain

Midbrain

Medulla

Parasympathetic

Cervical

Postganglionic fibers

Thoracic

Sympathetic

Lumbar

Parasympathetic

Sacral

Spinal cord

Eye
Lacrimal gland
Mucous membrane of nose and palate
Submaxillary gland
Sublingual gland
Mucous membrane of mouth
Parotid gland
Heart
Larynx
Trachea
Lungs
Esophagus
Stomach
Abdominal blood vessels
Liver and ducts
Pancreas
Small intestine
Large intestine
Adrenal gland
Rectum
Kidney
Bladder
Sexual organ
External genitalia

Figure 9-3 *Autonomic nerves and the organs they innervate. As shown, most organs receive fibers from both the sympathetic and parasympathetic portions of the system.*

nerves attached to the middle part of the spinal cord and is called the *sympathetic* division (Figure 9-3). The ends of most of its nerve fibers produce norepinephrine. This transmitter (and hormone) is generally associated with energy-releasing actions in the body. In stressful situations, norepinephrine prepares the body for emergency actions—the "fight or flight"

syndrome discussed in Chapter 8. A human being in such an excited physiological state can sometimes perform an astounding muscular feat such as lifting one end of an automobile from an accident victim lying beneath it.

Most of the viscera—internal organs—are innervated by both the sympathetic and parasympathetic systems. The function of this arrangement is to allow the action of an organ to be speeded up by one system and slowed down by the other.

To cite a few examples, the vagus nerve of the parasympathetic system slows down the rate of heartbeat, but the sympathetic system stimulates heart action. The vagus nerve causes the muscles in the wall of the stomach and small intestines to contract, but impulses from sympathetic nerves inhibit muscle action in these organs. Parasympathetic innervation stimulates the bladder and colon, whereas sympathetic nerves inhibit these organs. This type of control, with stimulation balancing inhibition, is necessary for the successful functioning of these organs, because there are times when they should be active and times when they should be inactive.

The actions controlled by autonomic nerves are often called *involuntary* because they are not usually subject to willful or conscious control by the individual. The rate at which the heart beats is closely adjusted to bodily activity and not to the wishes of the individual. But does this mean that an animal cannot *learn* to alter its visceral activities? Experiments performed with rats in recent years indicate quite the contrary. By subjecting rats to conditioning, the rats learned to increase and decrease their heart rates, blood pressure, and the contraction of the intestinal wall, to control the diameter of their blood vessels, body temperature, and to influence their rate of urine formation.

This conditioning technique sometimes called *biofeedback*, has been also applied with some success to human patients to modify high blood pressure, to slow down rapid heart rates, and to train epileptic patients to suppress abnormal brain waves. Experienced meditaters reportedly experience significantly reduced heartbeat and breathing rates while meditating. Whether this is due to the extensive relaxation of the body during meditation or to some unexplained influence of meditation on the viscera or autonomic nerves is a matter of debate. At any rate, visceral learning and meditation indicate that autonomic nerve action is not as entirely involuntary as once was believed.

Basic Concept

Autonomic nerves supply the viscera and are arranged in two antagonistic sets: sympathetic (with epinephrine re- leased at its synapses) and the parasympathetic nerves (with acetylcholine as its transmitter).

OTHER TYPES OF NERVOUS SYSTEMS

The organization of the nervous system discussed so far is typical of that found in the vertebrates, but is by no means the only type or even the most common type found in the animal world. A brief look at one of the other types provides an interesting comparison with the vertebrate system.

The most common type of nervous system in animals is the arthropod type found in insects and in crustacea such as the crayfish. These are complex animals with many sense organs and many intricate behavior patterns. One would expect to find a rather complex nervous system accompanying these features. Let us look briefly at the plan of this system, as found in an insect.

The basic plan of the arthropod system is that of a chain of ganglia extending from the head region into the abdomen (Figure 9-4). The anterior ganglia of the head region are usually fused into a *cerebral ganglion* termed a *brain.* Also present are large optic lobes connected to the large compound eyes typical of many arthropods. Extending from the brain are motor and sensory fibers that serve various structures in the head region. The brain is not comparable in its various parts to a vertebrate brain.

The brain is connected to another sizable body, the *subesophageal ganglion,* located on the underside region of the head. Its fibers innervate the mouthparts primarily. Extending from this ganglion are connectives (nerves) to the ventral nerve cord, a chain of paired ganglia lying in the thorax and abdomen. Each pair of ganglia sends nerves to the muscles and receives sensory information from sense organs located near the ganglion.

In addition to the regular neurons that help interconnect the ganglia of the ventral nerve cord are some large nerve cells called *giant fibers.* It has been suggested that the giant fibers function in quick reflexive movements such as leaping to escape a predator — or a fly swatter!

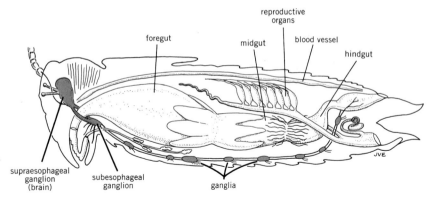

reproductive
organs

foregut

midgut

blood vessel

hindgut

supraesophageal
ganglion
(brain)

subesophageal
ganglion

ganglia

Figure 9-4 *Arthropod-type nervous system found in a grasshopper and many other invertebrates. How does its location in the body differ from that found in vertebrates?*

Finally, there is a ganglion that supplies many of the internal organs with nerve fibers—a visceral nervous system somewhat analogous to the autonomic system in vertebrates but far simpler.

Hence we find that the arthropod nervous system is radically different in its anatomy from the vertebrate type, is considerably simpler, contains far fewer nerve cells (100,000—200,000 compared with the billions in a vertebrate system), and yet manages to handle the complexities of flying, leaping, swimming, food gathering, courtship, mating, and complex social behavior. As a matter of fact, the complexity of arthropod behavior implies an extraordinary efficiency in neural organization and functioning—perhaps even greater than that of the so-called higher evolved vertebrate nervous system.

Basic Concept
The arthropod type of nervous system consists of a ventrally located chain of ganglia and giant nerve fibers. The brain is made of several fused ganglia.

THE NEURON

The building blocks of the nervous system are specialized cells known as neurons. The term *neuron* is derived from a Greek word meaning sinew or nerve, which probably reflects the fiberlike appearance of nerves in the body. We know now that a nerve is not a single neuron but rather a bundle of thousands of microscopically small individual neurons grouped like wires in a cable.

The function of neurons is to process and transmit information from one part of the body to another. Accordingly, even though they are only a few ten-thousandths of an inch in diameter, many of them are up to several feet long. In Chapter 3 we saw how neurons exemplify the form–function concept because of their extreme specialization for conducting nerve impulses. Not all neurons have the same form, however. Some, especially in the brain, are short and thick, with cytoplasmic extensions in many directions (Figure 9-5).

A neuron typically consists of a *cell body* containing the nucleus, numerous cytoplasmic projections called *dendrites,* and one long projection called the *axon* (Figure 9-6). Dendrites form a mass of short branched fibers around the cell body, and together with the cell body receive signals from other neurons. The axon transmits impulses from the dendrites and cell body to some other region of the nervous system. The end of the axon is split into a small set of branches, each ending in small swellings (synaptic knobs) that lie close to the dendrites of other neurons. Nerve impulses that reach the synaptic knobs may chemically affect these dendrites. In this manner information flows from neuron to neuron until reaching its destination.

The axon of most vertebrate nerve cells is sheathed by many layers of a fatty white coating called *myelin.* Myelin consists of the cell membrane of a *Schwann cell* wrapped around the axon many times (Figure 9-7). The sheath is interrupted at short (one millimeter) intervals along the axon called *nodes of Ranvier.* Myelin is somewhat comparable to insulation around a wire as it increases the efficiency of transmitting nerve impulses along the axon. The impulse jumps along the

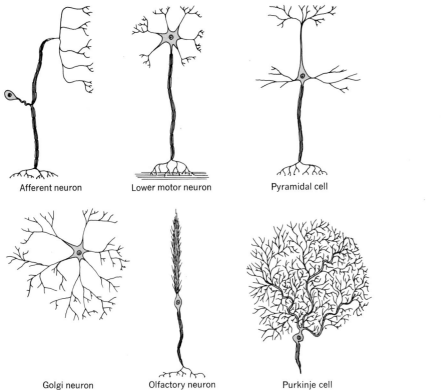

Afferent neuron

Lower motor neuron

Pyramidal cell

Golgi neuron

Olfactory neuron

Purkinje cell

Figure 9-5 *Neurons from different parts of the nervous system show a wide variety of forms. See if you can distinguish cell body, dendrites, and axon in each of the neurons.*

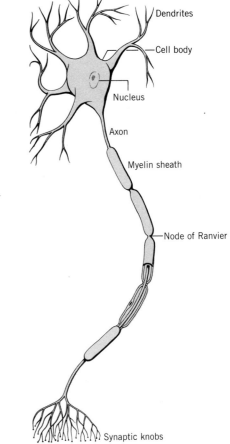

Dendrites

Cell body

Nucleus

Axon

Myelin sheath

Node of Ranvier

Synaptic knobs

Figure 9-6 *A generalized neuron and some of its major parts.*

axon, from node to node, thus increasing the speed of transmission. The importance of myelin is emphasized by the extremely debilitating disease, multiple sclerosis, which results when myelin is destroyed in the central nervous system. This produces muscle weakness, lack of coordination, and spasticity. Viruses are suspected as the cause of this degenerative condition.

Some nerve cells are specialized for detecting various changes in their environment; they are called *receptors.* They convert environmental changes (stimuli) into nerve impulses. Neurons conducting information from receptors into the central nervous system are called *sensory neurons.* From a sensory neuron, the impulse passes through one or more *association neurons,* a term applied to nerve cells between the sensory neuron and the final neuron. Nerve cells that transmit impulses to an effector organ such as a muscle are called *motor neurons* (Figure 9-8).

Basic Concepts

The nervous system is made of neurons, cells that are specialized for transmitting impulses.

A neuron usually consists of a cell body, dendrites, and an axon.

Sensory, association, and motor neurons are the major kinds of neurons found in nervous systems.

HOW THE NERVOUS SYSTEM FUNCTIONS

When a receptor is stimulated, it starts a series of *nerve impulses* into the nervous system. It is important to understand what these impulses are and what makes them move along a neuron. Another basic set of events occurs when impulses reach the ends of axons and encounter a *synapse,* the junction with another nerve cell. At the synapse, the impulse must be transmitted to other nearby nerve cells or to effectors such as muscles and glands. Because there are many synapses in a nervous system, *synaptic transmission* is of fundamental importance in nervous systems.

A third functional feature concerns the *reflex circuits* that underlie so much of the activity in the nervous system. In fact, the reflex circuit is often termed the

basic functional unit of a nervous system, just as the neuron is considered the basic anatomical unit.

The Nerve Impulse

Biologists were puzzled for a long time by the exact nature of nerve impulses because they are difficult to analyze without sophisticated electronic equipment. Because an impulse moves rapidly along a neuron, early investigators mistakenly considered the nerve impulse to be an electrical current flowing along a neuron. Later we will discuss the difference between the two.

In order to describe a nerve impulse, it is helpful to understand the physiological nature of a neuron when it is *not* transmitting impulses. As in all living cells the concentration of various kinds of ions inside the nerve cell differs from those on the outside in the tissue fluid surrounding the cell. In particular, there is a much higher concentration of potassium ions inside the nerve cell than outside. Conversely, there is a higher concentration of sodium ions in the tissue fluid around the cell than inside the cell. Because of the different concentrations of ions, the inside of the nerve cell membrane can be shown to be negatively charged compared with the outside. In fact, measurements indicate that there is approximately a 70 millivolt difference between the two sides. This is usually stated as a −70 mv difference since the inside is negative compared with the outside. This difference in electrical

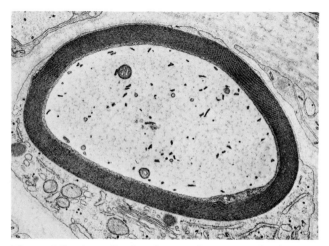

Figure 9-7 *A myelin sheath around an axon. The sheath consists of numerous layers derived from the cell membrane of a Schwann cell. Do you see its resemblance to insulation around a wire?*

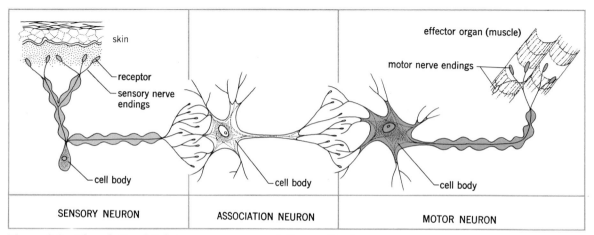

Figure 9-8 *The relationship between sensory, association, and motor neurons.*

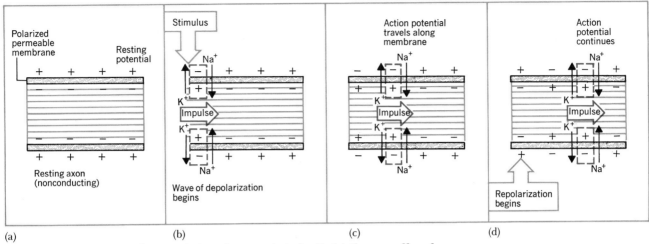

(a)   (b)   (c)   (d)

Figure 9-9 *Initiation and transmission of a nerve impulse. In* **(a)** *the nerve fiber shows a resting potential.* **(b)** *An impulse begins with a depolarization of the cell membrane as sodium ions move in and potassium ions flow out.* **(c)** *The wave of depolarization continues along the nerve fiber in a self-perpetuating manner.* **(d)** *After the impulse passes, the membrane pumps out sodium ions and the membrane repolarizes itself.*

charge is termed the *membrane potential* or *resting potential*. The membrane potential can be disturbed to produce an impulse (Figure 9-9*a*). Because the sodium and potassium ions tend to diffuse across the membrane, the nerve cell must continually extrude sodium ions and pull in potassium ions in order to maintain the resting potential. It is hypothesized that the cell membrane contains ionic "pumps" that perform these functions.

Normally, nerve stimulation originates in a sense organ. In the laboratory, impulses are usually obtained by applying a small electrical stimulus to an axon. In either case the result is the same: a sudden change in the permeability of the membrane to sodium ions, allowing them to pass in freely from the outside (Figure 9-9*b*). The mass movement of the positively charged sodium ions into the cell results in a positive charge on the *inside* of the membrane equal to about +30 millivolts (Figure 9-10). This reversal of charge is the impulse or *action potential*.

Potassium ions (also with a positive charge) now begin to flow out of the cell and counteract the effect of the inward movement of the sodium ions. The movement of the potassium ions initiates the process of *repolarization* (the return to the resting state). Repolarization is aided by the ionic pumps that move sodium out of the cell and reestablishes the potassium concen-

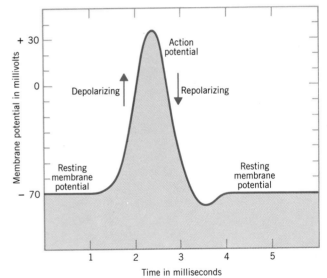

Figure 9-10 *A graph of the changes in potential between the inside and outside of a nerve cell membrane during a nerve impulse. The movement of sodium ions into the cell depolarizes it; its potential changes from about −70 millivolts to about +30 millivolts. Movement of potassium out of the cell repolarizes the membrane, changing the potential back to about −70 millivolts.*

tration inside (Figure 9-9*c,d*). Only then can the membrane be stimulated again. Every impulse is always the same strength in a neuron regardless of the intensity of the stimulus that produces it. This is called the *All or None Law:* A nerve fiber carries either a full-strength impulse or none at all. An analogy is the firing of a gun. How hard you squeeze the trigger determines whether or not the gun fires, but the bullet (nerve impulse) always has the same force.

This description concerns the events at only one point on the nerve fiber. How does the impulse move along the fiber? During the action potential, as described above, the outside of the membrane briefly becomes negatively charged. Adjacent to this point, the membrane is positively charged. This causes an electrical current to flow between the two areas because positive charges are attracted to negative ones. This current, in effect, stimulates the adjoining membrane to alter its permeability to sodium, which rushes into the cell and the impulse continues. In this manner the impulse is self-propagating and continues along the fiber for its whole length. A simplified analogy might be a row of dominoes standing on end. If the first domino in the row falls against the next one, its starts a self-perpetuating wave of falling dominoes.

Why is this movement of nerve impulses not the same as an electrical current through a wire? When an electrical current travels along a conductor, the conductor serves only as a passive carrier of electrons; these move in the same direction as the current with the speed of light. During transmission of a nerve impulse, ions rather than electrons move. Because axons are poor electrical conductors, the impulse must be continuously amplified to maintain itself. This requires time, thus an impulse travels relatively slowly. A nerve impulse retains its same strength only because of the active involvement of the nerve cell membrane in its propagation. Much of the research on nerve action involves measuring electrical changes under controlled conditions.

In one experimental technique, a nerve cell is placed in a container in which the conditions can be closely monitored. Electrodes are attached at one point on the fiber so that exactly measured stimuli may be applied to the neuron (Figure 9-11). At another point on the cell one tiny electrode is placed just outside the membrane and another just inside; these detect the electrical activity of the cell. The signals from these latter two electrodes are amplified and then displayed on an oscilloscope so that the activity may be

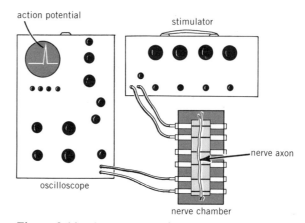

Figure 9-11 *An apparatus for viewing the electrical events of a nerve impulse. A segment of an axon is placed in a nerve chamber in which it may be kept active. Impulses, generated by an electrode, are amplified and displayed on the oscilloscope.*

watched and studied. You many wonder how it is possible to do such experiments on objects as tiny as individual axons. Researchers overcome this obstacle by using the giant fibers we mentioned earlier when discussing the arthropod nervous system. The giant axon from a squid, for example, is a millimeter in diameter. Much of our knowledge about impulses and their transmission was derived from experiments on structures such as squid axons. It is now known that mammalian axons function in a manner similar to invertebrates.

Basic Concepts

A nerve impulse is a self-propagating electrical event moving along a nerve fiber.

The All-or-None Law states that a nerve fiber carries a full-strength impulse or none at all.

An impulse involves electrical potentials but is not comparable to an electrical current flowing along a wire.

Synaptic Transmission

The term *synapse* means "joining" and, in the nervous system, refers to the junctions between two nerve cells. Synapses are concentrated in the gray matter of the spinal cord and the brain. There are also synapses in

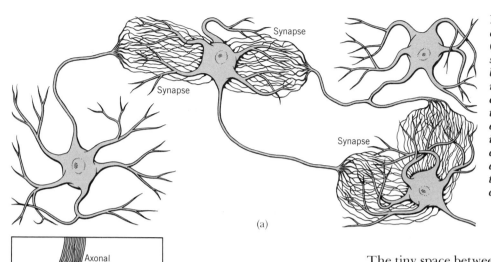

Synapse

Synapse

Synapse

(a)

Figure 9-12 *How neurons communicate with one another. (a) Numerous synaptic knobs surround the dendrites and cell bodies of adjacent neurons. Each neuron usually has synaptic connections with many additional neurons, thus providing alternate pathways for an impulse to travel. (b) A synaptic cleft between a synaptic knob and a dendrite. An impulse is transmitted chemically across the cleft.*

Axonal
ending

Mitochondrion

Axon
terminal

Synaptic
vesicles

Synaptic
cleft

Dendrite

(b)

autonomic ganglia and wherever motor nerve fibers end in muscle cells or glandular tissue.

The ends of most axons divide into numerous small branches, each terminating in many tiny swollen bodies called *synaptic knobs.* These lie close to the cell bodies and dendrites of nearby nerve cells (Figure 9-12*a*). By branching in this way, one axon can make synaptic connections with many additional nerve cells. It is convenient to speak of these cells as *presynaptic neurons;* the cells with which they make synapses can be termed *postsynaptic fibers.* The membrane of every postsynaptic cell is in contact with the synaptic knobs of many presynaptic fibers. As you see, this arrangement provides for almost innumerable possible pathways for impulses through the nervous system.

The tiny space between a synaptic knob and a postsynaptic fiber is called the *synaptic cleft,* or gap (Figure 9-12*b*). Although the cleft is a small, fluid-filled space, nevertheless impulses must cross it in some fashion or they would die out in the synaptic knob. This crossing, or *transmission* is brought about by chemical means. The knob contains tiny vesicles (sacs) of a chemical called a *neurotransmitter.* When an impulse reaches the synaptic knob, it stimulates some of the sacs to squirt their contents into the synaptic gap. The transmitter chemical diffuses across the gap, attaches to receptors, and alters the membrane permeability of the postsynaptic fiber (Figure 9-13). This allows sodium ions to rush in and depolarize the fiber. This event occurs simultaneously at many synaptic sites on the postsynaptic membrane—in fact, it is the sum of many synaptic events that finally sets off an impulse in the postsynaptic fiber. In many parts of the nervous system the transmitting agent released into the synaptic gap is *acetylcholine.* An enzyme, *acetylcholinesterase,* is present at all times in the gap. Its function is to break down the acetylcholine as it accumulates, preventing it from continually depolarizing the postsynaptic fiber.

The importance of this enzyme is seen when its action is blocked by another chemical. For example, organic phosphate pesticides such as malathion or parathion interfere with the action of acetylcholinesterase. As a consequence, impulses are transmitted continuously across the animal's synapses and its muscles go into uncontrolled spasms (convulsions).

At a synapse, only the presynaptic fiber is capable of secreting a chemical transmitter. This means that

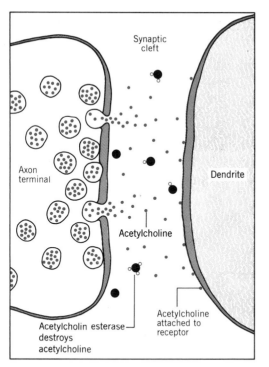

Figure 9-13   *Synaptic transmission. When an impulse reaches a synaptic knob, it causes the vesicles to release a quantity of transmitter chemical. The transmitter molecules attach to receptors on the postsynaptic membrane and then are deactivated by an enzyme.*

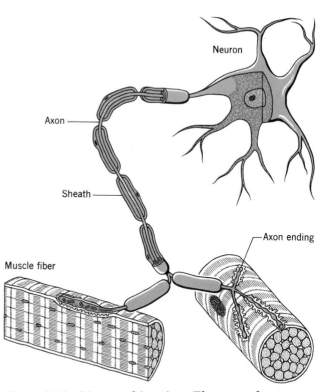

Figure 9-14   *Myoneural junctions. The axons of motor neurons terminate at synaptic junctions with muscle cells. Release of a transmitter chemical at these junctions initiates series of events that cause the muscle to contract.*

nerve impulses can travel in one direction only through the system. Should an impulse start in the middle of a neuron, it would travel to both ends of the nerve fiber, but only at the one end could it trigger off the release of a transmitter.

Up to this point we have been describing transmitter chemicals that are *excitatory,* in that they cause impulses to start in adjacent nerve cells. Studies of impulse transmission show that an *inhibitory* chemical is released at some synapses. When it reaches the postsynaptic membrane, the inhibitory chemical does not depolarize the membrane but, instead, *hyperpolarizes* it by increasing its resting potential. In any given set of synaptic junctions on a postsynaptic fiber, some are excitatory and some inhibitory. Whether an impulse is created in the membrane depends on the balance between its inhibitory and excitatory inputs.

When motor fibers enter muscle tissue, they divide into many small branches that end in swellings comparable to synaptic knobs known as *end plates.* These form synapses with muscle fibers known as myoneural junctions (Figure 9-14). At the junctions, the nerve impulses cause acetylcholine to be released from the end plates; it diffuses across the cleft and depolarizes the membrane of the muscle fiber, produces an impulse, and the muscle fiber contracts. Each motor axon, by branching, innervates many muscle fibers —as many as a thousand in some of the larger skeletal muscles in the thigh, or as few as five or six in the muscles that move the eyeball. A lower ratio of nerve endings to muscle fibers allows more precise control of the muscle as in the movement of an eyelid; a large leg or arm muscle does not require as great precision. The strength of the muscle contraction depends on whether many neurons or a few are "firing" simultaneously, providing another kind of control.

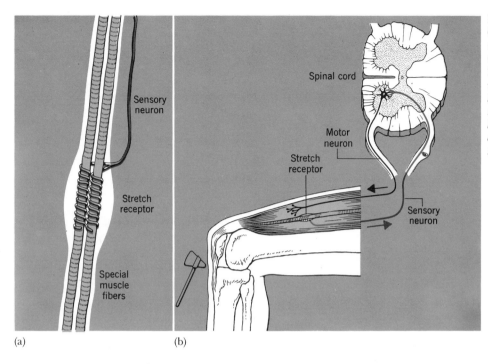

Spinal cord

Motor
neuron

Stretch
receptor

Sensory
neuron

Sensory
neuron

Stretch
receptor

Special
muscle
fibers

(a)   (b)

Figure 9-15 *Reflex action. At (a) stretch receptors attached to muscle fibers are stimulated when the muscle is stretched. (b) The knees reflex is set into action when the patellar tendon is lightly tapped (stretched) as shown. This is a simple neurological test for the functioning of certain nerve pathways in the spinal cord.*

The vital role of synapses in the nervous system is indicated by the number of chemicals and drugs that affect them. To name a few, some anesthetics work by inhibiting synaptic transmission; amphetamines act to *increase* the amount of neurotransmitter substance, thus accelerating certain functions in the nervous system. The deadly botulinum toxin prevents the production of acetylcholine, thus causing paralysis. Some snake venoms bind to acetylcholine receptors on the postsynaptic membranes to cause paralysis of the respiratory muscles. Curare, a strong plant poison, acts in the same manner.

Basic Concepts

At a synapse, nerve impulses cause the release of transmitter substances which are either excitatory or inhibitory in their action on adjacent neurons.

Enzymes in synaptic clefts deactivate some transmitter agents as they accumulate.

Impulses travel in only one direction because only axons release transmitter agents.

Reflex Arcs

As indicated previously, much of the functioning of a nervous system is by reflex action, that is, sort of an automatic stimulus—response reaction. The simplest reflex circuit involves a receptor device, a sensory neuron, a motor nerve fiber, and an effector (usually a muscle fiber).

An example is the stretch reflex of muscles. When a muscle is stretched, tiny sensory devices in it (muscle spindles) are stimulated (Figure 9-15a). Impulses travel into the spinal cord via sensory neurons, which synapse in the gray matter with motor fibers. The motor neurons conduct impulses to muscle fibers in the vicinity of the muscle spindles and the muscle contracts (Figure 9-15b). The muscle stretch reflex is the basis of some simple neurological tests that are often part of a physical checkup. A light tap just below the kneecap stretches the tendon attached to a large muscle on the front of the thigh. The muscle contracts slightly, and the lower leg jerks. In the ankle-jerk reflex, a tap on the back of the foot just above the heel normally causes the foot to move. There are many other muscle reflex actions throughout the body. The absence of one of these reflexes may indicate damage to the reflex pathway, possibly in the spinal cord.

All reflex circuits other than the muscle-stretch type contain additional neurons, association neurons (sometimes called *interneurons*) in the gray matter of the spinal cord. These neurons not only link the sensory and motor neurons, but they also often synapse with neurons entering other nerve tracts and other reflex circuits. In this way reflexes can be modified or coordinated with other reflexes. For example, if one limb meets a painful stimulus and flexes quickly to withdraw from it (the withdrawal reflex), the opposite limb *extends* itself reflexly. As someone once noted, without this coordination of opposite reflexes you might step on a tack, attempt to flex *both* legs, and end up sitting on the pain-producing object!

Reflex actions are valuable to the organism in that they require no decision-making by the brain, and, in fact, may not involve the brain at all.

Basic Concepts

In a reflex arc, impulses start in a receptor, travel into the
spinal cord along a sensory neuron, and are transmitted out through motor neurons to effectors. Association neurons may be interposed between the sensory and motor neurons in the spinal cord.

Reflex circuits make possible a number of stereotyped actions frequently used by an organism.

In this chapter we have looked mostly at the organization and functioning of the peripheral nervous system and the spinal cord. This part of the system is remarkable in its own right for its rapid, efficient handling of messages and for its seeming lack of fatigue, its precise directing of impulses to specific effectors, and its ability to integrate and coordinate a considerable variety of bodily actions by reflex circuits that do not necessarily involve the brain. To make a simplified analogy, so far we have looked at the telephone lines and how messages are transmitted. But as yet we have said little about the people who send the messages (the sense organs) or how the central office (the brain) handles them. These are the topics of the next chapter.

KEY TERMS

acetylcholine

acetylcholinesterase

action potential

all or none law

association neuron

autonomic nervous system

axon

biofeedback

central nervous system

cranial nerves

dendrites

gray matter

irritability

membrane potential

motor neuron

myelin

nerve impulse

neuron

neurotransmitter

nodes of Ranvier

parasympathetic nerves

peripheral nervous system

receptor

reflex circuit

repolarization

resting potential

Schwann cell

sensory neuron

solar plexus

spinal cord

spinal nerves

sympathetic nerves

synapse

synaptic cleft

synaptic transmission

white matter

SELF-TEST QUESTIONS

1. Name four major functions of the nervous system.
2. The nervous system works closely with what other major control system?
3. (a) What organs make up the central nervous system?
   (b) What parts make up the peripheral nervous system?
4. Where approximately does the spinal cord end?
5. Why is a portion of the spinal cord gray in color and a portion white in cross section?
6. What is the difference between a paraplegic and a quadriplegic?
7. What does it mean to say that a spinal nerve is "mixed"?
8. What parts of the body are served by spinal nerves?
9. If the dorsal root of a spinal nerve is cut, would it affect the motor or the sensory action of that nerve? Explain your answer.
10. (a) What parts of the body are served by the autonomic nerves?
    (b) How is the autonomic nervous system able to both stimulate and inhibit the actions of an organ?
11. Name the two major divisions of the autonomic system.
12. Describe the fight-or-flight syndrome.
13. How is the arthropod type of nervous system radically different in its anatomy from the vertebrate type?
14. What are the three major parts of a neuron?
15. Define myelin and its function in relation to neurons.
16. What is the relationship of myelin to multiple sclerosis?
17. What distinguishes sensory, association, and motor neurons from one another?
18. Describe the conditions in the tissue fluid around a nerve fiber at rest, and what is meant by the cell's resting potential.
19. Describe what happens when a nerve impulse begins in a nerve fiber.
20. What is meant by depolarization and repolarization?
21. In what way is an impulse self-propagating?
22. Give some reasons why a nerve impulse is not the same as an electrical current flowing through a wire.
23. How is a cathode-ray oscilloscope used to visualize a nerve impulse?
24. Explain the all or none principle.
25. Describe what happens when a nerve impulse reaches a synaptic knob.
26. In reference to a synapse, define the following terms: (a) synaptic knob; (b) synaptic cleft; (c) pre- and postsynaptic fibers; (d) chemical transmitter.
27. What is the function of special enzymes such as acetylcholinesterase in the synaptic cleft?
28. Explain the difference between inhibitor and excitatory transmitters.
29. How do you account for the observation that impulses travel in only one direction in a nerve fiber?
30. Describe a myoneural junction and tell what happens there.
31. Trace the passage of an impulse through a reflex arc, beginning with a receptor organ.
32. What is the basic advantage of a reflex action?

chapter 10

SENSE ORGANS
AND
THE BRAIN

chapter 10

questions to think about

What is the basic function of all sense organs?
How are the different sensory devices adapted to their specific functions?
In what respects are all vertebrate brains similar?
How is the brain studied?
How is a computer similar to a brain?

I think, therefore I am.

Descartes

In order to survive, all animals need to be constantly aware of events and changes taking place in their surroundings. One of the basic functions of a nervous system is to provide this kind of information continuously.

The perception of any environmental feature, such as light, requires a sensory device specifically adapted to that particular form of energy. To express the same idea a bit differently, an organism can respond *only* to those stimuli for which it has appropriate sense organs. Therefore, if an organism is to respond to 21 different kinds of stimuli, as does a human being, it must have a sense organ capable of responding to each stimulus.

This is one reason that biologists are usually skeptical when an individual claims to have "extrasensory" perceptive powers. The skeptic asks, "What and where are the sensory devices for receiving unusual stimuli such as another person's thought waves?" So far, no one has been able to find brain wave sense organs. A few years ago a popular magazine featured several individuals who appeared to demonstrate the ability to distinguish colors, while securely blindfolded, by rubbing their fingertips over an object or a printed picture. Investigators of this amazing fingertip vision eventually discovered the truth: What seemed to be fingertip vision was done by peeking from beneath the lower edge of the blindfold.

The stimuli to which most organisms respond are those most important to their survival, including light, sound, temperature, movement of the body and its parts, odors, touch, and pain. Some of these stimuli, such as those of body position, alert the brain to changes within the organism.

It is interesting to note that many animals live in a perceptual world far different from ours. For example, fish do not have auditory organs and therefore live in a world of silence. However, they have sense organs that we do not have, including a system of canals in the skin (the lateral line system) that enables them to perceive vibrations in the water. Also they have organs roughly comparable to our taste buds (chemoreceptors) but located on the outside of the body in the head area. These organs are extremely sensitive to chemical changes in their watery environment. This is how sharks and other fish detect substances such as blood from considerable distances.

A number of fish have evolved electricity generating systems that enable them to explore their environment, find prey, and even stun it with an electrical charge of impressive voltage. Mammals as diverse as porpoises and bats use echolocation, the emission of ultra high-frequency sound waves, as a means of communicating, for navigating, and for finding prey. Thus a bat can fly rapidly through a maze of obstacles in utter darkness in pursuit of a moth or other prey. Often we find that the bat's prey also has some sort of sensory device for detecting the pursuing bat and eluding capture. When biologists encounter a predator-prey system such as this, it leads them to believe that the two organisms have had a long evolutionary acquaintance!

One more point to be emphasized is that all receptors do the same thing: They convert the energy of stimuli into nerve impulses. And, if you recall, all nerve impulses are alike. Therefore, it is the task of various parts of the brain, rather than of the sense organs, to interpret the meaning of the impulses. Information (as impulses) from the eyes go to a specialized part of the cerebrum which interprets them as vision. Information from the auditory nerves goes to a special auditory area, and so on. In this sense we really "see" and "hear" with the brain rather than with eyes or ears.

In this chapter we will describe the major features of some of the principle sense organs, then follow with a look at the brain and how it processes this information.

Basic Concepts

Organisms can respond only to stimuli for which they possess the appropriate sense organs.

The kinds of stimuli to which an organism responds are those that are valuable to its survival.

Sense organs convert stimulus energy into nerve impulses which go to special areas of the brain where they are perceived as sensations.

SENSORY PERCEPTION

Light

Light is of such great importance to so many organisms that it is not surprising to find that nearly all forms of life have light receptors. These range from tiny bits of light-sensitive pigments in many unicellular forms to the complex image-forming eyes of arthropods and vertebrates (Figure 10-1). In addition to enabling animals to see, light also often plays a role in other body activities. In birds and a variety of other animals, reproductive cycles are tuned to *photoperiods,* seasonal changes in the amounts of light in a 24-hour time period. As the days become longer in the spring months, this change is noted by the animal's brain. The brain in turn stimulates the endocrine system to increase its output of reproductive hormones. The system is then turned off in the fall and winter by the shortening day length. In birds, migration movements are tied to photoperiodism by a similar mechanism.

Light or its absence has a strong influence on the activity cycles of animals. Many animals are active during daylight hours and inactive at night, but numerous other forms are adapted for a night (nocturnal)

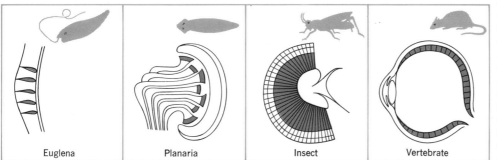

| Euglena | Planaria | Insect | Vertebrate |

Figure 10-1 *A variety of light receptors with their light-sensitive pigments shaded. Some receptors respond only to the presence of light (Euglena and Planaria), whereas more complex types, as in insects and vertebrates, form images. All have the common property of converting light energy into nerve impulses.*

life-style. Moths, bats, and many reptiles, amphibians, and some carnivorous mammals are nocturnal. As you might guess, such animals often have many adaptations that help them function better at night, such as large eyes that are very sensitive to dim light.

Basic Concepts

Light is of great importance in regulating the activity cycles of many animals.

Light receptors, found in virtually all animals, include spots of light-sensitive pigments, and complex photoreceptor organs.

Vision

The human eye, like that of most other vertebrates, is remarkable for the functions it serves in relation to vision. It regulates the amount of light that enters the eye, focuses images sharply on the retina, detects very minute movements of an object, and accurately registers a large range of colors. In fact this image-forming eye is sometimes called the *camera eye* because the light-sensitive retina is comparable to the film in a camera, the lens focuses the image on the retina, and the iris functions like a camera iris diaphragm in regulating the amount of light entering the eye (Figure 10-2).

The eyeball (Figure 10-2 top) is surrounded by a rather tough coat of connective tissue, the sclera. This tissue, forming the white of the eye that you can see when you lift the eyelid, helps to attach the eyeball in its bony protective socket. Six strap-shaped muscles attach to the sclera and move the eyeball in different directions. These six small eyeball muscles are remarkably alike in all vertebrates from fish to human beings and have changed little over a vast period of evolutionary time (Figure 10-3).

At the front of the eye, the sclera is modified into a transparent layer called the *cornea* (Figure 10-2). The cornea forms an important part of the visual system, because light waves passing through it are bent sharply inward. Behind the cornea is a small fluid-filled chamber whose floor is formed by the *iris* and the *crystalline lens.*

The iris covers the front of the lens except for a small circular opening called the *pupil.* By regulating the size of the pupil, the iris controls the amount of

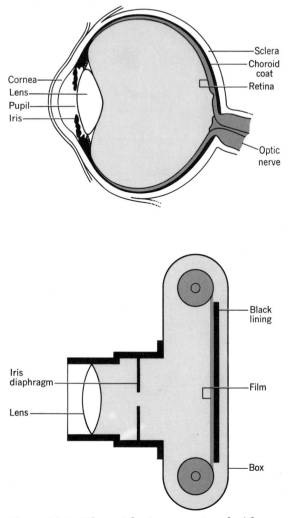

Figure 10-2   *The vertebrate eye compared with a camera. Can you match the parts that are similar functionally?*

light that reaches the retina. In bright light, the iris constricts. In dim light, the pupil enlarges and more light enters the eye. Dilation and constriction of the pupil is a reflex action under autonomic nerve control. Although their chief function is to protect the front of the eyeballs, the eyelids also help to regulate the amount of light striking the eyes.

The transparent lens is held in place by an envelope to which is attached *ciliary muscles.* By pulling on the envelope, these muscles change the shape of the flexible lens to help in focusing images on the retina. This process is known as accommodation.

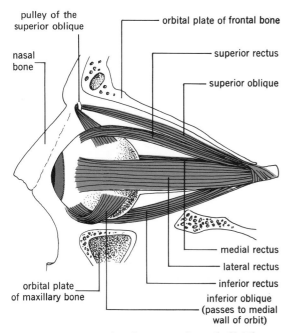

pulley of the
superior oblique

nasal
bone

orbital plate of frontal bone

superior rectus

superior oblique

medial rectus

lateral rectus

inferior rectus

inferior oblique
(passes to medial
wall of orbit)

orbital plate
of maxillary bone

Figure 10-3 *Muscles that move the eyeball. These are amazingly similar in all vertebrates.*

One of the properties of lenses is that they invert images. Hence our view of the world around us comes to the retina upside down. Early in life, however, our brain corrects this problem by interpreting upside down as right side up.

Much of the volume of the eyeball is taken up by the jellylike material (vitreous humor) that fills its large back chamber. The inside of this chamber is lined with layers of sensory cells and nerve cells that constitute the *retina.* The innermost of these layers consists of a field of light-sensitive cells called *rods* and *cones.* In between are many neurons that help "process" visual information. The outer layer of neurons, the optic nerve fibers, leave the eyeball as a unit to make up an optic nerve. There are far more rods and cones than optic nerve fibers, indicating that these fibers extract and bring together information from a number of receptor cells. This infers that the retina functions like a brain in sifting through information and deciding how much of it to transmit to higher neural centers. This is not such a far-fetched idea when we learn that the embryonic optic cup that forms each eye grows from a detached piece of the brain early in development.

At the site where the optic nerve leaves the eyeball (the optic disc), blood vessels also exit and enter. These vessels spread throughout the retina and are easily visible if one looks into the eyeball with a small light.

As you might have observed, doctors always examine the retina carefully with an instrument called an ophthalmoscope to observe the conditions of the retinal arteries and veins. A variety of diseases including diabetes and high blood pressure affects these vessels and can be detected from this examination (Figure 10-4).

Rods and cones, as their names imply, are elongated, specialized cells (Figure 10-5). They are modified epithelial cells rather than neurons. One end of the rod cell is packed with layers of membranes containing a light-sensitive pigment, rhodopsin. When light strikes rhodopsin, a chemical change in the pigment triggers an electrical response. This information is transmitted to the visual area of the cerebrum via the visual pathway.

Rods are sensitive to dim light and thus provide organisms with some degree of night vision. Cones function in brighter light and, unlike rods, are capable of color vision. There are evidently three types of cones, each of them sensitive to a different color of light: red, green, and blue. The brain puts together the information coming from these three types of receptors. As a result, the normal human organism is able to discriminate about 160 different spectral colors. Most of the cones are concentrated in a small central area of the retina called the *fovea,* which is about 2 millimeters in diameter. Our sharpest vision occurs in the fovea; consequently we do most of our reading and other visual tasks with this very small portion of the retina.

Nocturnal animals have considerably more rods than cones, and day-active (diurnal) forms have the reverse. In a human eye there are about 120 million rods and only 7 million cones. It is tempting to conclude, therefore, that human beings are intended to be nocturnal animals (a conclusion that might be reinforced by observing the increase in student activity on a college campus after dark). However, this conclusion is probably not valid. In fact, there are no nocturnal societies of people, especially among primitive tribes that live so closely integrated with their natural surroundings.

As you have probably observed, the eyes of many

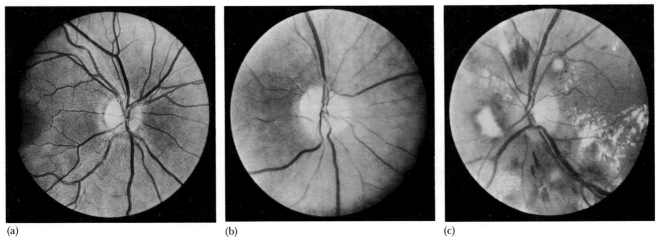

(a)
(b)
(c)

Figure 10-4 (a) *The retina as viewed through an ophthalmoscope to show a normal optic disc (center) and normal blood vessels. (b) The retina showing effects of high blood pressure. The optic disc is blurred as a result of swelling of the optic nerve, and some of the small arteries show irregular constrictions. (c) The retina showing effect of moderate diabetes. Extensive hemorrhages (blurred spots) are evident.*

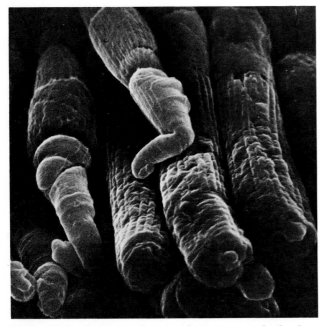

Figure 10-5 *Scanning electron photomicrograph of rods and cones (pointed bodies) in the retina of an amphibian.*

animals (not including humans) shine at night when caught in the beam of car headlights or a flashlight. There is a layer of light-reflecting cells, the tapetum,

behind the retina of nocturnal animals, which presumably helps the retina collect light more efficiently.

Judging from the numbers of people who wear glasses, eye defects must be exceedingly common. The most common problems are nearsightedness (*myopia*), farsightedness (*hypermetropia*), and *astigmatism*. In myopia, the eyeball is too long, so that when the lens attempts to focus for distant objects the image falls just in front of the retina. The consequence is that distant objects are blurred, whereas near vision is unaffected. In hypermetropia, the eyeball is too short, with the consequence that near objects are brought into focus behind the retina and are therefore blurred, but distance vision is unaffected. In astigmatism, the curvature of the cornea is not uniform, so that part of the image is blurred. All of these conditions are correctable with properly shaped lenses. Because eyeglasses have a number of practical drawbacks, eye specialists developed contact lenses, small lenses that are placed directly on the cornea. Although more expensive than glasses, contact lenses are in wide use today. They can be irritating to the eye, however, if worn continuously for too long a time.

Two additional eye defects that are also relatively common are cataracts and glaucoma. *Cataracts* occur when the normally transparent lens develops opaque areas. However, cataracts can also involve the cornea.

This is a slowly progressive condition that leads to blindness if not treated. The normal treatment is removal of the lens, after which the patient must wear rather thick glasses or a contact lens. Success has been achieved in some patients by replacing the diseased lens with an artificial one.

In the condition know as *glaucoma,* the fluid in the small front chamber of the eye does not drain off properly. Pressure builds up and eventually will cause blindness if not treated. Various drugs (including marijuana) may relieve the condition, although surgery is necessary in some instances to provide a drainage pathway for the fluid.

Basic Concepts

The structure and functioning of the vertebrate eye is generally analogous to that of a camera with its lens, iris diaphragm (iris), and film (retina).

Rods and cones are the basic visual units of the vertebrate eye.

Sound, Gravity, and Motion

Sound—vibrations of the air or of water—offers a basic means of communication for numerous animals. Noises are often produced during courtship behavior, in defending territories, for navigating (echo-location in bats, for example), and for various types of signals between animals. Sound also plays a part in sex identification.

Hearing. Human beings also use sound extensively for a variety of purposes, and our sound receptors are like those found in other mammals. Sound waves enter the external ear canal, causing the eardrum membrane to vibrate. The inner side of the eardrum is attached to a chain of three small bones located in the middle ear, and commonly called the hammer, anvil, and stirrup because of their shapes (Figure 10-6). These bones amplify the vibrations across the middle ear chamber and at the same time transmit them to the membrane of the oval window. The oval window lies between the *middle ear* and the fluid-filled inner ear chambers. As the oval window vibrates, it sets up waves of motion in the inner ear.

The *inner ear* contains two fluid-filled tubes wound into a space-saving spiral structure called the *cochlea*

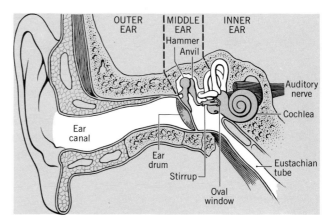

Figure 10-6 *Major structures of the ear.*

(from a Latin term meaning snail). The smaller of the tubes, the cochlear duct, contains the actual sensory device for hearing, the *organ of Corti* (Figure 10-7). The organ consists of long rows of *hair cells* (cells with tiny projections) that rest on a *basilar membrane.* The hair cells are overlain by a thin tectorial membrane in such a way that fluid vibrations push the basilar membrane containing the hair cells against the tectorial membrane. In some manner this triggers impulses in the hair cells. These impulses eventually reach the auditory area of the brain where they are translated as sound. In a rather amazing chain of events, the sound that began as vibrations of air is converted into waves in a fluid, and finally into nerve impulses.

The middle ear chamber, housing the delicate ear bones, is sealed off from the external ear by the eardrum. However, a membranous *eustachian tube* (Figure 10-6) connects the middle ear with the back of the mouth in such a way that whenever you swallow, the tube opens momentarily. This device serves to equalize the air pressure in the middle chamber with that outside the body. It is important whenever you ascend to high altitudes or dive underwater, because unequal air pressure within and without could otherwise damage the eardrum. Pain in the middle ear quickly tells you that the pressure is unequal and has to be dealt with. The eustachian tube has one drawback in that it allows harmful microorganisms to get into the middle ear when an individual has a thoat infection. A severe infection can occasionally damage the middle ear bones or the eardrum, causing deafness.

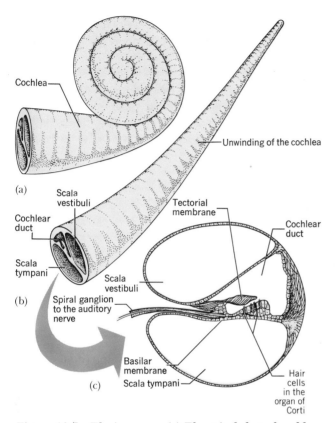

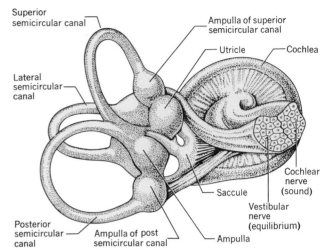

Figure 10-8 *Organs of equilibrium in relation to the cochlea. The fluid-filled semicircular canals, saccule, and utricle enable us to detect motion in many different planes in reference to the pull of gravity. Can you figure out which canal functions with which plane?*

Figure 10-7 *The inner ear. (a) The spiral-shaped cochlea; (b) The cochlea as it would appear straightened out. It contains three fluid-filled canals that join at the far end of the cochlea; (c) Location of the organ of Corti within the cochlear duct. Vibrations of the hair cells against the tectorial membrane produce nerve impulses that the brain interprets as sound.*

Equilibrium and Motion. The inner ear contains additional sensory devices and gives us a sense of balance (dynamic equilibrium) and the ability to sense the pull of gravity — that is, to tell down from up (static equilibrium). These devices are the liquid-filled *semicircular canals* and the *saccule* and *utricle* (Figure 10-8).

Each of the interconnected semicircular canals lies in a different plane, so that whether the head rotates up and down or from side to side, fluid in at least one of the canals ''sloshes'' in one direction. As a result, small tufts of hair cells at the entrance of each canal are stimulated as they are moved by the fluid. Nerve impulses from these cells are transmitted to the brain where decisions are made to activate the appropriate

skeletal muscles to maintain the body's balance.

Although the semicircular canals detect rotational motion of the head or body in different planes, they do not by themselves detect the pull of gravity or the movement of the body in a straight line. These senses are provided by two other organs, the saccule and the utricle, which are part of the same membranous system that makes up the semicircular canals. These organs contain hair cells covered by membranes containing crystals of calcium carbonate called *otoliths.* The otoliths rest on the hair cells, due to gravitational pull. Hence, any change in head position causes the otoliths to move, stimulating their hair cells. These sensations in turn enable us to distinguish whether we are standing upright, lying down, or standing on our heads. In addition, movement of the head also tends to displace the otoliths and thereby stimulates the hair cells. The saccule and utricle enable us to sense acceleration in one direction in addition to the rotational acceleration detected by the semicircular canals.

Unusual or excessive stimulation of the inner ear, as caused by the movements of a boat, airplane, or car, may produce the unpleasant symptoms of motion sickness. Some people are more prone to this complaint than others and may obtain some relief with certain drugs.

To provide a broad summary, the inner ear devices play a major role in the general orientation of the body in space. The nerve fibers of the inner ear go to various areas of the brain, including the cerebellum and cerebrum, and they also connect with nerve tracts into the spinal cord for coordination with body posture. The eyes provide important additional visual clues to our orientation in space.

An entirely different aspect of body orientation is that of *proprioception:* the awareness of the position of various parts of the body at any moment. This sense is due to the presence of proprioceptor organs in the tendons and muscles of all major joints in the body. Any movement of an appendage triggers nerve impulses in these sensory devices. The impulses eventually reach the cerebral cortex, making us aware that the appendage has moved and where it is at any particular moment.

Basic Concept

The detection of sound waves and the pull of gravity are performed by the closely related ear and semicircular canal organs in human beings and many other vertebrates. Both organs function by the movement of fluid against specialized groups of hair cells.

Chemical Sensing

The ability to detect chemicals in the environment by taste, smell, or a combination of the two, is common in the animal kingdom. It is essential for both finding and tasting food and for avoiding obnoxious chemicals. Although these senses are described separately here, they function closely together in many animals. Much of what we taste, for example, we simultaneously smell because the receptors for taste and small are located near each other and function in a similar manner.

In order for a substance to be detected by a chemical sensor, the substance must be in solution. Hence many water inhabitants have chemical sensors on the surface of their bodies. The catfish, for instance, can "taste" with its whiskers. However, in terrestrial animals, the chemical sensing organs are located in their mouths and nasal chambers where moist surfaces are available.

Taste. In humans, clusters of taste receptors called taste buds are found on the tongue, palate, back of the mouth, and on the flap (epiglottis) which closes off the voicebox (Figure 10-9a). A taste bud consists of a cluster of hair cells and supporting cells sitting in a tiny pit. To register as a taste, molecules of a substance

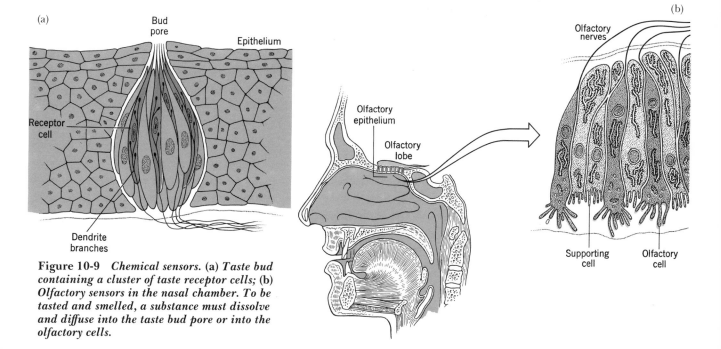

Figure 10-9 *Chemical sensors. (a) Taste bud containing a cluster of taste receptor cells; (b) Olfactory sensors in the nasal chamber. To be tasted and smelled, a substance must dissolve and diffuse into the taste bud pore or into the olfactory cells.*

must dissolve in the mucus of a taste bud and trigger off chemical action in the hair cells. Precisely how this works is not known.

There are four primary tastes and corresponding taste-bud receptors: sweet, sour, salt, and bitter. From these four tastes in various combination with their odors are derived all the numerous flavors we associate with foods.

Smell. In humans and other mammals, the receptor cells for odor detection are located together with numerous mucus cells in the roof of the nasal chamber (Figure 10-9b). To be detected as an odor, molecules of a substance must dissolve in the moist mucus and then interact in some way with the olfactory receptor cells to create nerve impulses in them. Although the human sense of smell is not nearly as acute as that of a dog or cat, many people are nevertheless able to distinguish several thousand odors. According to one hypothesis, molecules of different odors stimulate different receptor cells. The problem with this hypothesis is that all olfactory receptor cells appear identical in their structure.

Everyone has experienced the phenomenon of olfactory fatigue that occurs when an odor is continually present. After a brief period of time, the odor seems to disappear. This can be a fortunate adaptation for individuals who must live near paper mills and other odoriferous industries.

Basic Concept

The detection of chemicals by taste or smell occurs when molecules of a substance dissolve in the mucus surrounding specialized receptor cells and trigger nerve impulses in them.

Heat, Cold, Touch, and Pain

Skin forms the main barrier between the outside world and our bodies, so it is not surprising that it is sensitive to a variety of stimuli. These are touch and pressure, heat, cold, and pain. Obviously this is information that we need to know about in order to avoid injury from the environment.

The receptors for all these stimuli in the skin are naked nerve endings and nerve endings that are modified into various types of bulbs and other intricately shaped bodies (Figure 10-10). The naked nerve endings appear to function for any of the four stimuli, but the specialized nerve endings are stimulated by specific stimuli. For example, the end bulbs of Krause are cold receptors. The specialized receptors are unevenly distributed over the body, and some areas are more sensitive to certain stimuli than are others. The tip of the finger, for instance, has more touch receptors than pain receptors, whereas the back of the finger has many more pain than touch receptors.

Pain. Pain is of special interest because it always involves unpleasant or stressful feelings, it can be brought on by a variety of stimuli, and it has a strong emotional association. It is adaptive in the sense that it warns an organism of potentially damaging stimuli.

Pain has been studied extensively for medical reasons, particularly in reference to how it can be relieved by drugs or surgery. Some disease conditions may evoke pain so severe that medication will not relieve it. In these cases, drastic surgery may be used such as cutting nerve tracts in the spinal cord or brain.

Pain may be elicited from various parts of the body, including the skin, muscles, and internal organs. Pain in the superficial parts of the body is usually sharp and localized, although later it may give way to a dull, aching feeling. Deep pain as from the internal organs is more diffuse and frequently causes nausea.

Pain receptors are far less abundant in the viscera (the internal organs) than in the rest of the body. Visceral pain can be very severe, however, as you may recall if you have ever had a bad stomach ache or an attack of appendicitis. Very often, pain in an internal organ will cause the overlying body muscles to cramp severely. Thus, if the stomach or small intestines become distended with gas, the pain of the stretching causes the overlying abdominal muscles to contract vigorously and painfully.

An interesting sidelight of visceral pain is that it is often "referred" to another part of the body. An example that you have probably experienced is what seems like a dull toothache that actually turns out to have been referred pair from an infected sinus some distance from the seemingly aching teeth. Many examples of referred pain are known in medicine, and they can be used to help diagnose the actual source of certain kinds of pain. For example, a pain on the inner

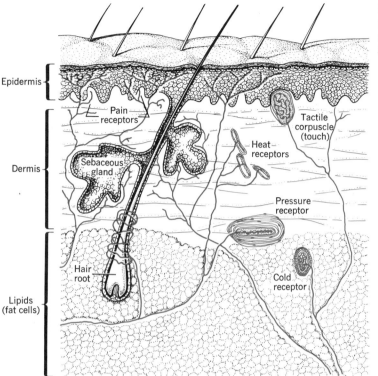

Figure 10-10 *Receptors for pain, heat, cold, touch, and pressure found in the skin. Are they more abundant in some skin areas than others? How do you know?*

Epidermis

Dermis

Lipids (fat cells)

Pain receptors

Sebaceous gland

Hair root

Tactile corpuscle (touch)

Heat receptors

Pressure receptor

Cold receptor

side of the left arm may be referred from the heart, and pain in the tip of the shoulder sometimes indicates an abnormal condition in the diaphragm. Diagnosing the real cause of referred pain is tricky, however, because referred pain does not always appear in the same location in different patients.

The most effective pain-killing drugs are the opiates, such as morphine. The drawback to these drugs is that patients risk becoming physiologically dependent ("addicted") on them after a period of time. Opiates are so effective because they attach to special receptor sites called *opiate receptors* found on certain brain cells. This blocks the receptor cells from producing a pain-causing neurotransmitter called Substance P. As bizarre as it sounded, it almost seemed as though the brain was adapted to using opiates for pain suppression. This "mystery" was solved when neurobiologists discovered two similar kinds of transmitters, *enkephalins* and *endorphins,* that associate with opiate receptors and prevent them from producing Substance P. These transmitters appear to be the

brain's "opiate." Attempts are being made to design a pain-killing drug based on the structure of these substances because they would presumably be nonaddicting.

Related to the discovery of endorphins is a possible explanation for the effects of acupuncture. This procedure consists of placing needles in the skin at various places around the body. Acupuncture relieves pain in many individuals, but until now there seemed to be no physiological reason for this procedure to work. The discovery of endorphins suggests that the placement of the needles may cause the release of endorphins, which in turn brings about the observed pain-killing effects.

Basic Concept

Touch, pressure, pain, heat, and cold are sensed in the skin by naked nerve endings or by modified nerve endings. Pain receptors are less common in the viscera than in the rest of the body. Visceral pain is often referred to another part of the body.

THE BRAIN

The brain is, in effect, the command post of the nervous system. It receives a constant flow of information from the sense organs, decides what to do with it, and directs various parts of the body to carry out any specific actions that are needed. This alone is a highly complex activity rivaling that of a computer. But this is only one of many functions of this most complicated of all organs. Some of its other functions will be discussed in the next few pages.

The brains of all vertebrates are similar in their basic plan, but the evolutionary modifications of this plan from group to group are extensive. Some of these modifications reflect an emphasis on a particular sensory device. The shark's brain has a huge olfactory bulb—a "smell" brain, so to speak. A bird, in contrast, has a "balance" brain with a large cerebellum. This part of the brain is responsible for the great muscular coordination required for flight. Mammals have the largest cerebrum—the "thinking" brain (Figure 10-11).

Other types of evolutionary modifications in the brain reflect the anatomical and functional complexity of the species. Mammalian brains, for example, are larger and more complex than those of amphibians. And the peak of complexity and size (for body weight) is seen, of course, in the human brain with its huge cerebral hemispheres covered with gray matter (the cerebral cortex). Many biologists believe it is this modification—the expansion of the cerebrum—that lifted humankind above the level of the rest of the animal world.

A great deal of knowledge about the structure of the brain has accumulated through the years, and more is probably known about the human brain than about any other vertebrate's brain. Even so, there are huge gaps in our understanding of how this 3-pound mass of trillions of cells carries out many of its functions. Memory storage and recall constitute two challenging "gaps" in our knowledge that we will discuss later.

Basic Concept

All vertebrates have the same basic brain plan that is modified according to the complexity of the organism and its particular specialized sensory devices.

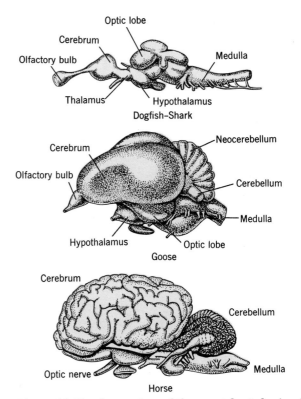

Figure 10-11   *Comparison of three vertebrate brains. Each brain is modified for specialized functions and reflects a different level of evolution. Which one appears to be the most complicated?*

Methods of Studying the Brain

The anatomy of the brain has been thoroughly studied, because the organ can be removed, preserved, and examined in many different ways. But how the brain works is not so easily studied, because it must be living and reasonably healthy in order to function properly.

One somewhat crude investigative technique is to remove or destroy a portion of an animal's brain and then observe what bodily activities are affected. The same type of information is sometimes derived in a clinical situation when people have suffered brain damage from an accident or a tumor. Useful information has also been obtained during some types of brain surgery. In many cases the patient can be conscious, because there are no pain receptors in brain tissue, and can describe what is experienced when various areas on the surface of the brain are electrically stimulated.

The cortex of the cerebrum has been thoroughly mapped with this technique.

With laboratory animals, a widely used experimental technique is to stick extremely slender electrodes into deeper parts of the brain and to observe the results of stimulation to these areas. When an electrode implanted in a certain part of a rat's hypothalamus, for instance, is stimulated, the animal becomes enraged enough to attack a cat. With this technique, many emotional centers of the rat's brain have been located —including (it is nice to note) a pleasure center. Enough evidence has been obtained from human volunteers to indicate the presence of similar centers in human brains.

Long ago it was found that there was a tiny amount of electrical activity associated with the brain's surface. With the development of appropriate electronic equipment, this electrical activity could be greatly amplified and recorded graphically, and the resulting *electroencephalograms* could later be analyzed and studied.

Studies have indicated that the electrical activity of the brain is generated in the cerebral cortex and occurs in fairly uniform pulses or rhythms. There are several types of these brain waves or rhythms, based on how often they occur per second and on the age of the organism. Thus, a human infant has characteristic waves that change in childhood to a different rhythm.

In adults, two kinds of brain waves are typical. *Alpha rhythms*, 8 to 13 cycles per second (cps), are recorded when an individual is relaxed with eyes closed. Any sort of sensory stimulation, or simply opening the eyes, abolishes the alpha rhythm. An adult who is awake and functioning normally exhibits the *beta rhythms* (18 to 32 cps). These rhythms give way to others when a person is sleeping or under anesthesia.

Making an electroencephalogram is a fairly simple matter of attaching electrodes to specific areas of the head and recording the signals (Figure 10-12). It is now a common procedure for diagnosing epilepsy and for locating brain tumors and hemorrhages. These conditions are often accompanied by abnormal rhythms from the area of brain damage.

An exciting new area in brain research deals with the discovery of a whole array (at least 30) of transmitter chemicals from brain tissue. All are peptides (made of amino acid chains) but they have a diverse range of

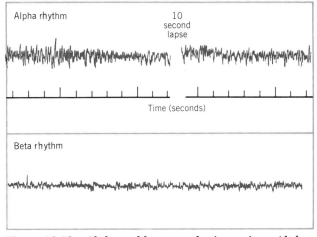

Figure 10-12 *Alpha and beta wave brain tracings. Alpha waves occur at a lower frequency than do beta waves.*

functions. Endorphins, enkephalins, and Substance P are part of this array of neuropeptides. Others such as vasopressin, insulin, and growth hormone, are also found other places in the body where they function as hormones. How all of these relate to one another in the functioning of the brain is yet to be worked out. It appears that the brain has its own fundamental chemical language for communicating within itself.

Basic Concept

Functions of the brain are studied by removing parts of it, stimulating surface areas and deeper areas electrically, analyzing its electrical activity, and by chemically analyzing its tissues.

Parts of the Brain

From the viewpoint of its development, the brain represents the greatly expanded and specialized front end of the spinal cord. It forms early in the life of an embryo from the neural tube, and for a long time it is by far its largest organ (see Chapter 14).

In an adult, the spinal cord seems to merge gradually into the brain stem, although the first pair of spinal nerves serves to mark the boundary between the two; this is also approximately where the spinal cord enters the opening into the skull.

Brainstem. The first part of the brain is known as the *brainstem.* It is composed of a series of interconnected parts: *medulla, pons, midbrain, thalamus,* and *hypothalamus* (Figure 10-13). The brain stem is often termed the "primitive" or "old" brain because this tissue makes up most of the brain of lower vertebrates such as fish and amphibians. In higher vertebrates such as mammals, the brain stem is overshadowed by the development of the cerebrum, or "new" brain. Even so, the brain stem is still highly important in mammals, as indicated by the fact that most of the cranial nerves attach to various parts of the brain stem, and most of the control centers for a variety of basic body functions are located there. These include centers for breathing, heartbeat, swallowing, sneezing, and vomiting. Injuries to the brain stem are especially dangerous because of the number of vital functions it controls.

The small but highly important structure, the *hypothalamus,* is the control center for many diversified and vital autonomic nervous system functions (Figure 10-13). One of these is temperature regulation (see Chapter 11).

The control centers for eating, drinking, and sexual behavior are also located in this part of the brain. Each center has two parts, one that inhibits the activity and one that increases it. If the inhibiting center for eating is destroyed experimentally in a cat, the animal will eat incessantly until it eventually dies. If the feeding center is destroyed, the animal refuses to eat and eventually starves to death. Normally there is a balance between these antagonistic centers such that the animal eats enough food to satisfy its caloric requirements and then stops. Obesity in human beings is occasionally related to a disease of the hypothalamus that affects the feeding center.

Perhaps the most striking function of the hypothalamus is its regulation of the activity of the anterior pituitary gland and, thereby, all the vital functions that the pituitary regulates. It does this by releasing a number of specific chemicals (neurohormones) each of which in turn influences the release of specific hormones by the pituitary gland. These chemicals are collectively called the *pituitary regulating factors.*

Cerebellum. The cerebellum (Figure 10-13) is a prominent part of the brain lying just above the medulla and pons. Like the cerebrum, it has an outer cortex of gray

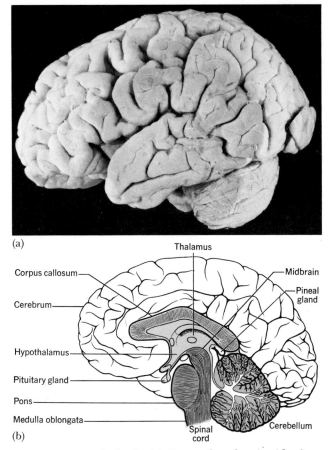

(a)

(b)

Figure 10-13 *The brain.* **(a)** *External surface in side view showing the extensive and convoluted surface of the left cerebral hemisphere. A portion of the cerebellum and brain stem are seen below the cerebrum;* **(b)** *Major parts of the brain in longitudinal section.*

matter and contains many folds and fissures. The cerebellum receives nerve tracts from the motor area of the cerebral cortex and has inputs from practically all sense organs in the body. It coordinates and refines all motor actions, including those concerned with posture and equilibrium, jointly with the motor cortex of the cerebrum. Together these two structures bring about the smooth, precise motor activities typical of a well-coordinated organism. Damage to the cerebellum does not abolish motor activity, but it does impair its precision. Injuries in the human cerebellum are reflected in a staggering walk, jerky movements, an inability to perform a simple feat smoothly such as touching an object with a finger, and slurred speech.

Cerebrum. The rest of our brain, an impressive 80 percent, consists of the two cerebral hemispheres (Figure 10-13). Their bulk alone is considerably greater than in other vertebrates, but the human cerebrum has an additional distinguishing feature of being covered with a 2- to 5-millimeter-thick layer of gray matter termed the *cerebral cortex.* Many other animals have gray matter, but not nearly so much.

Each hemisphere is divided by deep fissures into a number of subdivisions: an anterior *frontal lobe,* a lateral *temporal lobe,* and posterior *parietal* and *occipital lobes* (Figure 10-14a). In addition, each lobe has numerous smaller folds known as convolutions.

The human cortex has been studied extensively because the interpretation and integration of so many bodily activities is found here. Consequently, there are fairly detailed maps of the sites of these functions on the cortex. The back part of the frontal lobe, for instance, has a convolution called the primary motor area, which is the control site for the body's skeletal muscles. If this area is stimulated on a patient's exposed cerebrum, some part of the patient's body will twitch or jerk. Just behind this area is another convolution called the primary somatic sensory area. This area makes us aware of the skin senses such as touch and pain (Figure 10-14b).

The cortex of the temporal lobe serves as the interpreting center for hearing, taste, and smell, and also has centers for speech. A portion of the occipital lobe translates impulses from the optic nerve into vision.

These motor and sensory areas are well defined and occupy a relatively small part of the total area of the cortex. Damage to any of these areas severely affects the function it controls. The remaining areas of the cortex are termed association areas, because their functions are generalized.

The cerebral hemispheres communicate with each other by a large, compact band of fibers (about 200 million) called the *corpus callosum* (Figure 10-13b). Impulses are constantly exchanged between the two hemispheres at an immense rate, via these fibers, so that their functions are closely coordinated. What do you suppose would happen to the functioning of the hemispheres if the corpus callosum was cut? The an-

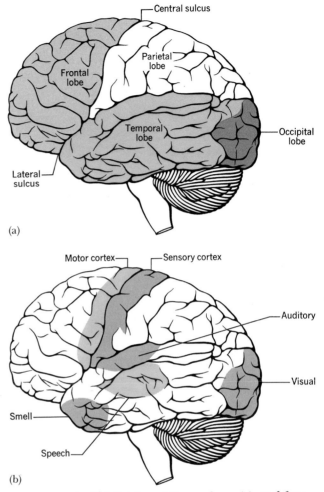

(a)

(b)

Figure 10-14 *Subdivisions of the cerebrum* (a), *and functional areas of the cortex* (b).

swer to this was provided first by experiments on animals and eventually in humans, in which the band was cut surgically to reduce the severity of epileptic seizures. In effect this surgery gives the patient two independent cerebrums—a split brain, so to speak. This condition does not noticeably alter the patient's behavior, but experiments with the patient show that each hemisphere functions quite well but independently of the other hemisphere. More surprisingly, these studies showed that the functions of the two hemispheres were not alike in all of their functions. For example, the left hemisphere is specialized for speech, language, and performing calculations. The right hemisphere is better able to identify objects by their form, to recognize faces and musical themes and to express emotion.

Basic Concept
The major parts of the brain are the brain stem, cerebellum, and cerebrum.

How the Brain Functions

Despite years of research efforts and numerous hypotheses, we still have only a hazy concept of how the brain functions. It is tempting in our world of complex machines and sophisticated electronics to compare brains to computers. Both process and store information, make decisions, work with electrical signals, and both consist of numerous small elements. The problem with this comparison is that the two carry out their similar functions by entirely different mechanisms.

The computer, with far fewer functional elements than the brain, operates extremely rapidly, often completing an operation in a millionth of a second or less. Its operations are determined by a series of instructions that specify each and every step. The instructions (a program) are typically written in a linear fashion in the manner of a recipe. The time required for a computer to perform a program is determined by the time required to perform the single operations or steps. Because computers are designed by humans, their functions are thoroughly understood; only in science fiction stories do they have mysterious functions of their own. Nevertheless, computers are fantastically efficient and useful devices that have enormously extended the neural capabilities of humans.

In the brain, neurons perform their functions more than a thousand times slower than computers and, moreover, use a system that involves both electrical potentials and chemical signals. Despite this, the work performed by a brain may rival that of a computer. This apparently is possible because the brain does not have to follow a linear program, that is, a step by step performance. The brain appears to consist of an immense number of "computing units" that operate simultaneously, in parallel, independently, and with a large amount of information-exchange with one another. Thus the performance of a brain tends to be limited by the number of computing units present and

not by their speed. Brains also contains hierarchical systems. This means that there are units which oversee and listen to the performance and output of units functioning at a lower level in the hierarchy. The system has been compared with a large corporation with different levels of management. In such organizations it is often those units that "make the most noise" which get noticed by upper management. Also, like such an organization, the nervous system is divided into a number of different units or departments that tend to talk to each other only at the higher (management) levels, and otherwise function relatively independently of one another.

The lower levels of organization in the brain are typically programmed (in a computer sense) to maintain basic vital functions of the body, like breathing, without supervision from higher levels. Other lower units are programmed to coordinate the basic patterns of motor activity such as walking or running. Commonly such units are controlled (from above) by being turned off or inhibited. When the higher levels relax their inhibitions, the motor pattern is expressed. An example is what happens when a chicken has its head chopped off: The poor creature runs about the barnyard rapidly (if briefly) instead of simply dying. This happens because removing the head also removes the inhibition of the "walking-running control unit."

The same principle is seen in the development of the nervous system. As an embryo develops, the lower levels of organization become operational before the higher inhibiting centers. An early activity is kicking and moving about in the uterus. Sometime later, and at a higher level of organization, one finds units that control many infantile reflexes such as sucking, and grasping objects placed in the palm of the hand. As still higher levels become functional in the maturing animal, these reflexes are inhibited for the life of the individual. Yet, in later life, if diseases damage these highest levels of organization, the infantile patterns of behavior may return.

The very highest levels of the brain appear to be able to focus their attention on only a limited amount of information. It is as though their efforts are directed towards the "job of the moment." For example, if a cat is placed in a cage with a metronome, electrical potentials can be measured on its scalp in response to each tick. If food is then placed in the cage, the response to the metronome disappears as the cat's attention shifts to a new, more important task.

As you can see, the function of the brain is determined by an incredibly complex and "busy" organization of nerve cells performing their assigned functions. Many neurons can, however, alter their performance based on prior experience. This plasticity is called *learning*. Many neurobiologists believe that learning is accomplished by altering the efficiency of synapses, or by neurons growing additional dendrites and branches so as to function in a new or different manner. Perhaps this is also the way that the nervous system compensates for the constant loss of neurons during our daily lives.

Basic Concepts

The brain evidently functions by the use of a hierarchial system of controlling units.

Learning is the alteration of performance of neurons based on prior experience.

KEY TERMS

basilar membrane	corpus callosum	inner ear	pain
brain stem	electroencephalogram	iris diaphragm	photoperiod
camera eye	endorphin	learning	proprioception
cataract	enkephalin	lens	pupil
cerebellum	eustachian tube	middle ear	retina
cerebrum	fovea	myopia	rods
ciliary muscle	glaucoma	neuropeptide	saccule
cochlea	hair cells	opiate receptor	semicircular canal
cones	hypermetropia	organ of Corti	utricle
cornea	hypothalamus	otoliths	

SELF-TEST QUESTIONS

1. What function do all sense organs have in common?
2. What determines the kinds of stimuli to which an organism can respond?
3. Explain how some animals may live in a perceptual world far different from ours.
4. What are some important functions regulated by light perception?
5. Explain how the eye is like a camera.
6. Name a function for each of the following: (a) sclera; (b) cornea; (c) crystalline lens; (d) iris; (e) pupil; (f) ciliary muscle; (g) rods; (h) cones
7. Why would you expect to find more rods than cones in the retina of a nocturnal animal?
8. Explain what each of the following terms mean: (a) myopia; (b) hypermetropia; (c) astigmatism; (d) cataract; (e) glaucoma
9. Describe how sound and gravity perception are brought about by fluids in the ear.
10. What is the basis for detection of chemicals by taste or smell?
11. Human skin contains receptors for which kinds of stimuli?
12. Discuss the nature of pain and what is meant by referred pain.
13. What is the function of endorphins and enkephalins?
14. How are all vertebrate brains similar in their basic plan?

15. What are neuropeptides?
16. List three ways of studying how the brain functions.
17. Name the major parts of the brain.
18. Discuss what is meant by a split brain.
19. How do the functions of the two cerebral hemispheres differ?
20. How is the brain (a) similar to a computer, and (b) unlike a computer in its functions?
21. Explain how it is thought that the brain functions by means of hierarchical control units.
22. How might learning be defined in terms of changes in neurons?

chapter 11

REGULATING THE INTERNAL ENVIRONMENT: HOMEOSTASIS

chapter 11

questions to think about

What are the basic elements of a self-regulating
 system?

Why is homeostasis necessary in the body?

Why is the kidney used as an example of a
 homeostatic organ?

Where is our biological "clock" located?

---

*The constancy of the internal environment is the
necessary condition of the free life.*

C. Bernard

As the previous chapters have probably made clear, a multitude of events take place inside the body at any given moment. All of these processes must be carefully regulated so that one does not interfere with another, and so that the best interests of the organism itself are always served. The body's internal environment, as it is sometimes called, is indeed far more stable than are its ever-changing surroundings. This internal stability is the outcome of the functioning of numerous complex internal regulatory mechanisms.

Take just one example—the regulation of the amount of blood sugar (glucose) in the bloodstream. Despite a variable intake of sugar in the diet and the continual removal of glucose from the blood by various organs, the glucose level in the blood fluctuates within carefully maintained limits. This regulation is brought about in part by the hormones *insulin* and *glucagon,* both produced by the pancreas. As the level of blood sugar rises it stimulates the pancreas to in-

crease its output of insulin. Insulin, in turn, causes many body cells to take in blood sugar, and the liver to convert blood sugar into a storage form called glycogen. If the blood glucose level decreases, secretion of the hormone, glucagon, causes glycogen to be converted back into blood sugar (Figure 11-1). Several additional hormones are involved in these conversions but insulin and glucagon are the major ones.

This system sounds simple enough, but think of what is involved. The pancreas must be able to secrete two different hormones, each at the appropriate time. In addition, the conversion of glucose to glycogen, or vice versa, involves enzymes in specific cells responding to the appropriate hormones. Therefore, the regulation of just one internal feature, blood sugar, involves a complex interaction of the body's glucose demands, hormone secretion, enzyme activation, and the function of organs. Add hundreds of additional equally complex control systems, and the internal environment suddenly seems awesomely complex. Indeed it is.

Some other internal features that require continual control are blood gases (oxygen and carbon dioxide), body temperature, rate of heartbeat, blood pressure, the water content of the blood and body tissues, the

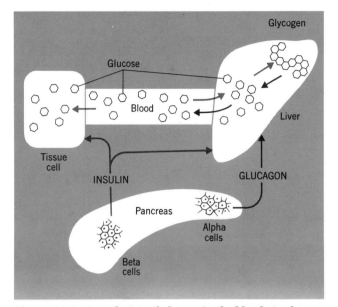

Figure 11-1 *Regulation of glucose in the blood. As the blood sugar rises, the pancreas secretes insulin, which in turn causes glucose to be taken up by body cells and to be stored in the liver as glycogen. When the level of blood sugar decreases, glucagon from the pancreas causes the liver and body cells to release glucose into the blood. In this manner, a relatively stable amount of glucose is maintained in the blood.*

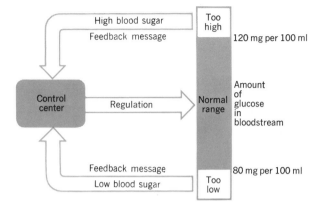

Figure 11-2 *Model of a homeostatic system for regulating blood sugar. The pancreas is the control center, and uses insulin and glucagon for regulation.*

proportion of electrolytes (ions) in the blood and tissues, and the excretion of body wastes. We will examine some of these in more detail later in the chapter. We will also take a closer look at the mammalian kidney, an organ that has a remarkable number of regulatory activities.

THE CONCEPT OF SELF-REGULATION: HOMEOSTASIS

The control and coordination of most internal features of the body are *self-regulating.* In other words, the control system corrects or adjusts inadequate conditions that would interfere with the function. Usually, a deficiency or an excess of the substance being controlled acts as a stimulus to some control center; and the control center, in turn, brings about some sort of

corrective measures. For example, a low level of glucose in the blood stimulates the pancreas to release glucagon. This type of mechanism is termed a *feedback system,* because information is fed back into the control center for appropriate action (Figure 11-2). Note that the system requires a controller (the pancreas in this case) that is sensitive to the feedback message (a shortage of glucose), and has a capability of responding to it (by producing glucagon). An excess of blood sugar is also a feedback message. See if you can trace the steps that are involved when that occurs.

This type of control mechanism with feedbacks is known as a *homeostatic system.* The combined action of all its homeostatic systems enables the organism to maintain a relatively uniform and beneficial condition within itself, a level at which the organism functions well. This level is often referred to as "normal." For instance, a "normal" blood sugar for an adult is 80 to 120 milligrams of glucose per 100 milliliters of blood. Over a period of time an individual's blood sugar may vary from the low end to the high end of this range, indicating that feedbacks are operating and that the system is regulating itself. That is, the system is constantly correcting for deficiencies and excesses of glucose in the blood. If a physician discovers that an individual maintains a blood sugar consistently above or below the normal range a defect somewhere in the homeostatic system is suspected.

Basic Concepts

The stability of the internal environment results from the functioning of numerous internal regulatory mechanisms.

Regulatory mechanisms involving feedbacks are known as homeostatic systems. Such systems are self-regulating.

EXAMPLES OF HOMEOSTATIC SYSTEMS

Chapter 8 describes a feedback system that functions in controlling vertebrate reproductive hormones. It is a well-studied system, which provides an excellent model of *homeostasis.*

As you read through the examples that follow, see if you can identify the feedbacks and control systems that operate in each. Note also how often each system involves chemicals (usually hormones), nervous control (usually through autonomic nerves), and osmoregulation (passage of materials through cell membranes). You may even find all three of these functioning in the same system!

Respiration

The vital act of breathing is a rhythmical one that takes place whether we are awake or sleeping (fortunately!) and is controlled by nerve cells in the reticular system in the brain stem. This center sends impulses in a steady pattern to the respiratory muscles to bring about the regular inhaling and exhaling actions described in Chapter 7. This breathing center communicates with the cerebrum to give us some degree of conscious control over how fast we breathe and even allows us to stop breathing for a limited time. Breathing is necessary mostly to control the amount of carbon dioxide and oxygen in the bloodstream, and it is these blood gases (especially carbon dioxide) that play a major role in the homeostatic control of respiration.

The mechanism works very generally as follows. As cells utilize oxygen in the body they simultaneously produce carbon dioxide. This of course depletes the oxygen and raises the level of carbon dioxide in the blood draining from the capillary beds that supply these cells. The increased level of carbon dioxide in the

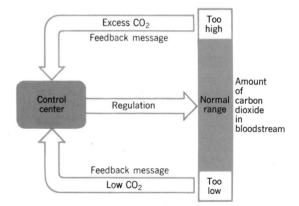

Figure 11-3 *Regulation of blood gases by using carbon dioxide as the feedback material. Where is the control center for this system located?*

blood stimulates chemoreceptors found in the walls of the carotid arteries, aorta, and in the brain stem. These receptors stimulate the respiratory center in the reticular system of the brain stem, which in turn activates respiratory muscles to increase the breathing rate and the heart to beat more rapidly. These actions propel more blood through the lungs and thus increase the exchange of the *respiratory gases.* Consequently, the blood oxygen level rises and the carbon dioxide concentration drops (Figure 11-3). Another aspect of this homeostatic system can be observed when a person hyperventilates (forces himself to breathe rapidly). As the carbon dioxide level decreases in the blood, heartbeat and blood pressure drop, sometimes drastically. From this description have you been able to determine the feedback mechanism? What is the control center?

The complete picture of respiratory control is more complex than what is described here. For example, oxygen can also stimulate the chemoreceptors in the carotid arteries and aorta but only when it is extremely deficient in the blood. Also, it is not actually the carbon dioxide that triggers the chemoreceptors but rather the hydrogen ions formed when carbon dioxide unites with water to form carbonic acid (H_2CO_3), and the carbonic acid partially breaks down to release hydrogen ions. Nevertheless, the end result is the same whether we generalize the process or try to include all of the details.

Body Temperature

Human beings, like other mammals and birds, maintain a stable body temperature which is often many degrees different from that of their external environment. This allows them to function within a wide range of temperatures and to inhabit a broad range of geographic areas.

Temperature regulation in the human animal has been studied intensively, because abnormal body temperatures (below 97°F or above 100°F) are often symptomatic of disease. Normal or average internal body temperature is 98.6° F, but the body temperature may fluctuate by many tenths of a degree throughout the day. This fluctuation suggests that a feedback control system is at work (Figure 11-4).

The major control centers for temperature control, sometimes referred to as the "body thermostat," are found in a portion of the brain called the *hypothalamus* (see Chapter 10). The skin contains temperature receptors, in the form of specialized nerve bodies, that inform the individual, via the hypothalamus, of the relative temperatures of objects that touch his body and whether his immediate environment is hot or cold. These skin sensations are relatively unimportant feedbacks in the temperature control process, however; the major feedback message is the temperature of the blood reaching the hypothalamus. If an individual is physically active or is exposed to a hot environment, his or her blood becomes warmer. Heat receptors in the hypothalamus detect the temperature change in the blood and send nerve impulses to various parts of the body. One response is dilation of surface capillary beds to allow more blood to reach the skin, where some of the heat is lost by conduction and radiation. At the same time, vasoconstriction may occur in the deeper regions of the body. Another response is an increase in breathing rate.

If these actions do not relieve the situation, additional impulses to the skin trigger sweating. The evaporation of these water molecules further aids in dissipating some of the body's heat into the atmosphere. On a hot, humid day all of these mechanisms may be called into action, and yet you may still feel very hot and uncomfortable. Even so, your internal body temperature will remain near 98.6°F unless you grossly (and foolishly) overexert (Figure 11-5).

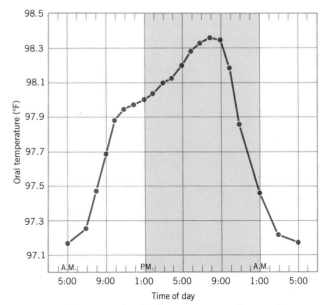

Figure 11-4 *A graph of body temperature fluctuations over a 24-hour time period. The data are based on 70 males. (Modified from* **Biological Rhythms and Human Performance***, W. P. Colquhoun, Editor, Academic Press, 1971.)*

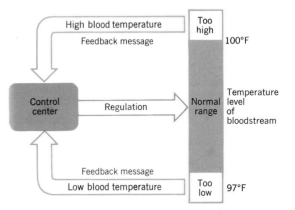

Figure 11-5 *Regulation of body temperature as a homeostatic system. Where is the control center for this system located?*

In cold weather, a more or less opposite chain of events occurs. The posterior part of the hypothalamus contains receptors that are stimulated by cooler-than-normal blood. When such a condition is detected, the hypothalamus sends out nerve impulses, this time causing constriction in the skin capillary beds. This

TABLE 11-1.   **Temperature Regulating Actions in the Body**

In Response to Heat:
   To increase heat loss—
      Vasodilation of skin capillaries
      Increase in breathing rate
      Sweating
   To decrease heat production—
      Lack of eating (loss of appetite)
      Inactive behavior
      Decreased secretion of thyroid stimulating hormone
In Response to Cold:
   To decrease heat loss—
      Vasoconstriction of skin capillaries
      Hair standing on end (goose bumps)
   To increase heat production—
      Shivering
      Hunger and eating
      Active behavior
      Increased secretion of epinephrine and norepinephrine

Source: Modified from W. F. Ganong, *Review of Medical Physiology,* 7th ed., Lange Medical Publications, 1973.

aids in keeping the blood deep in the body and, therefore, reducing heat loss. If this action does not retain sufficient body heat, then additional impulses trigger rapid muscle contractions (shivering) thereby increasing body heat. An increase in epinephrine and norepinephrine from the adrenal glands increases body metabolism. Appetite is also heightened. Table 11-1 summarizes the various activities that occur during temperature regulation in the body.

Fever—a homeostatic malfunction. Bacterial infections often trigger a chain of events that interferes with temperature regulation. White blood cells, in attempting to destroy the invaders, release a substance that seems to "reset" the hypothalamic thermostat for a higher body temperature. As the body temperature begins to rise, the usual mechanisms of surface vasodilation and sweating do not take place. In fact the skin remains cool enough (clammy) that its receptors may signal the posterior hypothalamus to warm up the body. The individual begins to shiver—he feels that he is having a "chill" and this causes an even greater rise in body temperature. This is such a consistent phenomenon that fever is virtually a sure indication of an infection.

Sometimes the control center regains its functioning at this point, the individual breaks out in a sweat, and the fever is "broken." If sweating does not occur and body temperature continues to rise, the individual's life is imperiled. Aspirin is a highly effective drug in such instances because it in some manner aids the control system to regain its normal functioning, possibly by resetting the hypothalamic thermostat. It has been suggested that fever may help the body overcome infections, but no mechanism for this action has been found.

THE KIDNEY AS A HOMEOSTATIC ORGAN

We now turn to a closer look at the kidney, which stands as a striking example of feedback mechanisms and intricate control systems all compressed into a small organ. Its overall function in any animal is osmoregulatory. In this role the kidney controls the amounts and kinds of substances in solution in the bloodstream, by removing excess materials from the blood. But a comparison of the anatomy, and functioning of kidneys in different groups of vertebrates is

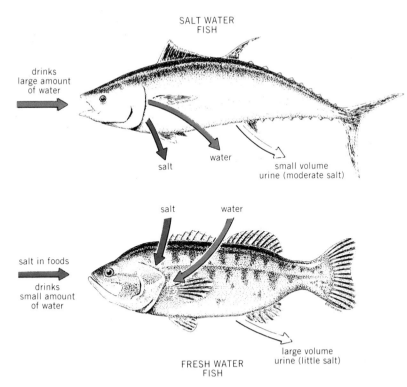

SALT WATER FISH

drinks large amount of water

salt

water

small volume urine (moderate salt)

salt in foods

drinks small amount of water

salt   water

FRESH WATER FISH

large volume urine (little salt)

Figure 11-6 *The osmoregulatory adaptations of freshwater and saltwater fish. The freshwater fish faces the problem of excess water entering the body through the gills. In contrast, saltwater fish tend to lose water through the gills.*

a fascinating study of adaptations and evolution. Consider for a moment how different are the environmental conditions faced by a freshwater fish, a saltwater fish, and a land-dwelling animal. The kidneys of the freshwater animal must constantly get rid of excess water molecules while retaining essential salts; the saltwater creature, on the other hand, must constantly excrete salts and retain water (Figure 11-6); and the terrestrial form must usually conserve most of the water while maintaining a proper balance of salts. Despite these challenging osmoregulatory problems, the kidneys of all groups of animals share certain anatomical and functional features.

The basic functioning unit in all kidneys is a very tiny tubular structure called a *nephron.* The first part of the nephron removes water, ions, nutrients, and waste products such as urea from the blood. Further along the nephron, most of these materials are returned into the blood. The material that remains in the nephron tubules undergoes the final and most delicate step of *osmoregulation,* a highly selective reabsorption of some of the material back into the blood. The remainder is

eliminated as urine. The reabsorbing process is adjusted to the needs of the organism at that particular moment and in its special environment. The organism may need to retain more water molecules at one time than at another. The nephron is able to make this adjustment and a variety of others by means of feedback controls that we will soon examine.

The human kidney consists of a tightly compressed mass of approximately one million individual nephrons. The nephrons make up the portions of the kidney known as the *cortex* and *medulla.* They drain their filtered fluids into the central region (*pelvis*) of the organ. From this general collecting area, the fluid (urine) flows through the ureters into the bladder (Figure 11-7).

Functioning of the Nephron

Blood reaches each kidney through a rather large renal artery leading directly from the abdominal portion of the aorta. This connection with the aorta is important, because it allows a high volume of blood

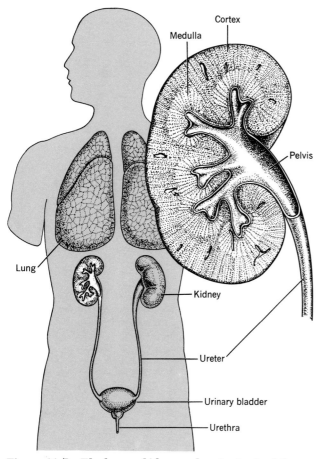

Figure 11-7 *The human kidney and major parts of the excretory system.*

under considerable pressure to reach the nephrons. Within the kidney the renal artery subdivides many times, eventually becoming small balls of capillaries known as *glomeruli.* Each glomerulus sits in a cuplike, hollow structure called *Bowman's capsule,* which forms the beginning of the nephron (Figure 11-8). Due to the high blood pressure, there is a massive flow and filtration of materials from the glomerular capillary bed into Bowman's capsule. This glomerular filtrate includes water, amino acids, salts, and glucose, as well as waste products such as urea. The only products that do not leave the glomerulus are blood cells and blood proteins, because they are too large to pass through the pores in the capillary walls. One would therefore never expect to find proteins or blood cells in urine except as a consequence of disease or injury to the glomeruli.

Because the amount of materials filtered into Bowman's capsule and the nephron tubule is so great, it is essential that most of them be returned to the blood. This massive tubular reabsorption takes place in the *proximal convoluted tubule.* Glucose, amino acids, and most of the salts are returned (primarily by active transport) to capillaries that surround the convoluted tubules. As these materials are reabsorbed from the filtrate, the concentration of dissolved materials in the blood rises. Thus, some of the water follows by osmosis and also returns to the blood. Because all of the glucose is normally reabsorbed at this point, none usually appears in an individual's urine. If it does, then a disease such as diabetes is suspected.

After this massive reabsorption, a small quantity of salts and urea and some water still remain in the filtrate. This fluid passes through the nephron into *Henle's loop* and the *distal convoluted tubule* (Figure 11-8). Here a complex set of events occurs, including *tubular secretion* and *selective reabsorption.* In tubular secretion, materials such as potassium ions, hydrogen ions, and ammonia are removed from the blood and added to the tubular filtrate. An important function of this event is to regulate the acid-base balance in the blood. Selective reabsorption removes certain substances from the filtrate and places them in the blood. What is reabsorbed, and how much of it, depends on the organism's needs at that moment. The selective uptake of water provides a good example of how this control system works (Figure 11-9).

The hypothalamus of the brain contains receptors (osmoreceptors) that are sensitive to the amount of water in the blood. Let us assume that the concentration of water in the blood is low. This imbalance triggers the osmoreceptors, which in turn stimulate the posterior lobe of the nearby pituitary gland to release vasopressin, also called *antidiuretic hormone (ADH).* When ADH reaches the kidneys via the bloodstream, it increases the permeability of the collecting tubule walls to water. As a result, more water leaves the tubules and enters the bloodstream, and the water balance in the blood is restored. If the blood contains too much water, on the other hand, the osmoregulators are not stimulated. Hence ADH is not released, and water in Henle's loop and the distal tubules is not returned to the blood but is eliminated instead.

By understanding the action of ADH, one can bet-

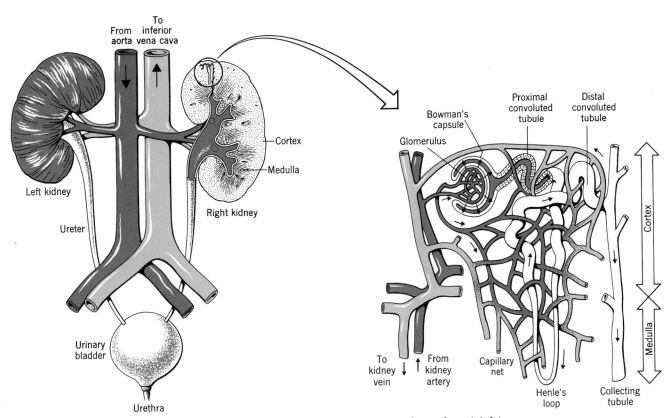

Figure 11-8 *Location of a nephron in the kidney (left) and the major parts of a nephron (right).*

ter appreciate the effects that beer and other alcoholic beverages have on the kidneys. Alcohol (plus the excess water intake involved in drinking) acts as a *diuretic.* That is, it inhibits the secretion of ADH and thereby increases the flow of urine. Caffeine also has a diuretic effect by increasing the filtration rate in the glomeruli. Cold weather has a similar result, but for a different reason. In cold weather the surface capillary beds constrict, blood pressure rises, and thus more blood is forced through the glomeruli.

In a rare disorder known as diabetes insipidus, too little ADH is available to be released into the blood. Consequently, water is not reabsorbed from the collecting duct but is lost from the body as urine. This loss may become serious enough to dehydrate the patient to a dangerous level.

The reabsorption of salts in the distal convoluted tubule is also subject to hormonal control. One of the steroid hormones from the adrenal gland tends to increase both reabsorption of sodium and excretion of potassium in the tubule. When either potassium increases or sodium decreases in the blood, more of the hormone is released from the adrenal gland. A return to normal salt concentrations results in a reduction in the amount of this adrenal hormone released.

The functioning of nephrons is actually even more complex than this description suggests, and it includes many additional control systems. But even this thumbnail sketch suggests why the kidney is considered the master homeostatic organ of the body.

The Artificial Kidney Machine

Diseases of the kidney are relatively common and may be life-threatening. Formation of urea is the body's way of converting highly toxic cellular waste substance (ammonia) into a less toxic form. However, urea in large amounts itself becomes toxic and, if not removed

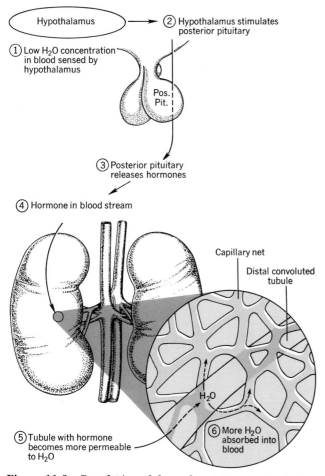

Figure 11-9 *Regulation of the reabsorption of water into the bloodstream. Can you figure out the control center and feedback signals in this system?*

from the blood, can cause death. Therefore, the design of a machine that could remove urea and other excess products from the blood was of great usefulness in treating kidney diseases. The artificial kidney machine accomplishes this purpose (Figure 11-10).

The machine contains two major components. One consists of many yards of small cellophane (dialysis) tubing arranged in a compact coil. The tubing acts as a semipermeable membrane, because it contains millions of tiny pores about the same size as those found in glomerular capillary beds. The other component of the machine is a tank containing dialysate (a solution)

in which the cellophane tubing is immersed. The chemicals in the dialysate are carefully adjusted to the needs of the patient who will use the unit.

Immediately before use, the tubing is filled with blood of the same type as that of the patient, plus an anticlotting chemical to prevent blood clots from forming in the tubing. A needle inserted in an artery in the patient's arm feeds blood into the tubing. As the blood circulates through the cellophane coils in the washing solution with the aid of a small pump, substances such as urea and excess salts and water diffuse through the porous tubular walls. Various kinds of small molecules and ions can also be put into the solution and thereby diffuse into the blood. After the blood has traversed the many coils of tubing and has been cleared of clots and air bubbles, it reenters a vein in the patient's arm. To properly clear the blood of toxic and excess substances takes many hours (four to six in many cases), because the diffusion process is slow and the blood has a long route to travel. Nevertheless the machine saves thousands of lives each year.

The machine is not really an "artificial kidney," because its method of functioning is based entirely on osmosis (or dialysis, as the event is often termed) rather than on blood pressure and selective reabsorption. The proper term for the machine is a *dialysis apparatus,* and the process of passing blood through the unit is called *hemodialysis.*

Basic Concepts

The kidney is an outstanding example of an osmoregulatory organ. Its nephrons remove excess substances and wastes from the bloodstream by a combination of actions that include filtration, massive reabsorption, osmosis, and selective reabsorption.

The artificial kidney uses the principle of dialysis to remove excess substances from the blood or to add materials to it.

BIOLOGICAL RHYTHMS

When discussing temperature control earlier in the chapter, reference was made to Figure 11-4 in order to show that body temperature fluctuates considerably throughout the day. Look at this figure again and see if

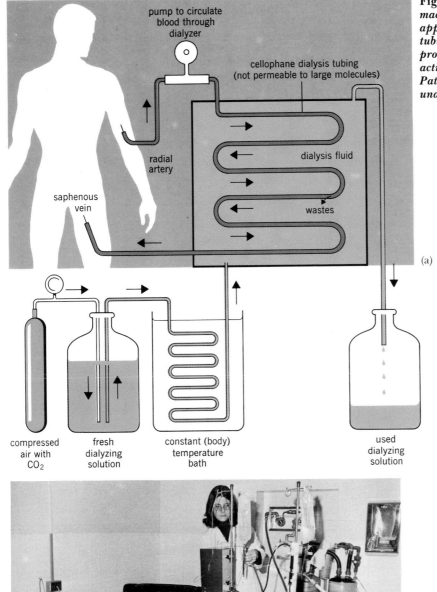

pump to circulate
blood through
dialyzer

cellophane dialysis tubing
(not permeable to large molecules)

radial
artery

dialysis fluid

saphenous
vein

wastes

(a)

compressed
air with
CO_2

fresh
dialyzing
solution

constant (body)
temperature
bath

used
dialyzing
solution

Figure 11-10 (a) *The artificial kidney
machine, more correctly called a dialysis
apparatus. Blood flows through the dialysis
tubing and is slowly cleaned of toxic
products and excess substances by osmotic
action.* (b) *A dialysis apparatus in use.
Patients can read or sleep while their blood
undergoes dialysis.*

(b)

the shape of the graph suggests another conclusion about body temperature. (Answer: It peaks at about 9:00 P.M. and is lowest early in the morning when measured over a 24-hour period.) This is an example of a biological rhythm, and because it is repeated approximately every 24-hours it is also termed a *circadian rhythm.* In a sense it is a larger scale rhythm imposed over the homeostatic system of temperature control. Many physiological events in humans and other organisms follow a circadian rhythm including blood pressure, pulse rate, the sleep/wake cycle, and even cell division in skin. Even some of the minerals in the body such as blood calcium and inorganic phosphate follow circadian patterns. A field of study— *chronobiology*—is growing out of the interest that scientists have in these internal body cycles.

The medical world is becoming interested in the implications of chronobiology because laboratory animals react differently to a variety of drugs in relation to the time of day they are administered. For example, if a group of rats is given a large dose of amphetamine at a certain time of day, most of them die. The same dose at a different time of day is lethal to only a small proportion of a group. The implication of observations such as these is that drugs given to humans for such ailments as high blood pressure should be taken in relation to their blood pressure circadian rhythm rather than on a regular schedule (such as every 4 hours). This type of timing could be especially crucial for cancer patients receiving high doses of strong anti-cancer drugs.

If biological rhythms exist, and abundant evidence indicates that they do, then where is the control center for these "biological clocks"? One hypothesis is that influences such as the day/night cycle, pull of gravity, temperature changes, and other external forces control the cycles. One drawback to this idea is that many cycles persist unchanged regardless of changes in the environment. Another hypothesis states that biological rhythms are controlled by an internal *biological clock.* This "clock" may be influenced or disrupted by external factors but even so, manages to maintain its rhythmical beat. Despite a considerable amount of research the actual location of the clock is unknown. Some evidence indicates that it may be in the hypothalamus of the brain together with many other vital control centers; additional evidence is badly needed.

The discussion of chronobiology should not be viewed as upholding the idea of "biorhythms" which found popular support a few years ago. According to this idea our lives are controlled by rhythms associated with our birth dates. By consulting the appropriate charts, an individual could predict when his or her biorhythms would be "up" or "down." This concept is not based on scientific evidence and thus lies more in the realm of astrology than in science.

Basic Concept
Many physiological events in the body follow a circadian rhythm.

KEY TERMS

antidiuretic hormone	diuretic	homeostasis	proximal convoluted tubule
biological clock	feedback system	hypothalamus	respiration
Bowman's capsule	fever	insulin	respiratory gases
chronobiology	glomeruli	kidney	selective reabsorption
circadian rhythm	glucagon	nephron	self-regulation
dialysis	hemodialysis	osmoregulation	
distal convoluted tubule	Henle's loop		

SELF-TEST QUESTIONS

1. What is meant by the body's internal environment?
2. Describe how insulin and glucagon function to regulate blood sugar.
3. List several internal features of the body that require continuous control.
4. Define the term *homeostasis*.
5. Name the basic components of a self-regulating system and tell how each works.
6. What is the control center for breathing?
7. What is the feedback mechanism in respiration?
8. Explain why holding your breath eventually triggers off the breathing reflex.
9. What feature of body temperature suggests that it is under homeostatic control?
10. Where is the body's "thermostat" located?
11. What is the feedback mechanism in temperature control?
12. List some actions that function to warm up the body on a cold day.
13. List some actions that function to cool off the body on a hot day.
14. Describe what happens to the temperature control system when a person has a fever.
15. What is the major function of the kidney?
16. List the parts of a nephron and tell their functions.
17. Tell what is meant by massive filtration, tubular reabsorption, tubular secretion, and selective reabsorption.
18. Describe how the kidney functions with the hypothalamus to regulate the amount of water in the blood.
19. Describe the general construction of an artificial kidney.
20. Tell how a dialysis apparatus differs from a real kidney in its functioning.
21. Tell what is meant by a circadian rhythm and give several examples.
22. Discuss the possible locations of the biological clock.

HOW TO MAINTAIN A LIVING MACHINE

chapter 12

questions to think about

Is the human body nothing but a complex living
machine?

What are the major protective devices of the body?

What are the five basic rules for maintaining a
healthy body?

Happiness lies, first of all, in health.

G. W. Curtis

Some years ago a popular human physiology text
called *Machinery of the Body** was in wide use in colleges.
The title implied that the body was comparable to a
machine with its various specialized parts; in order to
appreciate the machine, one had to understand the
functioning of each part. This simile of the body to a
machine is an apt one in many ways. Like the parts of a
machine, our organs and systems need to be maintained, they may break down and cease to function, or
they may simply wear out after a period of time. Sometimes we can repair or even replace defective parts so
that the body-machine can continue functioning.

At this point, however, our analogy ends, because
the body-machine has capabilities not found in manmade machines. Numerous internal homeostatic devices make it self-regulating, and it has the remarkable
ability to repair and heal its own injuries. Broken bones

mend, blood clots seal off cuts and gashes, and flesh
heals. Many additional built-in protective mechanisms
shield the body from a number of potentially damaging situations.

These mechanisms and how they function (or occasionally fail to function) make up most of the substance
of this chapter, which concludes with an attempt to
define health and to identify some basic conditions that
best help the body maintain its machinery in good
working condition.

Basic Concept

**The human body can be viewed as a sophisticated, self-re-
pairing machine containing a variety of protective mechanisms.**

PROTECTIVE MECHANISMS

Prior to birth, whether from inside an egg or from the
womb of its mother, an organism lives in a warm,
usually sterile, protective environment in which it is
provided with a nutritionally balanced diet. Upon
leaving this fetal haven, the young organism is immediately faced with a radically different and usually threatening environment. Hordes of microorganisms, some

---

* Anton Carlson and Victor Johnson, *Machinery of the Body*, third
edition. Chicago: University of Chicago Press, 1948.

harmful, infect its food, water, and air; larger organisms may inflict wounds and bruises on its body; the temperature of its surroundings often fluctuates widely; and the air rapidly dries out its body. As if these problems were not enough, the recently emerged animal must soon find or be provided with a new and adequate source of nourishment.

To withstand this unfriendly environmental onslaught, organisms must obviously have a variety of protective devices. For convenience of discussion, these are divided into the skin, blood system (including the immune response), and nutritional safeguards.

Skin

Most of us tend to regard our skin as just a simple body covering; we pay little attention to it except to apply cosmetics, shave or cut off excess hair, and patch minor cuts. Actually the skin is our largest organ (Figure 12-1), having a fairly complex anatomy and such a variety of functions that it constitutes one of the most important of the body's protective mechanisms.

Human skin is made of layers of specialized cells constituting the outer *epidermis* and an underlying portion called the *dermis* (Figure 12-2). The epidermis is especially important as a protective covering. New epidermal cells, continuously forming from a *basal layer,* slowly move to the surface of the skin. During this migration, they form a horny protein material called *keratin,* and slowly die. Thus, the surface of the skin is made of several layers of these dead, keratinized cells. The functions of the skin, mostly protective, include the following:

1. To form a protective shield against microorganisms, chemicals and sunlight. The keratinized outer layers of epidermal cells form a thin, flexible, relatively impervious barrier to microorganisms and many chemical substances (but not including the oils of poison ivy!). The pigment *melanin,* produced by cells in the epidermis called *melanocytes,* shields the deeper layers of skin cells from the harmful untraviolet radiation in sunlight. Increased exposure to sunlight causes the melanocytes to produce extra granules of melanin, and thus a darkening of the skin. Every spring and summer, great hordes of light-skinned people take part in a ritual called sunbathing in order to obtain a tanned skin. From a biological viewpoint this is an unfortunate

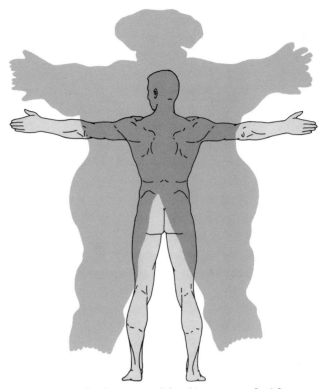

Figure 12-1 *Surface area of the skin compared with actual body size. A 6-foot man has about 15,000 square inches of skin area.*

custom for two reasons: Prolonged exposure to sunlight causes the skin to become leathery and wrinkled. Even more seriously, ultraviolet light is carcinogenic; repeated exposure of the skin to sunlight heightens the chance of developing skin cancers later in life.

Although the outer epidermal cells function as a barrier to microorganismic invasion of the skin, they also provide a dwelling place for considerable numbers of bacteria, fungi, and one microscopic animal, the hair follicle mite. All of these tiny inhabitants live primarily on organic materials produced by secretions from the skin. The most abundant are bacteria, which are found in greatest density on the skin of the face, the armpits, and the groin area (Figure 12-3). Although most skin bacteria are harmless, two varieties can cause infections if they get beneath the outer epidermal cells. *Staphylococcus aureus* may cause boils and pimples, and *Cornybacterium acne* infects hair follicles to produce the curse of many teenagers, the facial blemishes known as acne.

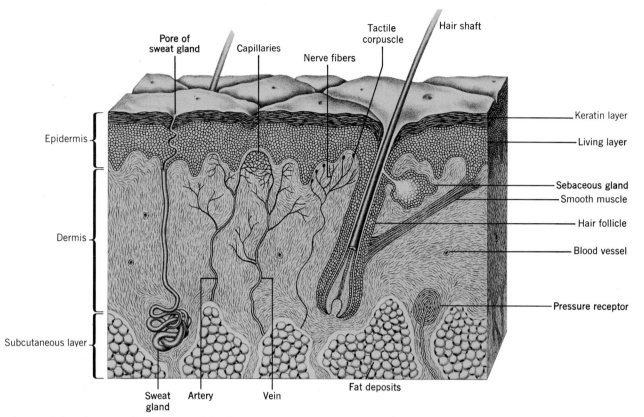

Figure 12-2   *Layers and structures of the skin. Can you determine from this diagram why skin is called an organ rather than a tissue?*

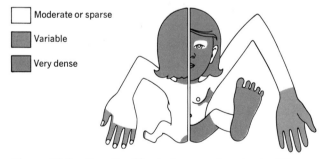

Moderate or sparse

Variable

Very dense

Figure 12-3   *Density of bacteria on various parts of the body.*

A New Zealand bacteriologist, Mary Marples, considers the microscopic life on the skin equivalent to an ecological system of larger organisms. She proposes a picturesque analogy, likening the forearm to a desert, and armpit to a tropical forest, and the scalp to cool woods. She notes that the resident skin inhabitants occupy most of the suitable living places, making it difficult for new arrivals, including the potential pathogens, to establish themselves. One inference from this observation is that the American obsession for keeping the skin clean by frequent bathing and use of cleaning creams may be unwise, because it can leave the skin vulnerable to be colonized by harmful alien bacteria and fungi. Thus, one might say: Bathe infrequently and protect your friendly skin bacteria!

Another protective aspect of skin, although it is not of major importance in human beings, is the presence of hair. Hair is a derivative of the epidermis. The short stiff hairs in the nose and outer ear openings help keep out dirt particles and perhaps an occasional insect. Scalp hair shades the top of head from sunburn. The remainder of body hair is evidently functionless in human beings, a holdover from our ancestry.

2. To cool the body and to regulate blood pressure. As a cooling organ the skin contains two devices, capillaries and sweat glands (see Figure 12-2). Upon receiving signals from autonomic nerves, arterioles that lead into capillary beds in the dermis open up and allow blood to flow through. Heat can then radiate from the blood through the skin. Simultaneously, sweat glands may secrete their slightly saline solutions onto the surface of the skin. The evaporation of sweat also aids in dissipating heat from the body. In dry air we are usually unaware of this evaporation process. But in humid air, sweat remains on the skin and is less effective in the cooling process.

Resistance to blood flow, caused by shunting blood into capillary beds, increases blood pressure. Should it become too high, blood-pressure sensors signal the arterioles to divert their flow from capillary beds directly into venules (Figure 12-4). This shunting procedure, which occurs throughout the extensive expanse of dermis, acts somewhat as a safety valve in reducing blood pressure.

3. To aid in wound healing. As a barrier between the external world and the body's tissues, the skin is vulnerable to a variety of damaging events such as abrasions, scrapes, cuts, and burns. Fortunately, the skin has effective regenerative powers that take care of such damage. In the event of a wound, the blood coagulates to form a *fibrin clot* and thus seals off the injury (Figure 12-5). Special cells called fibroblasts then move into the area and begin to form strands of connective tissue. Eventually they fill the wound with scar tissue and the healing process is complete. In cases of deep or extensive wounds, it is common medical practice to hold the cut surfaces together with clamps or stitches until healing can occur.

4. To initiate the synthesis of vitamin D. As described in an earlier chapter, vitamin D is important in the transport of calcium from the intestines, and a deficiency in this vitamin leads to poor bone formation. A chemical forerunner of vitamin D, provitamin D, is found in the skin. This compound must be activated by ultraviolet light, as in sunshine, before the liver and kidneys can make it into the active form of vitamin D.

5. To provide sensory information. In Chapter 10 we noted that the nerve receptors for touch, pain, cold, and heat were located in the skin. They provide sensory information about the conditions of its environment.

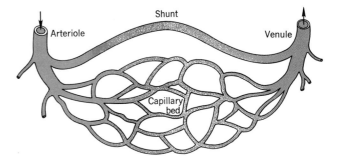

Figure 12-4 *Arteriole capillary shunt. Blood that normally passes through capillary beds can be diverted directly into venules on certain occasions. How does this action reduce blood pressure?*

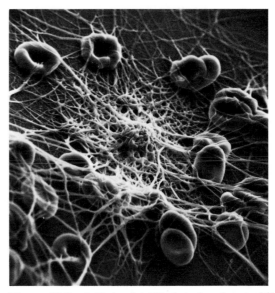

Figure 12-5 *A fibrin clot. The interlaced mat of fibers helps to close off injured blood vessels. The round objects are red blood corpuscles.*

6. To make an organism distinct. Skin identifies an organism by its body form, facial features, or distinctive markings such as fingerprints. This latter feature with its patterns of loops and whorls can even distinguish identical twins.

Basic Concepts

Skin is a complex, protective organ consisting of layers of

cells organized into an outer epidermis and an inner dermis.

The major functions of skin are to shield the body from the external world, to heal wounds, to cool the blood, to regulate blood pressure, to provide vitamin D, to provide sensory information about our surroundings, and to render an organism distinct.

Blood

Blood is the body's strongest line of defense against injuries and against foreign objects that get past the skin barrier. In the case of injuries, *blood clotting* efficiently prevents excess loss of blood and seals off the injury from bacterial invasion. The formation of a blood clot is such a complex reaction (at least ten different factors are involved) that only a few details can be provided here.

Located in the blood are numerous tiny cell fragments termed *platelets*. These fragile bodies contain several of the substances necessary for the clotting reaction. When an injury occurs, platelets accumulate at the site. They then begin to break apart, releasing their contents. These substances, including calcium and potassium, enter into reactions that eventually convert a blood protein, fibrinogen, into a fibrin *clot* (see Figure 12-5). The clot seals the wound and prevents additional bleeding. Eventually an enzyme (plasmin) dissolves the clot so that it will not hinder blood flow through the area.

The importance of blood clotting is emphasized by the hereditary disease, *hemophilia*, which occurs when one of the clotting agents is not present in the blood. The victims, usually males, bleed excessively from relatively minor cuts or injuries. Blood transfusions or injections of a purified preparation of the missing clotting factor are effective treatments in many cases.

Clotting, although essential for protection against wounds, is involved in one relatively common but serious medical disorder. A clot (*thrombus*) can form spontaneously within a blood vessel and block the blood supply to tissues or organs. The consequences are especially severe if the thrombus occurs in blood vessels supplying the wall of the heart (a coronary thrombosis), the brain (cerebral thrombosis), or the lungs (pulmonary thrombosis). Small pieces of a thrombus (*emboli*) may also break free and move elsewhere to stop up a vessel in another part of the body.

Another protective function of blood lies in the ability of some of its white cells to engulf bacteria. Two types of white blood cells, *neutrophils* and *monocytes,* are capable of *phagocytosis.* If bacteria enter the body, neutrophils (the most common of the several kinds of white blood cells) congregate around the bacteria and begin to engulf them, somewhat in the manner of an amoeba feeding on a smaller organism; monocytes help in this process but are present in far fewer numbers. Inside the neutrophil, enzymes destroy the entrapped bacteria. But in doing so, they also eventually destroy the neutrophil itself. These and other dead cells accumulate in the wound site, forming pus. Even this process is beneficial because the damaged tissue cells release *histamine.* This chemical, together with *kinins* in the blood and tissue fluid, cause the capillary walls in the area to become more porous to newly arriving neutrophils and monocytes. In addition, kinins attract additional white cells to the area which in turn produce the redness (inflammation) observed around an infection or wound area.

If the infection continues for a time or is widespread, substances released from the dying neutrophils affect the temperature control center in the hypothalamus, "resetting" the body's thermostat (see Chapter 11). An elevated body temperature—a *fever*—results. This is such a consistent phenomenon that a fever almost invariably indicates a bacterial infection somewhere in the body. Similarly, a pronounced increase in the number of white cells, as determined from a white blood cell count, also indicates an infection.

Antibiotic drugs are widely used for treating serious bacterial infections. Examples of these drugs are streptomycin, tetracycline, and penicillin. They act by inhibiting the protein-making process within bacterial cells. Without this vital process, bacteria cannot grow and reproduce. The term *antibiotic* should not be confused with the term *antibody* described in the immune system.

Basic Concept

Blood protects the body by clotting to seal off wounds, and destroying microorganisms through phagocytosis.

The third type of protection provided by the blood is the *immune system.* The functions of the immune system are to recognize invading microorganisms, cells from another organism (as from a transplant), or other substances such as large molecules, and then to take action to destroy or neutralize them. The system is remarkably sensitive to these "foreign" objects in the body, and is adept at distinguishing between *self* (normal body constituents) and *not-self.* Very little was known about this unusual and intricate system until recently. But because of its great importance in combating diseases, possibly including cancer, the study of immune systems has become a major area of research. Details of its functioning are rapidly emerging.

The immune system is based on small white blood cells called *lymphocytes* and on antibodies, the chemicals that some of them produce (Figure 12-6). Many lymphocytes and antibodies circulate in the blood, where they more or less patrol the body. Considerable numbers of lymphocytes accumulate in lymph nodes, the thymus gland, and the spleen.

Current studies indicate that there are two basic types of lymphocytes in terms of their functions, although all originate in bone marrow (Figure 12-7). Some lymphocytes, as part of their development, pass through the thymus gland (and hence are termed *T cells*). They specialize in combating cancer cells, bacteria, fungi, and cells or tissues that are introduced into the body from another organism. T cells carry out their protective functions by forming cell-killing chemicals, and by attracting other white cells that can engulf

the foreign body. They do not form antibodies, however.

Other lymphocytes that do not go through the thymus gland as part of their developmental cycle, also circulate through the bloodstream and lymphatic system. These lymphocytes, termed *B cells*, if stimulated will divide and form plasma cells that give rise to huge numbers of large protein molecules called antibodies.

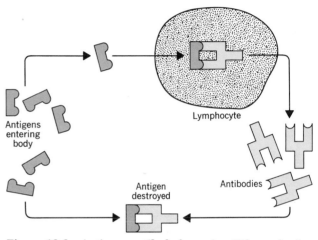

Figure 12-6 *Antigen–antibody formation. When a foreign material (antigen) enters the body, lymphocyte blood cells respond by forming antibodies. The antibodies inactivate or destroy the antigen.*

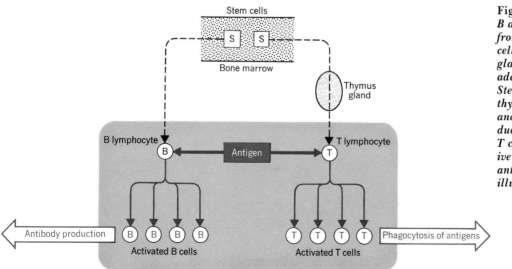

Figure 12-7 *Formation of B and T lymphocyte cells from stem cells. Some stem cells migrate to the thymus gland to become T cells adapted for phagocytosis. Stem cells that bypass the thymus are called B cells, and they specialize in producing antibodies. B and T cells together are effective in combatting the same antigen as shown in the illustration.*

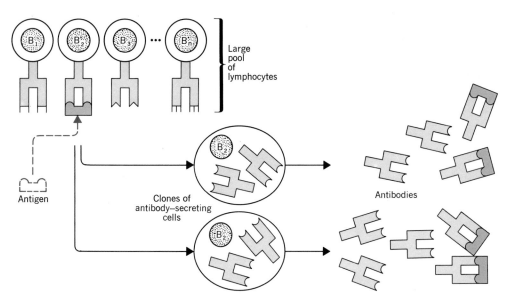

Antigen

Clones of
antibody–secreting
cells

Antibodies

Large
pool
of
lymphocytes

Figure 12-8 *Clonal selection concept. Each of the B cells in the pool at the top is modified to respond to a specific antigen. When an antigen enters the body, it binds to a B cell that has the appropriate pattern (epitope) on its surface. This is cell B in the illustration. Cell B_2 multiplies (makes clones of itself) and the clones in turn produce antibodies as shown. In this manner, B_2 cells quickly make large quantities of antibodies.*

If antibodies meet a substance that is not normally a part of the bloodstream (an antigen), they attempt to destroy it. This event is called an antigen–antibody reaction.

How B lymphocytes recognize a substance that is foreign to the body and also make antibodies has puzzled biologists for a long time. Now the recognition system is known to depend on specific patterns of foreign molecules called *epitopes*. B lymphocytes have a small amount of antibody on their cell surface. When one of these B lymphocytes meets with an antigen that fits with the antibody on the surface, it will divide and differentiate into plasma cells which produce that antibody in tremendous quantity. Because there are at least a million different epitope patterns, there are evidently a million different matching B lymphocytes with corresponding antibody (Figure 12-8).

Upon identifying an antigen, some of the B lymphocytes multiply to produce numerous copies of themselves—*clones* of identical cells. This system greatly increases the amount of antibody produced when we are re-exposed to that foreign substance. The cloned cells that do not form antibodies seem to retain a "memory" of the event so they are available to respond quickly if the same antigen invades the body at a later time. This is the reason that the body often responds more vigorously and rapidly to a second

infection by the same kind of microorganism (Figure 12-8).

Most inoculations or vaccines operate on this principle. By using dead or disabled disease organisms they make the body prepare antibodies against a specific disease such as polio, smallpox, or typhoid. Because of this acquired immunity, the body reacts as it does to a second infection if these diseases are ever encountered later.

For reasons not well understood by immunologists, lymphocytes and antibodies do not normally react against the body's own cells, or against each other for that matter. This phenomenon is termed *self tolerance* and is evidently "learned" by the immune system during embryonic life.

Sometimes the immune system works so well that the body becomes hypersensitive to such things as the poison in bee and ant stings, antibiotics, and plant pollens. What evidently happens is that when the immune system recognizes these substances, the resulting antibody–antigen reaction causes a massive release of histamines from white blood cells. This substance attracts additional neutrophils to the area, and as they are destroyed they release enough lysosomal enzymes to inflame and damage nearby tissues. This explains the reddened eyes, sneezing, red skin welts, and other symptoms that accompany these reactions. *Anti*hista-

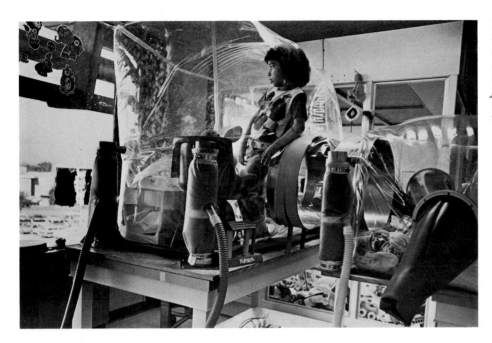

Figure 12-9 *This small child is destined to spend his life inside a plastic isolation chamber similar to the one shown here, because he would soon die of infections if exposed to microorganisms in the surrounding environment.*

mine drugs alleviate the reaction by inhibiting the action of the excess histamines.

Another type of malfunction of the immune system involves a variety of *autoimmune*—literally immunity to self—*diseases.* These include several forms of nephritis in which kidney nephrons are damaged, allowing proteins to pass in the urine (a diagnostic test for nephron damage), and rheumatoid arthritis. Occasionally a baby is born with an immunodeficiency disease, that is, without an effective immune system. Sometimes this condition can be remedied by transplants of bone marrow or thymus tissue. If this is not successful the unfortunate infant must be kept in a germ-free environment. This usually takes the form of a clear plastic tent or bubble into which only sterile air, food, and water are admitted (Figure 12-9). Without these precautions, the infant would soon die of an infection that would be minor for a normal individual.

T cells have an extremely important function in protecting against viruses, intracellular parasites, and tumors. Also T cells control the amount of antibody produced by B cells and other immune functions.

Because T cells are so efficient at what they do, they cause problems when surgeons transplant organs or graft skin on a patient from an other individual. The T cells cause the body to attempt to reject the alien tissues. To reduce this danger, tissue compatibility tests are performed and closely related family members are tested as donors. An identical twin makes an ideal donor! If these attempts fail, treatments are used to suppress the patient's immune system. These include radiation and a variety of immunosuppressant drugs. The problem with using such suppressants is that they leave the body vulnerable to a variety of infectious diseases and malignancies.

The final protective function of the blood involves *interferon* and *natural killer* (NK) *cells.* Interferon is a protein produced by several types of white blood cells (including T cells) in response to a virus infection; it appears to help the body fight off such infections. Currently, interferon is being studied as a possible anticancer drug. Natural killer cells are a group of lymphoid cells recently identified in the circulatory system that are especially efficient in attacking cancer cells. Their activity in this regard is enhanced by the presence of interferon.

Basic Concepts

The immune system consists of lymphocytes and antibodies produced by lymphocytes. Antibodies recognize antigens by the epitopes on their surface membranes and then attempt to destroy the antigens.

Vaccines cause the immune system to produce antibodies against specific disease-causing organisms.

An individual's immune system must usually be suppressed with drugs or radiation before his or her body will tolerate organ or tissue transplants from another individual.

T cells, interferon, natural killer cells, macrophages, and B cells help the body resist infections and the growth of cancer cells.

Nutritional Safeguards

Another type of protective mechanism relates to the body's nutritional requirements. As discussed in earlier chapters, the body requires specific nutrients in certain amounts in order to maintain health. Yet animals (and people) probably rarely eat a balanced diet, day by day. Still they seem to function quite well. Obviously the body's chemistry must be highly adaptable, and able to adjust to considerable fluctuations in its demand and supply of nutrients.

The greatest and most constant nutritional demand is for an energy source. In Chapter 5 we noted that the basic raw materials for energy conversion in a cell were simple organic substances such as the three-carbon molecule called pyruvate. Also the point was made that such molecules may be derived from carbohydrates, lipids, or proteins. The most abundant and easily obtained supply is from carbohydrates, but when this source is lacking the body can turn to other sources.

The survival value of this mechanism has been demonstrated many times by people who were forced to live on a limited diet for long periods of time. An example is provided by the experience of an English couple, the Baileys, whose sailboat sank suddenly in the Pacific Ocean in 1973.* The couple drifted in a raft for 118 days before being rescued—the longest survival time at sea ever documented. During most of this period, their sole food was flesh from turtles, fish, and an occasional sea bird, all eaten raw. Thus their diet was unusually high in protein. A medical examination shortly after their rescue showed that each had lost about 40 pounds, both were anemic and showed some symptoms of deficiencies of vitamins B and C, and both suffered a number of minor ailments such as swollen

limbs. But they survived the ordeal, even though forced to subsist on a severely limited and monotonous diet. Their bodies obtained sufficient energy by converting turtle and fish proteins into small organic molecules.

In terms of obtaining vitamins and minerals, the body has several alternatives when dietary intakes are low. Some of these items are stored in tissues to some extent and can be pulled out and used if necessary; the fat-soluble vitamins A, D, E, and K are examples. Calcium, iron, and phosphorus are "borrowed" from organs such as bones and the liver when low amounts are in the diet. Normally these debts are paid back when dietary levels of these nutrients rise again. Only when dietary deficiencies are of some duration does the body begin to show symptoms of nutritional diseases.

This adaptability is of great survival value to the body. However, it often misleads people about the value of sound nutritional practices. College students, and many other people, know from personal experience that they can live for some time on a diet of Big Macs, Cokes, and snack foods without becoming ill. Therefore, why bother with a balanced diet? One answer is provided by the Bailey's experience: They survived, but symptoms of dietary deficiencies eventually began to appear.

Even in total starvation the body can survive for several months by utilizing its own tissues—not only the fat reserves but also its proteins. The consequences are not pleasant (emaciation, lethargy, and mental depression) but life can be preserved.

Basic Concept

The body adapts to fluctuations in its nutrient supplies by obtaining organic molecules for energy from protein, lipids, or carbohydrates, by storing certain minerals and vitamins for later use, and by borrowing nutrients from various organs.

WHEN PROTECTIVE MECHANISMS FAIL: CANCER

Despite the efficiency of immune systems and other protective devices, some diseases may overwhelm the

* Maurice and Maralyn Bailey, *Staying Alive.* New York: David McKay Co., 1974.

body and destroy it. One of these is the unrestrained cell growth commonly called malignant tumors or cancers. In reality there are over 100 types of cancers that affect various tissues of the body and exhibit many different symptoms. But all cancers have several common characteristics. One is unrestricted cell division (although not necessarily a *rapid* rate of division, as is commonly believed). The consequence is a cell mass that enlarges until it crowds surrounding tissues, takes over their nutrient supply, and eventually destroys the organism. Another characteristic is that the mass of cancer cells has very little order or internal organization. And finally, all cancers are capable of metastasizing, that is, spreading via the blood and lymph systems to other parts of the body, where they start new cancers.

Cancers can be divided into four broad categories. The most common type, in human beings, are the *carcinomas:* solid tumors that may originate in any part of the body where cells form surface membranes or layers. These include the skin, the linings of the digestive, respiratory, urinary, and reproductive tracts, the coverings around parts of the nervous system, and the tissues of the breast. Slightly over 85 percent of human cancers fall into this category.

Another type of solid tissue tumor is the *sarcoma,* found in muscle, bone, cartilage, and other connective tissues. This type is uncommon in people (about 2 percent of human cancers) but is often found in laboratory animals and cell cultures.

The second most common type of cancer in human beings (5.4 percent) is the *lymphomas,* in which abnormal numbers of lymphocyte white blood cells are produced by the spleen and the lymph nodes. Hodgkin's disease is an example.

Leukemias constitute about 3.5 percent of human cancers. Leukemias result when abnormally large numbers of white blood cells are produced from bone marrow. The remaining 4 percent of human cancers cannot clearly be classified in any of the above categories.

Cancers account for about 20 percent of all deaths in the United States, second only to cardiovascular diseases. Nearly half of these deaths are due to cancers of the lung, large intestine, and breast (Figure 12-10). One of these, lung cancer, would almost disappear if cigarette smoking was eliminated. Because cancer is

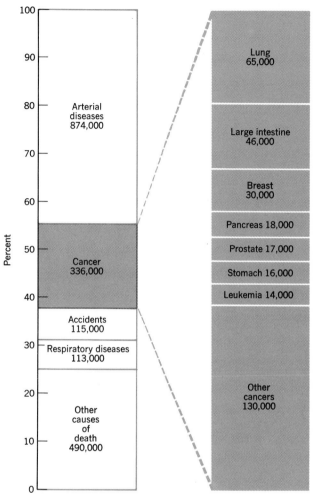

Figure 12-10 *Leading causes of death in the United States. Cancers are responsible for about one-fifth of all deaths, and most cancer deaths result from a few common forms of the disease as indicated on the chart.* (*Modified from "The Cancer Problem" by John Cairns,* **Scientific American,** *November 1975.*)

one of our leading causes of death, a tremendous amount of effort and resources has been devoted to early cancer detection, methods of treatment, and search for the agent or combination of agents that cause uncontrolled cell growth.

A variety of treatments may be used after a cancer has been detected. For localized cancers, removal of the cancer by surgery and treatment with high energy radiation are frequently used. Radiation damages cells that are dividing; therefore cancer cells are affected. In the case of widespread cancers, radiation and chemotherapy are frequently employed together. In chemotherapy a variety of chemicals are available that attack cancer cells. Most have strong and unpleasant side effects but are nonetheless immensely helpful in controlling the growth and spread of tumors.

A fourth and highly promising type of treatment is that of *immunotherapy*—the use of chemicals that help the body's immune system attack and destroy cancer cells. Some immunologists believe that the body often performs this function when cancer cells first appear but becomes ineffective if the tumor reaches a certain size. Immunologists are seeking a way to turn the immune system back on so that a tumor can be rejected naturally. Another immunological technique that appears promising for cancer diagnosis and treatment involves the production of very specific antibodies. It has been found that B cells, producing a single type of antibody, can be fused with a myeloma cancer cell. The fused cell is called a *hybridoma*. Hybridoma cells can be cultured to form colonies of identical cells (i.e., clones of themselves). In this strange "partnership" the cancer cells stimulate rapid cell reproduction, and the B cell produces its antibody. The advantage of this arrangement is that a culture of these hybridoma cells produces a quantity of only one kind of antibody. This is called a *monoclonal antibody* because it is produced by a single strain of cells. A single "pure" antibody such as this has many uses in medical diagnostic tests. Also, by obtaining a quantity of monoclonal antibody against a particular cancer such as leukemia, it is hoped that a potent new anticancer treatment will become available. Some success has already been obtained with laboratory animals and in a few human cases.

Despite all of the complex treatment procedures, about half of all cancer victims die within five years after the cancer is diagnosed and treatment begins. The success of treatment hinges mostly on detecting the tumor before its metastasizes. After it has spread, the prognosis for survival is usually dim.

Research on the cause or causes of cancers has been frustrating and unrewarding. Some researchers believe that many cancers may be caused by viruses that infect the DNA of cells and thereby gain control of the cell's chemical machinery. If such viruses could be identified, then treatments such as vaccines or drugs might be devised to counteract them. Many animal cancers are known to be caused by viruses, but only one unusual type of human cancer has been linked with these ultra-tiny bodies.

Another major school of thought believes that cancers are induced by external agents or chemicals (*carcinogens*) which alter the normal genetic control of cell division. These could also include viruses. There is considerable evidence for this concept. For instance, there are well-established links between cigarette smoking and lung cancer and between the amount of fiber in the diet and cancer of the colon. Many additional environmental agents such as high-energy radiation, sunlight, soot, asbestos fibers, benzene, arsenic, and vinyl chloride, to name a few, cause cancer in laboratory organisms and in the human body. It has been stated that we may live in a veritable "sea" of carcinogens and that removal of these substances from our air, food, and water would significantly reduce the incidence of many cancers.

If this viewpoint is correct, some of the considerable amounts of money and time now devoted to finding cancer treatments (cures) could be diverted to identifying and removing carcinogens from our surroundings. This is being done only on a small scale at the present time.

Basic Concepts

The term *cancer* is applied to masses of cells showing unrestrained cell proliferation, having little if any internal organization, and having the ability to metastasize.

Cancers are treated by removal (surgery), radiation, chemotherapy, and, in the future, immunotherapy.

One theory maintains that at least some cancers are caused by viruses, whereas another (the environmental theory), holds that many, if not all, cancers result from a number of substances that are known to interfere with normal cell division.

THE HUMAN QUEST FOR HEALTH

As almost everyone has experienced, you do not really appreciate being healthy until you become ill. Then you yearn to be free of the headache, pain, nausea, or other unpleasant feelings that the illness involves. It is difficult to define *health* or a *healthy state* precisely. To many people it means soundness of body and mind and the absence of disease. From the biological or physiological point of view, health might be defined as the physiological condition that prevails when an organism's major homeostatic systems are functioning properly. Workers in the health sciences often view health and illness as part of a continuous scale. Thus everyone lies somewhere along this continuum, never being completely healthy or completely ill. Which of these ideas fits *your* concept of health?

In recent years Americans have become increasingly interested in physical fitness and health. One reason is probably the frequent reminders in the news media of the rapidly increasing incidence of heart attacks and some forms of cancer. We often hear that we are a nation of overweight people. Also there has been an increasing interest in sports and, perhaps, a desire to look fit and athletically trim. For whatever reasons, many people now pay closer attention to their nutrition and participate in exercise programs such as jogging and bicycle riding.

All of this activity is beneficial, of course, if done properly, but it has also given rise to many questionable and sometimes fraudulent businesses that claim to sell "health" in one form or another. Because health is primarily a state of being, it is ridiculous to think that it could be purchased. Nevertheless, so-called health food stores flourish, and many people are led to believe that the "organic" foods, vitamin and mineral supplements, and other products they sell are endowed with special health-giving qualities. It is even claimed that the consumption of these products will cure specific ailments. These claims are not supported by any reputable scientific evidence, but many people unfortunately do not know this. Ironically, many of the food items sold in health food stores can be purchased at far less cost in any supermarket.

Because of the public's interest in physical fitness and weight control, another business venture called the *health spa* has also been quite successful. For a fee, individuals may use the various exercise facilities of the spa and receive a certain amount of instruction about its programs and special diets. The problem with the health-spa approach is that it emphasizes building muscle tone—"firming up" the body. However, the most beneficial exercises from a health standpoint are activities that improve cardiovascular fitness. The only way to obtain this kind of benefit is from regular, *vigorous* activities such as jogging, running, swimming, and bicycling. These exercises are inexpensive and can often be performed in one's own neighborhood.

TO MAINTAIN A HEALTHY BODY

Although health may have many meanings to different people, and may be sought in various ways, an essential prerequisite to health is avoiding illness. Medical scientists have analyzed data from many sources in an attempt to find the most important factors in keeping people disease-free throughout all or most of their lifetime. As you would probably guess, they have found no single magical ingredient but a few basic factors do seem to emerge whenever studies are made on a large sample of people. These are the major ones:

1. Heredity. If your parents and grandparents led long, healthy lives, the chances are greatly in favor of your doing likewise. We obviously have no control over our ancestry, but it is valuable to know the medical histories of our parents and close relatives in order to be alert to the risk of problems such as heart disease and cancer. If they are present, it is wise to avoid certain activities related to them (smoking, for example).

2. Exercise. Many studies have been conducted on the relationship between exercise and the incidence of heart attacks in various groups of people. The evidence does not unfailingly indicate that exercise *always* lowers the rate of heart diseases. But it does show strongly that the proper type of exercise almost always increases cardiovascular fitness. That is, the body improves its ability to take in oxygen and to utilize it; and the heart is able to pump more blood with less effort. These changes in turn impart a number of health benefits to the body, including resistance to disease of the lungs, blood vessels, and heart. The chance of surviving those diseases is greatly increased if good cardiovascular fitness is maintained.

As indicated before, vigorous exercises are best for attaining cardiovascular fitness. Many people obtain good results from 20 to 30 minutes of continuous physical exertion, such as running or swimming, every other day. Individuals who follow a regular exercise program faithfully, make it part of their life-styles, derive additional benefits such as better weight control and a heightened feeling of well-being.

3. Weight control and diet. Many studies show that obese people have more illnesses and die considerably earlier in life than do nonobese people. *Maturity-onset diabetes,* for example, is a disease found in older, overweight people; its symptoms often disappear if the person takes off the extra pounds.

Several aspects of obesity were discussed in Chapter 5, including the basic remedy for it: reducing the intake of calories. Exercise is also helpful in that it may decrease the desire to eat, but exercise without diet is not effective. We are continually bombarded with new diet schemes in the news media, but in the long run they usually fail because they are too monotonous for individuals to make part of their life-styles. This is probably fortunate, because many of these fad diets have serious nutritional shortcomings.

4. Alcohol. Alcohol is humanity's most widely used and abused drug. About 80 percent of the American population use it to some extent. Alcohol's effects on health and longevity are well documented. Generally speaking, a small and sporadic intake of alcohol evidently does no harm because the liver effectively breaks it down into acetaldehyde (at a rate of about one-fourth ounce per hour), which in turn is eventually metabolized into energy, carbon dioxide, and water (Figure 12-11). Alcohol in a small amount can actually serve as a relaxant and appetite stimulus.

But any health benefits from using alcohol are greatly overshadowed by the damage it does to the lives and health of the great numbers of people who consume it in large amounts, both sporadically and over long periods of time. Alcohol is a powerful neural depressant affecting nearly all brain functions, including motor coordination. This means that most individuals who drink over 4 or 5 ounces of alcohol cannot reliably operate machinery or vehicles, as evidenced by the high correlation between auto accidents and drunkenness. A widely used (and correct) statistic is that a drinking driver is involved in one-half of all fatal auto accidents. The intake of 10 or more ounces of alcohol in a relatively short time interval can cause death by inhibiting vital control centers, such as those that control heart rate and breathing, in the brain stem.

Another physiological danger occurs when alcohol is used in conjunction with other drugs. A drinking individual is unusually sensitive to the effects of many drugs, so the usual effect of the drug is much stronger and lasts considerably longer. This happens because alcohol interferes with the liver's normal ability to inactivate many drugs. The combined use of alcohol and drugs such as tranquilizers and barbiturates ("downers") causes a considerable number of deaths.

Dependence on alcohol and its overconsumption over a period of time—alcoholism—has become a serious and major public health problem in most of the western world. Over 5 percent of the American population (6–8 million people) are dependent on alcohol to the extent that it impairs their health, makes it difficult for them to remain employed, and generally makes them incapable of functioning rationally in day-to-day life. The emotional and financial burden of these individuals on their families and on society is tremendous. The only remedy—to stop using alcohol—is, unfortunately, not acceptable to many alcoholics and hence the disease continues unabated.

There are two major types of damage caused by alcoholism: malnutrition and liver disease. In continued use, alcohol inflames the stomach, pancreas, and small intestines. This damage in turn impedes the digestion and absorption of food. The breakdown product of alcohol, acetaldehyde, interferes with the activation of vitamins in the liver. In addition, the energy provided by alcohol (7.1 calories per gram) is sufficient to decrease an individual's appetite so that he or she often eats only sporadically. All of these factors combined may cause malnutrition.

Even on an adequate diet, however, it has been found that changes in the liver occur rather quickly,

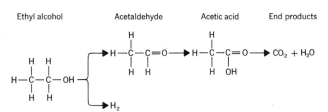

Figure 12-11 *Oxidation of ethyl alcohol in the liver. Numerous enzymes are involved. In excess amounts, acetaldehyde can interfere with the activation of vitamins in the liver and is also thought to damage liver cells.*

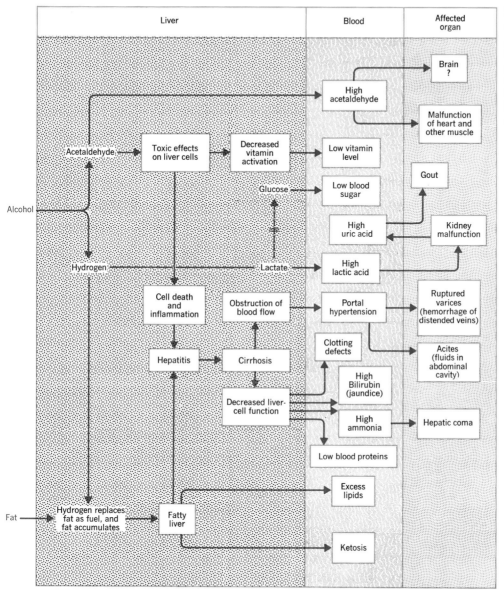

Figure 12-12 *Numerous complications may result from excessive alcohol consumption over a long time period. (Modified from "The Metabolism of Alcohol" by Charles S. Lieber, Scientific American, March 1976.)*

which may lead to liver disease. At first, fat accumulates among the liver cells. If a heavy intake of alcohol continues, liver cells begin to die and be replaced by fibrous scar tissue. This process may continue until most of the liver's normal functions are greatly diminished. This is the irreversible and often fatal liver disease called *cirrhosis.* It is thought that the damage to the liver results from its long-term exposure to large amounts of acetaldehyde. The chart in Figure 12-12 indicates the numerous complications that can occur. The health message here is obvious: Alcohol is very dangerous and should be used in moderation.

5. Tobacco. The last item that medical statisticians relate to health is the use of tobacco, particularly cigarettes. Evidence that cigarette tars are carcinogenic is overwhelming. The rise in incidence of human lung

TABLE 12-1. Total Cancer Deaths Expected in the United States in 1981 for Sites Associated with Smoking and Projected Percentage of These Deaths Attributed to Smoking

	Males		Females	
Site of Cancer	Estimated Deaths	% Attributed to Smoking	Estimated Deaths	% Attributed to Smoking
Lung	77,000	97	28,000	74
Mouth	6,300	78	2,850	46
Esophagus	5,800	83	2,300	50
Pancreas	11,500	28	10,500	22
Larynx	3,100	99	600	57
Bladder	7,300	28	3,300	22
Kidney	4,900	28	3,200	22
All Tumors	227,500	41	192,500	14

Source: "The Causes of Cancer" by Arnold E. Reif, *American Scientist,* July–August 1981, p. 440.
For males, the figures include fatalities attributed to smoking cigars and pipes, as well as cigarettes.

cancer is directly proportional to our increased use of cigarettes (Table 12-1). The cigarette industry at one time disputed this evidence on the grounds that lung cancer in female smokers occurred far less often than in males, suggesting that there was no correlation between cancer and smoking per se. Now, however, the incidence of lung tumors in female smokers is rapidly catching up to that in males. The older studies did not take into account that females had simply not smoked cigarettes as long as had males.

Cigarette use is also known to aggravate a variety of other serious conditions, including heart diseases, hypertension, and emphysema. The 1982 Surgeon General's report on smoking states that tobacco kills an estimated 340,000 Americans each year as a consequence of these diseases. The report, understandably, states that smoking is the most important public health issue of our time and the chief preventable cause of death. The message from all of these depressing findings and statistics is again clear and simple: *Stop smoking* and improve your chances of living a healthy life.

Basic Concept

The basic factors that enhance health and longevity are heredity, vigorous exercise, weight control and diet, moderate use of alcohol, and avoidance of tobacco.

The identification of these five major factors in the maintenance of health is based on long-term statistical studies of large groups of people. Among other things, this means that the findings do not necessarily apply to every individual. Some few people eat poorly, drink heavily, never exercise, are obese, and yet manage to live long lives; others may observe all of the "health rules" and yet suffer a coronary in their early forties. Despite these exceptions, however, the odds of leading a healthy life are greatly in your favor if you exercise often, eat wisely, do not smoke, and drink lightly. There are other factors that relate to health, of course, such as stress and emotional states, but their precise roles are not as well documented as the ones discussed here.

It seems almost perverse that, in the living world, only human beings possess such great knowledge about themselves. And yet, in many instances we deliberately choose to ignore it!

HEALTH CARE—SOME BIOETHICAL PROBLEMS

In Chapter 1, the field of bioethics was defined as the application of ethical decision-making processes to various types of biomedical problems. The prevention and treatment of diseases, that is, health care, frequently involves this type of decision making. A few examples are presented here for you to consider and evaluate.

What type of medical care should a nation attempt to provide for its citizens? Should it be based on a fee system as in the United States or should every citizen have guaranteed access to adequate medical care? If the latter, how will it be financed? Is health care a social right or an earned commodity? Is an individual who knowingly ignores good health practices by abusing drugs, overeating, not exercising, and so on, be entitled to health care as a social service? These and many more problems face our society as medical services become increasingly expensive. For this reason, many individuals participate in various kinds of health insurance plans; indeed, some form of national health insurance for all citizens seems destined to come about eventually. Medicaid and Medicare already serve this function to some degree for tens of thousands of people.

Another area that frequently involves moral-ethical decisions is that of experimentation on human subjects. Suppose a drug has been developed that is highly effective against a certain type of cancer in laboratory animals, although it has many dangerous side effects. As a medical researcher, what kind of moral-ethical decision would you make about trying the drug on a human cancer patient? For example, would you require permission of the patient and thoroughly explain the effects of the drug? Would you administer the drug to a comatose dying patient? Suppose the patient is a small child? Obviously, each of these situations requires a different set of judgments. Many "bioethicists" believe that experimentation should be used only with full permission and understanding of the patient, should not be painful or life threatening, should not involve institutionalized patients such as prisoners even if they volunteer, and should at all times respect the dignity and sanctity of the human body. What do you think?

KEY TERMS

antibiotic drug	dermis	immune system	monoclonal antibody
antibody	embolus	immunotherapy	natural killer cell
antigen	epidermis	interferon	neutrophil
antihistamine	epitope	keratin	phagocytosis
autoimmune disease	fever	kinins	platelet
B cell	fibrin clot	lymphocyte	skin
basal cell layer	hemophilia	melanin	T cell
cancer	histamine	melanocyte	thrombus
carcinogen	hybridoma	monocyte	vitamin D
clone			

SELF-TEST QUESTIONS

1. In what sense is the body like a machine?
2. How is the body superior to a machine?
3. Describe the general anatomy of the skin.
4. How does the skin form a protective barrier against sunlight?
5. Why is it detrimental to expose the skin to excessive amounts of sunlight?
6. Under what circumstances are bacteria beneficial to the skin?
7. How does the skin help cool the body and regulate blood pressure?
8. What is the relation of vitamin D to the skin?
9. How does skin give uniqueness to an organism?

10. Describe the general course of events that results in blood clotting.
11. Discuss the dangers of a thrombus in the body.
12. Why is a bacterial infection often accompanied by pus formation and redness around the infection site?
13. Name the two main functions of the immune system.
14. What are the functions of the B cell lymphocytes and T cell lymphocytes?
15. How does the immune system recognize substances that are foreign to the body?
16. Define the difference between antigen and antibody.
17. How does a vaccine work in building immunity in the body?
18. Describe the mechanism of an allergic reaction.
19. What is meant by "autoimmunity"?
20. What are natural killer cells?
21. Discuss the function of interferon.
22. In what ways does the body adapt to fluctuations in its nutrient supplies?
23. What basic features do all cancer cells have in common?
24. List four broad categories of cancers.
25. Name three methods commonly used in treating cancers.
26. Define hybridoma and monoclonal antibody.
27. Discuss the two major schools of thought about the cause of cancers.
28. Give two definitions of health.
29. List five basic rules to help maintain a healthy body.

ABOUT BEGETTING: SEX AND REPRODUCTION

chapter 13

questions to think about

Why is reproduction necessary?

Why is reproduction really a cellular event?

In what ways are the male and female reproductive cycles similar?

Why are most birth control methods devised for females rather than for males?

Should individuals have freedom of choice in relation to abortions?

The thing that takes up the least amount of time and causes the most amount of trouble is sex.

John Barrymore

It is of some interest to note that most of the adaptations found in our bodies, like those of other animals, center around two organ systems: the digestive system and the reproductive system. (Name a few adaptations and see if you agree!) The two systems have nothing in common except that both involve survival: The organism must have nutrients to survive from day to day and its species must reproduce in order to survive from generation to generation. We previously looked at some of the numerous ways that organisms obtain and process food materials; in this chapter we want to look at a number of adaptations involved in reproduction.

Reproduction has several basic functions. For one, it bridges the gap between generations of organisms, thereby, in effect, counteracting the death of the individual. In doing this, it also insures the genetic continuity of each species through long periods of time. Reproduction also maintains the population levels of species, and often introduces genetic variation into populations so that they can better survive changing environments.

These basic functions are as true for humans as for the rest of the living world but with two differences. Human beings frequently utilize sex for enjoyment only and not for the purpose of reproducing. We seem unique among animals in this regard. In addition, sex and reproduction have assumed tremendous cultural significance in most human societies—perhaps more than they warrant at times. For example, we find that nearly every aspect of our daily lives is affected by some type of emphasis on the sexual differences between males and females. Even in infancy, babies are dressed in blue or pink clothing to signify their sex, and soon the young child is taught the roles that it is expected to assume—little boys are expected to be assertive and stoical, whereas little girls should be submissive and emotional, for example. As teenagers and adults we continue to be bombarded with sex-related stimuli, sometimes in highly irrelevant situations. For example, the advertising media routinely link sexuality with products that have nothing to do with sex, such as soft drinks, alcoholic beverages, automobiles, airlines,

Figure 13-1 *Example of how sexuality is exploited in order to help sell nonsexual products. Are you influenced by this type of advertising?*

food, and TV sets (Figure 13-1). To any educated person this blatant exploitation of sex appears ridiculous. Yet, advertisers can prove that sexy advertising sells more of their products.

Sex most assuredly does play a prominent role in the biology of our bodies, in our behavior, and in our emotions. It is, therefore, important for people to understand these roles in order that they may view sexuality in its proper perspective—as a normal biological trait that can greatly enrich human life but does not necessarily have to dominate it.

Basic Concept

Reproduction forms the link between generations of organisms, maintains their population levels, and prolongs their genetic continuity through time. In humans, reproduction is also of great cultural significance.

SEXUAL AND ASEXUAL REPRODUCTION

All the varied methods of reproduction found throughout the living world can be grouped into two basic categories: *sexual* and *asexual*. In asexual repro-

duction, a portion of an organism grows into a duplicate of itself; the process does not require special sex cells or separate male and female parents. Asexual reproduction is accomplished in a variety of ways. An organism may divide into two or many organisms by fission, as do most one-celled organisms. The parent's body may grow a *bud* that gradually becomes another organism and then breaks free, as in plants such as yeasts and animals such as *Hydra* (Figure 13-2*a*). Pieces of the parent may split or fragment from the parent and grow into new individuals, as happens in sponges and sea anemones (Figure 13-2*b*).

Asexual reproduction has the advantages of simplicity, because it requires only one parent. It also has the potential for rapidly producing many offspring. A disadvantage is that the offspring show virtually no variability, because all are identical to their parent. This can be a liability in an evolutionary sense. Most environments are slowly changing, and survival favors the *variants* that possess the right combination of adaptations at the right time. A population that was all identical, however large, would be highly vulnerable to environmental changes.

Sexual reproduction involves the union of specialized sex cells, specifically *eggs* and *sperm*. The product of this

(a)

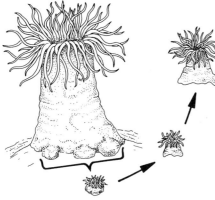

(b)

Figure 13-2 *Budding in Hydra.* **(a)** *The buds, formed by an outgrowth of the parent's body, eventually break free and become separate organisms. (CCM: General Biological, Inc., Chicago)* **(b)** *Fragmentation in a sea anemone. A small piece of the animal splits off and grows into an adult organism. Why would you expect these offspring to be identical to their parents?*

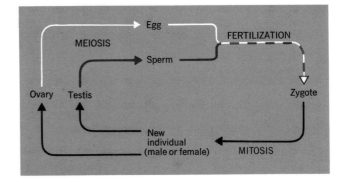

Figure 13-3 *The sexual reproduction cycle.*

union is a fertilized egg, or *zygote*. The production of sex cells is usually preceded or accompanied by a special type of cell division termed *meiosis*, which usually occurs in separate male and female parents (Figure 13-3). Quite obviously, sexual reproduction is much more complex than asexual reproduction. Its advantage is that it promotes variability among offspring by uniting genetic material from two different parents. This variability is important as the source of adaptability in populations of organisms.

You have only to look at children from the same parents to observe the phenomenon of variability in offspring in humans. It is well established that people, like other organisms, vary from each other not only in appearance but also in their internal anatomy, body chemistry, and heredity.

Humans have been more than successful in increasing their numbers over time. In fact there are evidently so few natural restraints on human reproduction that our species has grown to almost unmanageable numbers in some parts of the world. Even with famine and starvation the numbers keep increasing rapidly. One of humanity's major problems is how to slow down this tremendous population increase. A major part of the solution to this problem lies in finding ways to *prevent* human reproduction that are reliable, safe, simple, inexpensive, and acceptable to large masses of humanity. As you read through the next sections on the anatomy and physiology of sex, keep a mental note of the different ways in which one might interfere with the reproductive process without harming the human organism in any way.

Basic Concepts

Asexual reproduction requires only one parent and produces multiple offspring quickly. The offspring show little variability, an evolutionary drawback in a changing environment.

Sexual reproduction involves the union of an egg and a sperm to produce a zygote. The offspring vary in genetic makeup from each other, the major source of evolutionary adaptability.

CELLULAR REPRODUCTION

Although we refer to the reproduction of organisms, all reproduction in reality begins at the cellular level. Cells multiply to make more cells; some other cells specialize into eggs or sperm that in turn fuse and grow into new organisms. Hence, to comprehend and appreciate the wonderous process of reproduction requires a knowledge of what happens at this level.

There are two basic types of cell division (or multiplication): mitosis and meiosis. Both involve a series of orderly events occurring within the nucleus, as described below, but the end results are quite different. Mitosis is the universal way in which cells reproduce more cells of like type throughout the body. Meiosis, by contrast, occurs only in reproductive tissues (ovary or testis) and is involved in the production of eggs or sperm.

Mitosis

As noted in Chapter 3, mitosis constitutes part of the life cycle of a cell (the M stage in Figure 3-15). The nuclear events that constitute mitosis are arbitrarily defined as prophase, metaphase, anaphase, and telophase (Figure 13-4). The events that occur in each phase are generally as follows. In *prophase* the nuclear material condenses into tiny, threadlike bodies called *chromosomes.* Each chromosome consists of two strands of hereditary material (DNA) and protein. The nuclear membrane disappears, and microtubules from the cytoplasm form a basketlike structure, the spindle, enclosing the chromosomes. The chromosomes then attach to the spindle fibers near the middle of the spindle.

In *metaphase* the chromosomes are clustered across the expanded middle region of the spindle. This is a particularly useful stage for the cell biologist, because chromosomes are in their most compact form and thus can be seen and counted more easily than in other stages. The drug, colchicine, is often used for studying chromosomes because it interferes with spindle formation and halts the mitotic process at this stage. When chromosomes are counted; it is found that there is a consistent and typical number of them in each cell of any given organism. Thus, human cells typically contain 46 chromosomes at this stage of mitosis.

Anaphase is the stage in which the spindle fibers attached to the chromosomes evidently contract in such a manner as to pull each chromosome apart lengthwise into two equal strands. Because this happens to all of the chromosomes more or less simultaneously, equal amounts of chromosomal material (including the DNA) move to opposite ends of the spindle. At the end of anaphase, a set of *daughter chromosomes* (single-stranded) is located at each end of the spindle.

In *telophase* the spindle disappears and a nuclear membrane forms around each of the daughter chromosome sets. The cytoplasm of the cell begins to constrict until it pinches the cell into two daughter cells of approximately the same size and containing identical amounts of DNA. This latter event, involving the constriction of the cytoplasm, is called *cytokinesis* (Figure 13-5).

Basic Concept

Mitosis consists of four recognizable stages: prophase, metaphase, anaphase, and telophase. As a consequence of the process, the resulting two daughter cells normally contain identical amounts and kinds of chromosomal and DNA material.

Meiosis

As defined previously, sexual reproduction involves the fusion of two specialized sex cells, an egg and a sperm. Meiosis is the mechanism that prevents the nuclear material from doubling in amount each time this fusion takes place. The meiotic process precedes (or sometimes accompanies) the formation of the sex cells; therefore each sex cell contains only half as much hereditary material as the parent cell from which it

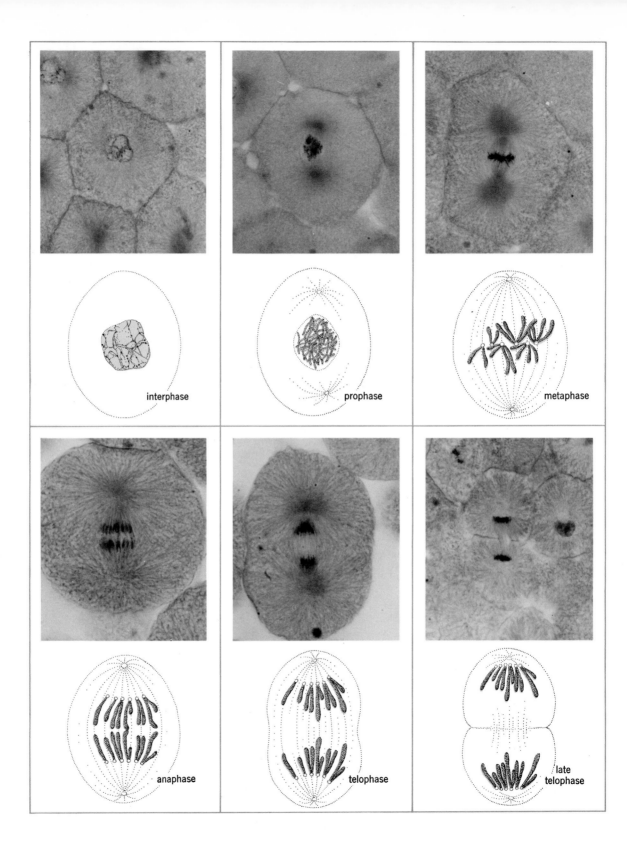

interphase

prophase

metaphase

anaphase

telophase

late
telophase

Figure 13-4 *(opposite) Mitosis in whitefish blastula cells. Interphase is not a stage of mitosis. (Photographs and line drawings by Richard A. Boolootian, Science Software Systems, Inc.)*

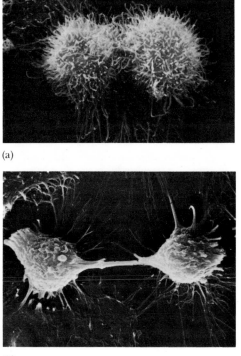

(a)

(b)

Figure 13-5 *Cytokinesis as observed with the scanning electron microscope.* **(a)** *Early telophase showing constriction of the cell;* **(b)** *Late telophase with the daughter cells almost separated.*

came. Thus when two sex cells combine in fertilization, the resulting cell (called a zygote) has the normal amount of hereditary substance. For example, human body cells contain 46 chromosomes (23 pairs); human eggs and sperm contain 23 chromosomes each (one member of each pair) as a consequence of meoisis. We refer to the reduced number as the *haploid* number and to the full number of chromosomes as the *diploid* number.

The complex process of meiosis takes place generally as follows: As in mitosis, the nuclear material (chromatin) resolves itself into a discrete number of threadlike *chromosomes* (23 pairs in the nucleus of a human cell). As the chromosomes become shorter, thicker, and rodlike, each one locates itself adjacent to the other chromosome like itself (its homologue) and the two become closely entwined (Figure 13-6). While entwined, some of the paired chromosomes break and exchange parts with each other. This is called *crossing over* and has considerable hereditary importance, as discussed in later chapters.

While these events are underway, the spindle envelops the chromosomes and eventually a spindle fiber attaches to each one at a site on the chromosome termed the *centromere.* The paired-up chromosomes then orient themselves on the middle (metaphase plate) of the spindle. The microtubules attached to each centromere then pull the paired homologous chromosomes apart in such a way that one group of 23 chromosomes moves toward one end of the cell and the other group of 23 moves in the opposite direction. The two chromosomes constituting each pair always go to opposite ends of the cell, but which one will go to which end is a matter of chance. Thus, with 23 different chromosome pairs, the number of chromosome combinations is vast. This fact has great hereditary significance because it gives rise to almost any combination of traits in any given reproductive cell. Each of these groups of 23 chromosomes now constitutes a *haploid nucleus* (haploid meaning half the number of chromosomes found in the body cells of the organism).

In the second part of meiosis (Figure 13-6), another spindle forms in each of these haploid nuclei and the chromosomes again orient themselves toward their respective metaphase plates. It now becomes evident that each chromosome consists of two distinct strands of material; indeed, each strand now has its own separate centromere. (This double-strandedness occurred much earlier in the meiotic process but it now becomes important.) The microtubules shorten and pull the chromosomal strands apart. At this point the original "mother cell" has given rise to four haploid daughter cells. Later in the chapter we will relate meiosis to the actual formation of egg cells and sperm cells.

Basic Concept

Meiosis is a complex set of nuclear divisions that results in the formation of haploid daughter cells.

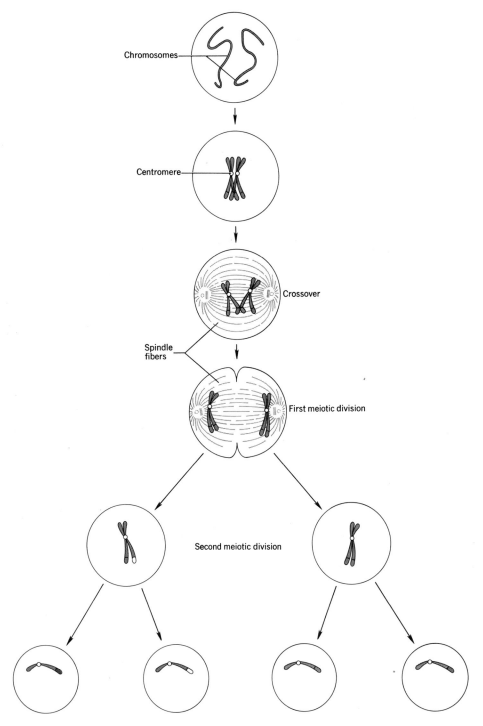

Figure 13-6 *Meiosis shown with only one pair of chromosomes for clarity. Early in meiosis (top) homologous chromosomes pair and become closely entwined. Crossing over is shown in the middle figures. Two daughter cells result from the first meiotic division, and each is haploid as indicated. In the second meiotic division, each chromosome is pulled apart into two strands as shown at the bottom, and each strand goes into a separate daughter cell. The result of this intricate set of cell divisions is that one diploid cell forms four haploid daughter cells. These undergo further changes to become functional reproductive cells.*

STRUCTURE AND FUNCTION OF THE REPRODUCTIVE ORGANS

Human organisms are sexually *dimorphic*. This means that mature males and females are distinguishable by means of external features of various kinds. A sexually mature female typically has enlarged mammary glands (breasts), relatively little body hair, a relatively high-pitched voice, broad hips and narrow shoulders and external genitalia covered by pubic hair.

A sexually mature male typically has underdeveloped but visible mammary glands, a considerable amount of body hair (but far less than most other male mammals), a relatively deep voice, narrow hips and broad shoulders, and external genitalia surrounded but not hidden by pubic hair. Males are also much more muscular than females. All of these features in both sexes are termed *secondary sex characteristics*, because they are not present at birth but develop later (at puberty) under the influence of the sex hormones (see Chapter 8). Infants are, of course, distinguishable at birth as to sex but only because of their external genitalia. Table 13-1 lists additional differences between males and females.

Sexual dimorphism is not unique to *Homo sapiens*, although it is more pronounced in humans than in most mammals, including other primates. One exception is the male baboon which differs from the female in body size and in having large canine teeth and a hairy mane.

THE FEMALE REPRODUCTIVE SYSTEM

The human female reproductive organs consist of the *ovaries, Fallopian tubes, uterus, vagina,* and *vulva* (external genitalia) (Figure 13-7). The ovaries, each about the size of a pecan, are located in the pelvic cavity, held in place by ligaments attached to the uterus and pelvic wall. Each ovary has a dual function in that it produces both eggs and hormones (estrogens and progesterone).

Adjacent to each ovary, and partially curled around it, is a Fallopian tube about 4 inches long. The tiny opening into the tube is surrounded by a fringe of fingerlike projections. The functions of the Fallopian tubes are to take in eggs as they are ovulated from the

TABLE 13-1. Some Sexually Dimorphic Features in Males and Females[a]

Male	*Female*
Life expenctancy shorter	Longer-lived
Death rate higher at all ages	Lower death rate
Sex determined by Y chromosome	Sex determined by absence of Y chromosome
Y chromosome has no structural genes	X chromosomes have genes
Sex cells (sperm) produced from puberty to death	Sex cells (ova) not produced after about mid-forties
Larger body size	Smaller body size
Heavier bones, muscles, heart, lungs, salivary glands, kidneys, and gonads in proportion to body weight	Heavier brains, livers, spleens, adrenal glands, thymus glands, stomach, and fat deposits in proportion to body weight
Basic metabolic rate higher	Basic metabolic rate lower
More red blood corpuscles and hemoglobin	Proportionately larger complex of bone marrow, spleen, thymus, lymph node and lymph tissue. A greater immunological competence.

[a] *Source:* Modified from A. Glucksmann, "Sexual Dimorphism in Mammals," *Biol. Rev.* 49:423–475, 1974.

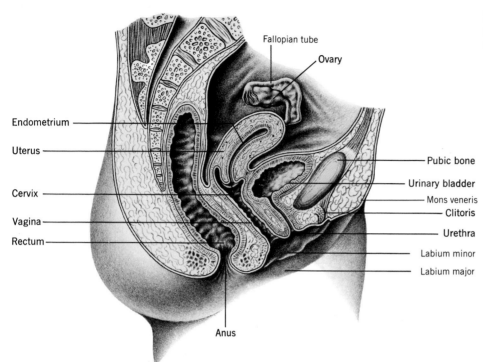

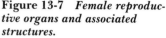

Figure 13-7 *Female reproductive organs and associated structures.*

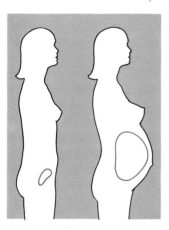

Figure 13-8 *Enlargement of the uterus during pregnancy.*

ovaries, transport them to the uterus, and to provide the site for the union of eggs and sperm.

The *uterus* (*womb*) is a muscular organ about the size of a human fist, richly supplied with blood vessels. Its function is to provide a suitable area in which a fertilized egg can implant and grow. During pregnancy the uterus is capable of remarkable distention, stretching to many times its normal size to accommodate the growing embryo (Figure 13-8). But in the absence of a

fertilized egg the vascular uterine lining (*endometrium*) breaks down about once a month and passes from the body in menstruation. After each menstrual period the uterine lining again builds up and becomes prepared to accept a fertilized egg. The lower portion of the uterus where it joins the vagina is called the *cervix*. The cervical canal into the vagina normally remains closed except during the menstrual period and during birth.

Uterine tumors are fairly common and often necessitate removal of the uterus (*hysterectomy*). In addition, cancer of the cervix is the second most common type of cancer in females (breast cancer is the most common). Thus, it is highly advisable for women to have a Pap test when recommended by her physician. In this test, named after its originator, Dr. Papanicolaou, a physician rubs a cotton swab on the cervical opening and smears some of the clinging material onto a microscope slide. A pathologist can then examine this material for cancer cells.

The *vagina* is a thin-walled muscular tube about 4 inches long and capable of considerable stretching. It not only accommodates the enlarged penis of the male during intercourse but also becomes the birth canal during delivery of a baby. The vagina opens into the

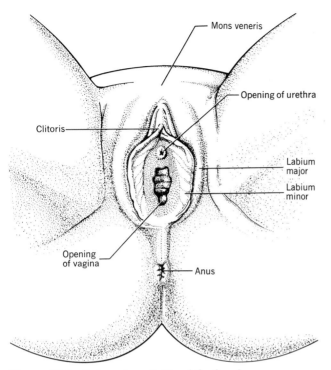

Figure 13-9 *External genitalia of the female.*

vulva (the external genitalia) and may be partially closed, at least in young women, by a thin membrane, the *hymen.* One of the many misconceptions that people have about sex is that the hymen remains intact until the first act of intercourse, and that one "proof" of virginity is, therefore, an unbroken hymen. In reality, there are a variety of experiences other than intercourse that may destroy the hymen.

On each side of the vaginal opening are the *labia,* or lips (major and minor) of the vulva that serve partly to enclose and protect the vaginal opening (Figure 13-9). Just above the vaginal opening is the entrance of the urethra from the uninary bladder. This orifice is closed except during urination. Within the vulva the labia minora enclose a small protuberance, the *clitoris.* The clitoris, partially covered by a small foreskin, consists of very sensitive tissue and functions in sexual excitement. The labia major merge, at the upper angle of the vulva, into the *mons veneris* ("mountain of Venus"—the Roman goddess of love), a mound of fatty tissue lying over the pubic bone.

During sexual excitement, various events take place in the uterus, vagina, and external genitalia. The va-

gina secretes a lubricating fluid, the clitoris enlarges, and the labia undergo characteristic changes. Additional physiological events occur during intercourse and at the time of *orgasm.* Orgasm in females, as in males, is an intensely pleasurable physical and emotional event. The orgasm often takes place in several steps or stages accompanied by rhythmic contractions in the vagina and uterus. Strong muscle contractions may occur in other parts of the body as well. Heartbeat, blood pressure, and breathing rate increase markedly during orgasm. So far as can be determined, the event is as intense in females as in males even though there is no ejaculation of semen. In addition, females may experience multiple orgasms or prolonged orgasms that greatly add to the enjoyment of their sexual experience. In a sense, an orgasm in either sex is a sudden and explosive release of sexual tensions and usually leaves an individual in an extremely relaxed physical and emotional state. For students who wish further information, there are a number of excellent paperbacks now available. One of these is *Sex and Human Life* by Eric Pengelley.*

The Formation of Egg Cells

An ovary contains about a half million potential egg cells (primary *oocytes*) but only 400–500 will ever mature. Each oocyte is surrounded by a layer of follicular cells. At puberty, FSH (*follicle stimulating hormone* from the anterior pituitary gland) stimulates further development of the eggs and their follicles—the process of *oogenesis*—but in a highly selective manner. Usually only one oocyte develops into an ovum each 28 days. Under the stimulus of FSH, a follicle enlarges tremendously to leave the extremely tiny oocyte suspended in a mass of follicular fluid. Meanwhile the oocyte enters the first half of its meiotic cycle: The chromosomes pair, arrange themselves on a metaphase plate, attach to spindle fibers, and separate into two haploid sets. The spindle at this time is oriented so that one of the chromosome sets actually leaves the egg cell proper to become a small mass known as a *polar body* which eventually degenerates.

* Eric T. Pengelley, *Sex and Human Life,* 2nd edition. Reading, Massachusetts: Addison-Wesley Publishing Company, 1977.

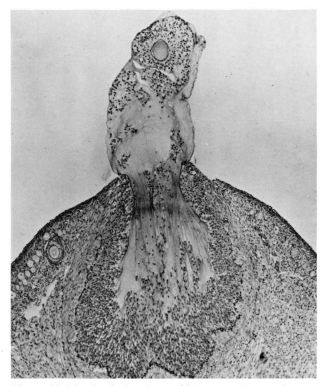

Figure 13-10 *Ovulation in a rabbit ovary. Can you locate the oval egg surrounded by follicle cells? The remains of the follicle seen here in the ovary become the corpus luteum.*

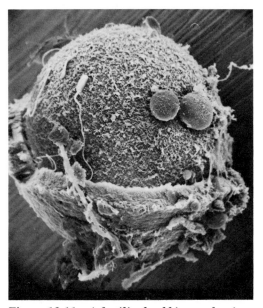

Figure 13-11 *A fertilized rabbit egg showing two polar bodies (×1600). Do you see the sperm cell on one side of the egg? The resemblance of this illustration to one of the muppets is coincidental!*

The chromosomes in the egg cease their meiotic activity and remain in metaphase. Ovulation is brought about by LH (*luteinizing hormone*), which causes a follicle to move to the surface of the ovary and rupture, releasing the ovum to the outside. Following ovulation the oocyte, with a gelatinous mass of follicular cells, oozes out of the follicle (Figure 13-10) and enters the adjacent Fallopian tube. The follicle itself remains in the ovary, becoming the *corpus luteum.*

If fertilization (union with a sperm) occurs, another spindle forms in the egg nucleus. Now the double stranded chromosomes line up on the metaphase plate and separate into two masses. As before, one of these masses leaves the egg as another polar body; neither of the polar bodies has any further role in reproduction (Figure 13-11). The final result of meiosis is a single egg with one haploid nucleus. When it is joined by the sperm nucleus in fertilization, the full diploid (double) chromosome number is restored.

Unlike men, whose reproductive cycle is potentially lifelong, women cease to ovulate sometime in their mid- or late forties due to a decrease in the amount of anterior pituitary gonad-stimulating hormones. This period, called *menopause,* signals the end of a woman's reproductive cycle but not necessarily the end of her sex life. With no fear of pregnancy and no need to use contraceptives, a couple may find their sex life even more satisfying than before.

Basic Concept
Eggs are haploid cells produced after the initiation of meiosis. Ovulation ceases during the mid- or late forties in a female, a period termed menopause.

The Female Reproductive Cycle
The female cycle is activated by hypothalamus releasing factors (acting on the pituitary gland) at around 10 to 12 years of age. This onset of puberty also marks the development of the female secondary sex characteristics.

After the cycle is set in motion, it functions generally as follows. Following the end of a menstrual period, the pituitary gland releases increasing amounts of the follicle stimulating hormone, FSH. As noted previously FSH causes immature follicles (egg-containing bodies) in the ovary to increase in size. Some of the cells making up the follicle begin to produce estrogen hormones. One of these, *estradiol,* causes the uterine lining to grow thicker. As a part of a feedback mechanism, the estrogens also modify other portions of the control system. For example, they affect the pituitary in such a way that the amount of FSH released is gradually decreased and luteinizing hormone, LH begins to be secreted. LH causes the follicle to mature and rupture from the ovary. The remains of the ruptured follicle in the ovary form a glandular mass of cells called the *corpus luteum.* This in turn produces still another hormone, *progesterone,* which helps prepare the lining of the uterus to receive an egg. The events up to this point in the ovary are termed the *ovarian cycle* (Figure 13-12).

If the egg is not fertilized, it passes into the uterus and disintegrates. The corpus luteum ceases to produce progesterone, and eventually the uterine lining (endometrium) breaks free and menses occurs. After this the endometrium begins to build back up and the cycle starts again. The events in the uterus constitute the *uterine* or *menstrual cycle* (see Figure 13-12). The combined ovarian and uterine cycles take approximately 28 days.

In terms of time, the ovarian and uterine cycles are coordinated as follows. The first day of the 28-day cycle can conveniently be defined as starting with the first day of a menstrual period. Menses lasts three to five days and then FSH (from the pituitary) promotes the growth of a follicle in the ovary. Follicle cells in turn produce estradiol and the uterine lining begins to grow. Between the eighth and fourteenth days, FSH decreases and LH increases in the bloodstream. The result is ovulation, which occurs on approximately the fourteenth day, or half-way through the cycle. For the next seven days, the corpus luteum produces enough progesterone to both maintain the endometrium and inhibit secretion of FSH. After the twenty-first day, the corpus luteum begins to deteriorate and the amount of progesterone decreases. Finally, at about the twenty-eighth day the endometrium begins to break loose and

a new menses (and cycle) begins (see Figure 13-12).

As most women know, the cycle may vary considerably. Menstrual periods may occur sooner or later than every 28 days, and they may be prolonged or brief. Twenty-eight days is the *average* time for a large sample of females. Some females experience cramps (*dysmenorrhea*) on the first day of their periods. These are usually relieved by moderate exercise and often disappear entirely after birth of the first child. On occasion birth control pills are prescribed to relieve dysmenorrhea.

Biologists are always curious about consistent or characteristic cycles such as this one because they signify some adaptive advantage to the species, either now or in its evolutionary past. What could be adaptive about a 28-day cycle in humans? There is no clear answer to this question. Perhaps an advantage is that the female is capable of becoming pregnant every 28 days. Because humans typically bear only one offspring per pregnancy and have a relatively long pregnancy time, this continual sexual receptiveness would have aided primitive human beings in maintaining their numbers.

Basic Concept

The human female reproductive cycle involves a complex of hormonally controlled events taking place in a 28 day cycle.

THE MALE REPRODUCTIVE SYSTEM

The male reproductive system consists of a pair of *testes,* the *vas deferens,* three accessory glands (*prostate, seminal vesicles,* and *Cowper's gland*), the *urethra,* and the *penis* (Figure 13-13).

The two testes are located in a sac, the *scrotum,* suspended beneath the penis. The penis and scrotum constitute the external genitalia of the male. The location of the testes, which dangle from the body in the scrotum, leaves them exposed to possible injury. A safer anatomical arrangement would locate the testes within the body cavity, as the ovaries are in females. But the biological problem with this arrangement is that the relatively high body temperature in mammals is detrimental to sperm. Suspending the testes outside

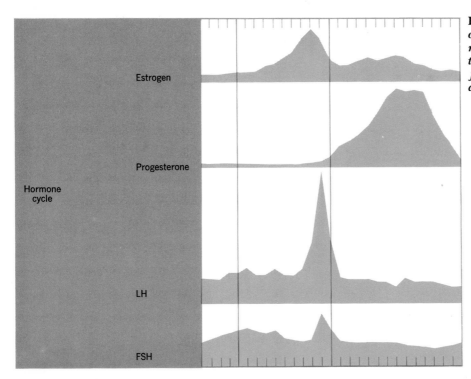

Figure 13-12 *Major events that occur in the ovary and uterus in relation to the various reproductive hormones. See if you can follow these events as they are described in the text.*

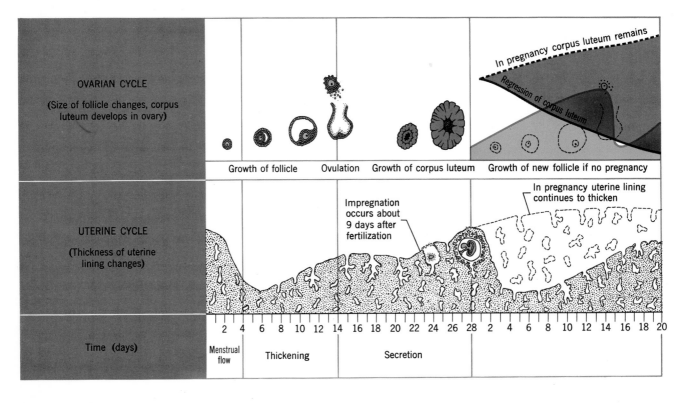

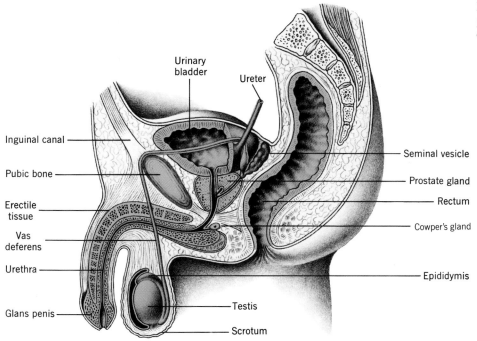

Figure 13-13 *The male reproductive organs and associated structures.*

Urinary bladder

Ureter

Inguinal canal

Pubic bone

Erectile tissue

Vas deferens

Urethra

Glans penis

Seminal vesicle

Prostate gland

Rectum

Cowper's gland

Epididymis

Testis

Scrotum

the body cavity avoids this problem. In other mammals, such as mice and elephants, the testes are usually located within the body but descend to the outside during the mating season. Even in human males, the scrotum draws up during cold weather, whereas warmth causes it to relax and become longer. The human testes, however, remain in place.

Each testis is a dual-purpose organ because it produces both sperm and testosterone hormone. Internally, a testicle is divided into numerous compartments containing tiny, tightly coiled tubules. These are the sites of sperm formation. As sperm cells form in these tubules, they move into a larger collecting tube, the *epididymis,* and eventually into the main sperm duct, the vas deferens. The vas deferens on each side passes up into the pelvic cavity of the body via an *inguinal canal.* This passageway, located in the lower abdominal wall, is also of interest for another reason. This region of the lower abdominal wall is thin and weak in many individuals. Not uncommonly, a portion of the intestines pushes into this weak region, creating an inguinal *hernia.* Corrective surgery is often the only remedy.

Each vas deferens, after entering the pelvic cavity, loops around the urinary bladder to join a tube leading

out of a seminal vesicle gland. The combined tube, the ejaculatory duct, runs through the prostate gland to empty into the urethra (see Figure 13-13). These glands add a fluid secretion, the *seminal fluid,* to the sperm cells to aid their transport through the urethra. A pair of small glands—*Cowper's glands*—opens into the urethra just beyond the prostate. They also secrete a small amount of fluid. The prostate gland often becomes troublesome in later life because it may swell and obstruct the urinary passage. Also, cancer of the prostate is not uncommon.

The urethra is a tube that extends from the bladder to the outside of the body via the penis. It serves a double function, transporting urine and semen, but usually not at the same time. At the moment of orgasm, rhythmical muscular contractions transport semen from the ejaculatory duct into the urethral tube and through the penis. This is termed *ejaculation.* Additional contractions occur in the prostate gland and penis. Because the orgasm is such a pleasurable event, it indeed becomes the primary reason that most people engage in sexual intercourse. But people frequently seem to ignore the fact that this aspect of sex is the prelude to reproduction—that is, that orgasm is an adaptation to promote and encourage reproduction. Obviously then, a realistic researcher in methods of birth control must find ways to prevent the union of sperm and egg that do not diminish the pleasurable aspects of the sex act itself. It is of some interest to note that the role of orgasm in nonhuman animals is not known. Other female mammals give no outward signs of experiencing an orgasm during copulation, and males very little.

The average adult penis is 3 to 4 inches long when not erect. Some anatomists claim that the human penis is the largest in relation to body size of any mammal. During sexual excitement, or as the result of almost any gentle stimulation, the penis elongates to an impressive 6 inches or more, which constitutes an *erection*. This unusual size increase is caused by blood flowing into the three elongate bodies of spongy tissue that fill the body (shaft) of the penis and the enlarged *glans* at the end (Figure 13-13). There is no muscle tissue or bone within the human penis, but the tough connective tissues surrounding the spongy tissue stretch tight and hard during an erection. The slang expression, "having a hard-on" is an aptly descriptive phrase. The function of an erection is to enable the penis to enter the vagina. A male can experience orgasm and ejaculation without a full erection but cannot engage in sexual intercourse. The term *impotence* refers to the condition in which a male cannot have an erection, or at least cannot maintain one long enough to have intercourse.

The skin of the penis, necessarily loose on the nonerect penis, forms a cover, the *foreskin* or *prepuce*, over the glans. Frequently the foreskin is removed from the newborn infant in an operation termed *circumcision*. This is thought to create better hygenic conditions for the glans.

The Formation of Sperm Cells

The origin of *spermatozoa* (sperm cells) and their development is called *spermatogenesis*. It begins with unspecialized cells in the testes and involves a complex of intracellular events that does not actually begin until the male reaches puberty. At that time, under the influence of the pituitary hormone FSH, cells in the tubules of the testes begin to multiply by mitosis. These cells become *spermatocytes*, which are destined eventually to differentiate into mature sperm.

Each spermatocyte goes through the process of meiosis described earlier in the chapter. The consequence is that each spermatocyte with its diploid number of chromosomes produces four haploid cells called *spermatids*.

The spermatids then go through a process of specialization to become motile spermatozoa (Figure 13-14). In each spermatid a tiny structure, the centriole, grows a long filament that protrudes from the spermatid and eventually becomes the central filament of the sperm's long tail. Most of the cell's cytoplasm is discarded, leaving the condensed nucleus to become the head end of the spermatozoan. The head is partly enveloped by a sac called the *acrosome*, formed by Golgi apparatus. The acrosome contains enzymes that aid the sperm in entering an egg. Just below the head of the sperm a small amount of cytoplasm remains, enclosing a series of mitochondria. This section is the *middle piece* of the sperm. Mitochondria produce the ATP energy necessary for the sperm's rapid motion. Beyond the middle piece is the long tail, consisting of bundles of microtubules in an enclosing sheath. Mature sperm move gradually into the vas deferens and ejaculatory duct. One ejaculation normally contains about 500 million sperm, estimated to be about 72 days old. It is believed that spermatozoa older than this disintegrate. Beginning at puberty, sperm formation normally continues throughout the life of the male, and in astounding numbers. For example, many males are thought to produce in excess of 300 million sperm per day.

Basic Concept

Sperm are highly motile haploid cells produced as a consequence of meiosis and specialization of the cytoplasm of the cell and nuclear condensation.

The Male Reproductive Cycle

As in the female, the male reproductive cycle is dormant until puberty. At this time, a hypothalamus releasing factor causes the pituitary gland to stimulate the testes into action. The testes respond by starting to form spermatozoa and by secreting male sex hormones called *androgens*, primarily testosterone. Androgens bring about development of the typical male secondary sex characteristics. As androgens build up in the blood,

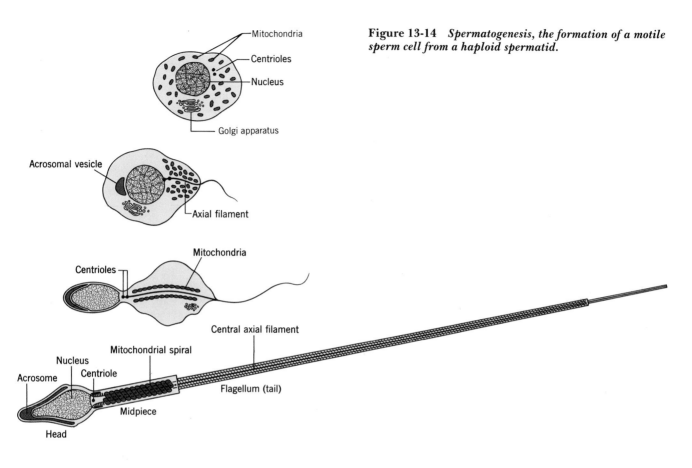

Figure 13-14 *Spermatogenesis, the formation of a motile sperm cell from a haploid spermatid.*

they eventually interact with the hypothalamus in a feedback control system (see Chapter 8) which maintains hormonal levels for efficient function of the system.

Unlike the female reproductive system, the male system functions throughout the life of the male. In this respect it resembles the reproductive cycle of most other mammalian males.

FERTILIZATION

When the male reaches orgasm during intercourse, a huge number of sperm are deposited near the cervix of the uterus. Many of these sperm work their way up the uterus, possibly by moving up mucus strands, until they attain the Fallopian tubes (Figure 13-15). Many sperm do not get this far, so their numbers are greatly diminished by the time the egg is reached. Even so, enough sperm remain to surround the egg and attempt to enter it. While in the reproductive tract, the sperm interact with substances that prepare the acrosome to release its enzymes. This process is termed *capacitation* and is evidently necessary for successful fertilization.

As sperm surround an egg, their acrosomes release enzymes that start dissolving a pathway through the mass of follicular cells around the egg and the egg membrane (the *zona pellucida*). Eventually one sperm reaches the egg itself (Figure 13-16). Once this happens, the egg's zona pellucida in some manner becomes impenetrable to further sperm cells.

Soon after entering the egg, the haploid nucleus of the spermatozoan fuses with the haploid egg nucleus to restore the full diploid number of chromosomes. If the sperm contributes an *X* sex chromosome, then the fertilized egg develops into a female. If the sperm contributes a *Y* sex chromosome, the zygote grows into a male.

The fertilized egg, still in the Fallopian tube, begins to divide, enters the uterus, and embeds there to finish its development. These events will be discussed more fully in the next chapter.

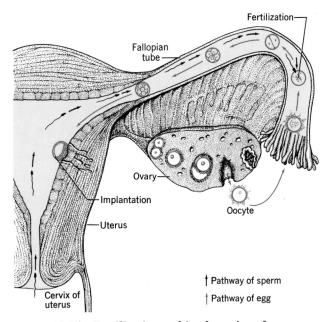

Figure 13-15 *Fertilization and implantation of an egg. The egg unites with a sperm in the Fallopian tube and then undergoes division (cleavage) as it passes into the uterus and implants there.*

Several times we have stressed the importance and means of preventing fertilization. But what about the opposite problem, infertility? The male may be infertile because of insufficient numbers of sperm in his ejaculate or because the sperm are abnormal in some way and cannot get to the egg. In this case a couple may still have children by means of *artificial insemination.* Sperm are obtained by the couple's physician from a donor who remains unknown to the couple. The physician places the sperm in the wife's vagina at approximately her time of ovulation. A fertilized egg and a subsequent pregnancy may result. This technique, termed *AID* (for artificial insemination donor), is commonly used and results in about 20,000 births a year.

If the male does not produce enough sperm but they seem normal otherwise, it may be possible to collect several ejaculates and have these used in place of donor sperm. This procedure is called *AIH*, for artificial insemination husband. It is, of course, preferable to AID, because the baby is then genetically related to both of the parents rather than just to the female.

A woman may be sterile for a variety of reasons, including lack of ovulation, adhesions (closures) in the oviduct, or other abnormalities of her reproductive tract. Ovulation can sometimes be stimulated by injections of FSH and LH; in fact multiple births may occur following this treatment. In some instances, surgery can be used to correct anatomical problems.

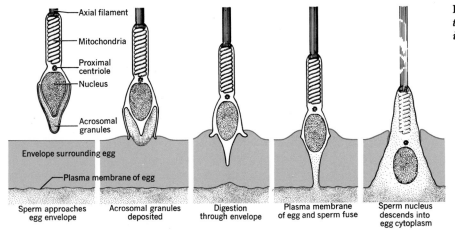

Figure 13-16 *How a sperm penetrates an egg with the aid of enzymes in its acrosome.*

Basic Concept

Union of egg and sperm nuclei is termed fertilization. Capacitation of the sperm and release of acrosomal enzymes are necessary for successful fertilization. Artificial insemination may be used in cases of male sterility; hormonal treatment can be used to induce ovulation in females.

CONTRACEPTION

How to Prevent Union of the Sperm and Egg

Many biologists and world political leaders believe that it is important to find ways to slow down human reproduction in order to lower the rate of world population growth. How can this be done in a manner that is acceptable to most cultures?

The anatomy of the male and female systems, as described above, should suggest several methods. Beginning with the male, the problem is to keep the sperm from getting into the vagina. One method of doing so is to withdraw the penis just prior to orgasm. This is *coitus interruptus,* an ancient technique but not a very reliable one, since droplets of sperm may leave the penis prior to orgasm. The technique also offers little satisfaction to either sex partner. Another old but much more effective technique is to sheath the penis in a covering that retains the sperm and semen after ejaculation (Figure 13-17). The *condom* (or "rubber") has an interesting history, having once been made of linen cloth, animal skin, or even large seed pods. Most are now made of latex rubber. The condom is a reasonably reliable, cheap method of birth control that has the additional advantage of protecting against venereal diseases. A disadvantage is that it can be used only when the penis is erected. The condom has a failure rate of 15–20 percent.

Far more reliable than the condom is a *vasectomy*—a simple and relatively painless operation that consists of cutting and tying off the sperm ducts on each side at the base of the scrotum. However, this operation is rarely performed on younger men who do not have children as it is considered irreversible. Researchers are trying to develop techniques for reversible vasectomies, such as inserting tiny valves in the sperm ducts, but these are still in the experimental stages.

With reference to the female, there are several

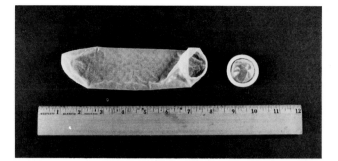

Figure 13-17 *A condom in its rolled up and unrolled form. When used properly, the condom is a reliable contraceptive device that also offers protection against sexually transmitted diseases for both sexual partners.**

methods available for keeping the sperm from the eggs. A rubber cap called a *diaphragm* can be fitted over the cervix to prevent the entry of sperm into the uterus. This protection is reinforced with a layer of spermicidal jelly (Figure 13-18). The diaphragm has a slightly higher failure rate than the condom, but can be inserted sometime prior to intercourse. Spermicidal *foams, douches,* and *suppositories* are also available, but again these substances must be carefully applied at a generally inopportune time. They also have high failure rates.

In recent years, *intrauterine devices (IUDs)* of various sizes, shapes, and materials have come into wide use (Figure 13-19). These inexpensive devices, usually made of plastic or plastic-coated wire, are inserted in the chamber of the uterus by a physician and remain there until removed or accidentally expelled. Why the IUD prevents implantation of the egg in the uterine wall is not known, but it works very reliably for millions of women. Its failure rate in actual use is about 1–5 percent. Drawbacks to the technique include bleeding and uterine cramps in some individuals, danger of puncturing the wall of the uterus, and the fact that the IUD can be inserted only by trained medical personnel.

---

* A new device, the contraceptive sponge, became available for public use after this chapter was prepared for publication. It consists of a small sponge containing a spermicidal chemical. The user places it in her vagina prior to having intercourse and may leave it in place for several days. It is less expensive than a diaphragm, does not require a prescription, and is claimed to be highly effective.

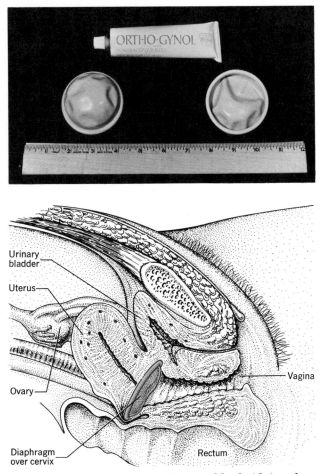

Figure 13-18 *A diaphragm (front and back sides) and a tube of spermicidal jelly. This flexible cap is placed over the entrance into the uterus just prior to having intercourse, and then covered with spermicidal jelly. The drawing shows the diaphragm in place.*

In a technique analogous to vasectomy, it is possible to cut or seal off the Fallopian tubes permanently. An instrument called a *laparoscope* is inserted through an incision in the abdominal wall and manipulated so as to sever or seal the tubes with an electric current. Unlike a vasectomy, this operation must be performed in a hospital. It is also considered irreversible.

Basic Concept

A variety of physical techniques and procedures is available for use by males and females for preventing the union of eggs and sperm. These include the condom, diaphragm,

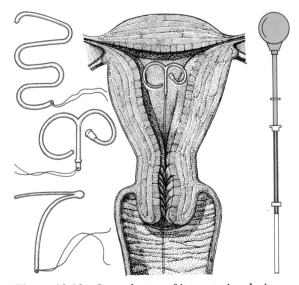

Figure 13-19 *Several types of intrauterine devices and the instrument for inserting them. The strings enable the user to know that the device is in place, or can be used to remove it. The drawing on the right shows an IUD in the uterus.*

spermicidal materials, the IUD, vasectomy, and closing the Fallopian tubes.

Interfering with the Reproductive Cycle

The Rhythm Method. This method is based on avoiding intercourse during the time of ovulation. In a perfectly regular cycle the "unsafe" period for having sex would be from day 10 to day 17, the time period that should safely encompass ovulation. The problem is that the ovarian cycle in many females is not sufficiently regular or predictable to allow the unsafe period to be reliably known. Some individuals use the basal body temperature method to more accurately determine ovulation time. This is based on the observation that body temperature rises measurably on the day following ovulation in many females and remains elevated during the rest of the cycle. To use this technique, the body temperature must be taken and recorded daily upon waking. After the rise in temperature is observed, intercourse is avoided for several days. A further refinement in the rhythm method involves observing the nature and quantity of mucus found in the vagina (but produced by the cervix of the uterus) during the female's cycle. Following the end of a menstrual period, there are several days in which the vaginal wall feels relatively dry to the touch (a safe period for intercourse). Mucus then begins to be produced and increases in quantity until the vaginal wall feels slippery and stringy, an indication that ovulation is pending. Several days following ovulation, the mucus becomes cloudy and tacky to the touch and intercourse may be resumed.

Even by carefully using the above techniques, the rhythm method remains unreliable: however, it does have the advantage of requiring no artificial devices and is the only method officially accepted by the Roman Catholic Church.

The Birth Control Pill. The hormones that function in reproduction have been investigated in considerable detail in humans. This knowledge resulted in the eventual development, about 20 years ago, of the *birth control pill.* Pioneers in its development included Gregory Pincus; a colleague, Min-Cheuh Chang; and John Rock. They were originally not interested in birth control but were attempting to regulate problems associated with the menstrual cycle. The basic idea was to interfere with the feedback portion of the female's hormonal cycle in such a way as to prevent ovulation and thus produce a less severe menstrual period. After considerable research and experimentation with laboratory animals, Pincus and his colleagues found that a combination of an estrogen and progesterone would effectively stop ovulation. The estrogen-progesterone combination evidently inhibits the hypothalamus from producing the neurosecretion that in turn stimulates the anterior pituitary to produce FSH and LH. In the absence of these hormones, follicle formation occurs but eggs are not ovulated. Thus Pincus and his colleagues discovered a technique of birth control. When "The Pill" finally came into use for humans, it was immediately highly successful in preventing pregnancy.

The oral birth control pill commonly used today usually contains a mixture of a synthetic estrogen and progestin, a progesterone-like substance. The approximately 35 different brands on the market vary in the amount of estrogen contained. A monthly supply usually consists of 21 pills. The user starts the pill on the fifth day after the onset of her menstrual period and takes one a day until the supply is gone. Then a menstrual period occurs (or else a 7-day time period elapses) and she commences another series of 21 pills. If all goes well, this schedule allows a menstrual period to occur as it would normally. Hence the user is assured that she is not pregnant. A woman at first may have difficulty in finding a pill with an estrogen dosage appropriate to her particular cycle. Too much estrogen may prevent the menses from taking place at all, and too little will cause bleeding at intermittent times during her 28-day cycle. Also some women experience side effects when taking the pill; hence it is important for users to be under the supervision of a physician.

Failure rate for the birth control pill approaches zero if the pill is used properly. Its popularity is reflected by the number of users — approximately 8 million women in North America and over 50 million throughout the world. Females who suffer side effects from estrogen may find that they can use the *minipill* that contains progestin only.

Basic Concept

Interference with the female hormonal reproductive cycle has proved to be an efficient birth control technique. The pill, estrogen plus progestin, prevents ovulation.

The bulk of research on human reproduction still concerns better means of contraception, especially in relation to the female's cycle. It offers more promise for improved contraception because it is more intricate than the male's and thus provides more potential ways of interfering with the reproductive events.

The search for a reliable male contraceptive pill also continues. Progestin is known to suppress sperm formation by inhibiting secretion of anterior pituitary gonadotropins, but the safety and side effects of its action are not well known. The chemical 5-thio-glucose is effective in preventing sperm formation in mice. It can be given orally, it does not appear to have side effects, and it allows the return to complete fertility when no longer administered. Whether a "thiosugar pill" can be developed for men is not yet known.

A prominent expert on contraception, Dr. Carl Djerassi of Stanford University, believes that it will be well into the 1980s before such entirely new methods as the once-a-month pill and the male contraceptive pill will be developed, if at all. Part of the reason for his pessimism is the great expense and long time period required to develop new drugs. The Food and Drug Administration requires that any new contraceptive be tested thoroughly on a variety of laboratory animals, including rats, dogs, and monkeys, for up to 10 years before it will consider releasing the drug for use by the public. Any serious side effects, such as carcinogenic properties, would of course be unacceptable. Because of the expense, often millions of dollars, and the long testing period, new contraceptives appear slowly and the prospects that a dramatic new device will reach the market soon are dim.

INDUCED ABORTION—A BIOETHICAL DILEMMA

The term *abortion* refers to the expulsion of a fetus (human embryo older than 8 weeks) from the womb before it can survive outside its mother's body. It may occur spontaneously (a miscarriage) or be deliberately caused in various ways—an induced abortion. In recent years there has been a worldwide trend toward legalized abortion to the extent that approximately 65 percent of the countries of the world now have liberalized abortion laws. In the United States alone there are more than 1.5 million induced abortions per year; worldwide, more than 50 million per year. Clearly, induced abortion is widely used for controlling pregnancy and may be considered as one of the foremost methods of birth control.

The U.S. Supreme Court ruled in 1973 that states cannot prohibit a female from having an abortion during the first three months of pregnancy. In effect this ruling gave women the right to decide whether or not to continue a pregnancy, that is, to have control of their own body. The demand for induced abortions rose rapidly and continues to increase despite vigorous opposition by antiabortion groups. Because abortion can be performed legally by trained medical personnel, the risk of death or illness from the procedure has dropped drastically: in fact is about one-seventh the risk of dying from pregnancy and childbirth. Also, most abortions are now done in abortion clinics on an outpatient basis at far less cost to the patient than going to a hospital. Thus, it has become relatively safe and inexpensive to have an abortion.

Most abortions are performed by suction curettage. In this procedure, a physician pushes a small tube into the uterus and then removes the embryo or fetus with a suction device. The procedure does not take long and causes little or no discomfort to the patient. The fetus of course is destroyed as a consequence of the above events. Suction curettage is used through the initial 16 weeks of pregnancy, the most favorable period for inducing abortion safely.

The Supreme Court's decision to "liberalize" abortion, that is, to allow an individual free choice in the matter, has led to an enormous amount of controversy. Antiabortion forces contend that abortion is equivalent to taking a human life and therefore should never be allowed. Proabortion adherents argue for the benefits of reducing the number of unwanted children and protecting the rights of women to make their own decisions regarding reproductive matters. Many individuals find that they do not like the idea of aborting a living fetus but at the same time support the concept of individual freedom of choice in the matter. Which of these viewpoints do you support?

Most of the arguments over abortion center around whether a fetus is a person: When does "humanhood" actually begin? A contemporary ethicist, Daniel Callahan, sums the arguments in three broad viewpoints. One is the contention, maintained by most antiabortionists, that humanhood begins at conception with the fusion of an egg and a sperm. The developing embryo is a defenseless human that must be protected (unless this action endangers the life of the pregnant woman). A second viewpoint is that the developing embryo is a potential human being and should be aborted only under carefully specified (usually medical) conditions. The third view is that a fetus is not a human being while in the womb and that the individual carrying the fetus has complete control over its destiny, including the choice of destroying it.

As in all bioethical matters there are no absolute rights or wrongs in the matter of abortion, but there are certainly many values to be considered. Why don't you examine your own value system in relation to this difficult issue and see where you stand?

KEY TERMS

abortion

acrosome

anaphase

androgens

artificial insemination

asexual reproduction

birth control pill

capacitation

centromere

cervix

chromosome

circumcision

clitoris

coitus interruptus

condom

corpus luteum

Cowper's gland

cytokinesis

daughter cell

diaphragm

diploid

dysmenorrhea

ejaculation

endometrium

epididymis

erection

estradiol

fallopian tube

fertilization

follicle stimulating hormone

haploid

hernia

hymen

hysterectomy

impotence

inguinal canal

intrauterine device

labia

luteinizing hormone

meiosis

menopause

menstrual cycle

metaphase

mitosis

mons veneris

oocycle

oogenesis

orgasm

ovarian cycle

ovary

ovulation

Pap test

penis

polar body

progesterone

prophase

prostate gland

rhythm method

scrotum

secondary sex characteristic

seminal vesicle

sexual dimorphism

sexual reproduction

spermatids

spermatogenesis

spermatozoa

telophase

testes

urethra

uterine cycle

uterus

vagina

vas deferens

vasectomy

vulva

womb

zona pellucida

zygote

SELF-TEST QUESTIONS

1. Name three basic functions of reproduction.
2. Give an example of how sex and reproduction are of great cultural significance to human beings.
3. What is the difference between sexual and asexual reproduction?
4. Give some examples of asexual reproduction.
5. List an advantage and a disadvantage for sexual and for asexual reproduction.
6. Name the two basic types of cell division.
7. In your own words describe the major events of mitosis.
8. What is the major end result of mitosis?
9. (a) Where does meiosis occur?
   (b) What is the function of meiosis?
10. Describe the events of meiosis.
11. (a) What does sexual dimorphism mean?
    (b) Give some examples in human males and females.
12. List the major female reproductive organs and at least one function of each.
13. Describe how egg cells are formed.
14. What happens to the female at menopause?
15. Describe the major events that occur in the female reproductive cycle.

16. Name the major organs of the male reproductive system and a function of each.

17. Describe how sperm cells are formed.

18. Are mature sperm haploid or diploid? Explain your answer.

19. How does the male reproductive cycle work?

20. How and where does fertilization take place?

21. What do the initials AID and AIH stand for?

22. What are three means available to males to prevent sperm from entering the vagina?

23. Which of the above means is the most reliable?

24. Describe three means available to females to prevent sperm from uniting with eggs.

25. Describe what is meant by the "rhythm" method of birth control.

26. Tell how the birth control pill works.

27. Discuss abortion as a bioethical problem.

chapter 14

DEVELOPMENT
AND GROWTH

chapter 14

questions to think about

How does an embryo form?

How is the fetus nourished?

Which organ system develops first?

How does birth take place?

What are the stages of the human life cycle?

A baby is an angel whose wings decrease as his legs increase.

French Proverb

Probably the most astonishing and complex process in all of biology is the series of events involved in the development of an organism. The process is astonishing because it begins with a single, undifferentiated cell—that is, a fertilized egg—and concludes with an exceedingly complex organism, often consisting of trillions of cells. Furthermore, even though these trillions of cells are identical in their genetic makeup (DNA), they become greatly different in appearance and function during development.

The very general events of development from egg to organism are similar in all animals, but there are almost innumerable variations in details. In mammals, including human beings, the egg is fertilized in the oviduct (Fallopian tube) and begins immediately to divide as it moves through this tube. Several days later the dividing egg, now a tiny ball of cells, enters the chamber of the uterus. Within a day or so the cell mass embeds in the spongy wall of the uterus, where the rest of its complex development into an embryo and fetus takes place. As the embryo grows and develops, it is nourished by means of the placenta. In humans, approximately 9 months later the fetus has become sufficiently developed to carry out its vital biological activities on its own; birth then occurs.

This fascinating unfolding from the simple egg to the complex organism has intrigued people for centuries. Embryology, also called developmental biology, is one of the older branches of biology and still commands a considerable amount of study and research.

FORMATION OF THE EMBRYO

Fertilization

Fertilization occurs when an egg and a sperm unite to form a zygote as described in the previous chapter. This event is significant in several ways, the foremost being that it combines the hereditary material, DNA, from two sources (the parents) who differ genetically. Hence, the zygote contains the genetic blueprint for a unique individual. Secondly, the penetration of the

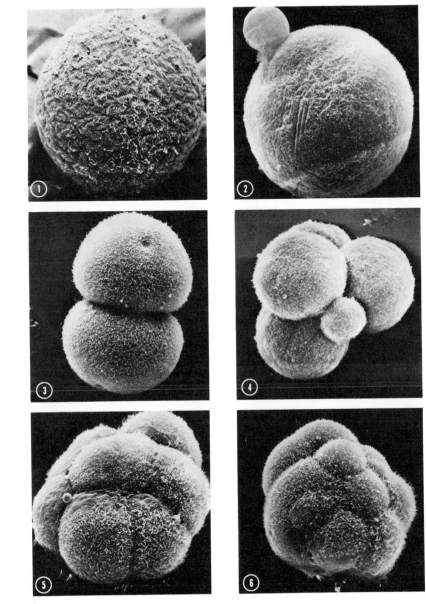

Figure 14-1 *Early cleavage stages in a mammalian (mouse) egg. (1) Unfertilized egg just after fertilization. (2) Fertilized egg just after fertilization with a polar body. (3) Two-cell stage with cells showing numerous microvilli. (4) Four-cell stage and third polar body (small round object). (5) Eight-cell stage. (6) Solid ball of cells just prior to becoming a blastocyst.*

sperm stimulates the egg to complete meiosis. The egg then embarks on a brief series of cell divisions called *cleavage* (Figure 14-1).

Cleavage and Blastula Formation

In cleavage the zygote divides into 2 cells, which divide again into 4, then 8, then 16 and so on, until a hollow ball of cells, the *blastula,* is formed. Blastula is a general term, and in mammals the blastula stage is called a *blastocyst* (Figure 14-2). This stage is reached about five days from the time of fertilization. Formation of the blastocyst (or blastula) signifies the end of cleavage.

During the process of cleavage, the total amount of cytoplasm remains about the same as it was in the zygote. That is, the cleaving cells become smaller with each cell division. Consequently, the many-celled blastocyst is extremely tiny.

The initial events of cleavage are similar in all vertebrates. However, the eggs of vertebrates are not all alike. Some, such as amphibian and bird eggs, contain considerable amounts of nutrient yolk for the embryo, whereas others, such as mammalian eggs, have practically none. The nature of the egg influences cleavage to a considerable degree. When little or no yolk is present, as in mammalian eggs, the cleavages produce cells of approximately equal size. But when larger amounts of yolk are present, as in bird or shark eggs, only the upper portion of the egg cleaves, so that the embryo develops as a disc on the upper surface of the yolk (Figure 14-3). The amount of yolk is related to how the embryo is to be nourished. The mammalian embryo is destined to be nourished soon by the mother, whereas the bird embryo must be sustained entirely by the yolk contained in the egg.

Basic Concepts

Cleavage of the zygote produces a small, multicellular blastula or blastocyst.

Mammalian eggs have relatively little yolk. The amount of yolk in an egg determines the type of cleavage it undergoes.

Gastrulation and the Germinal Cell Layers

As described above, the blastocyst is a tiny ball of cells with a fluid-filled center called the *blastocoel*. Soon after it forms, a mass of cells grows into the cavity from one side of the wall. This group, called the *inner cell mass,*

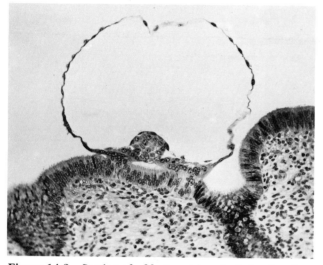

Figure 14-2 *Section of a blastocyst (monkey) showing the inner cell mass (small clump of cells) and blastocoel (clear area).*

Figure 14-3 *Cleavage patterns in different kinds of vertebrates. The amount of yolk in an egg determines its type of cleavage.*

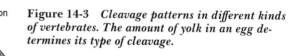

becomes increasingly distinct from the outer cells of the blastocyst (see Figures 14-2 and 14-3). In subsequent stages the inner cell mass forms the embryo itself, whereas the outer layer takes part in the formation of the placenta.

The blastocyst with its inner cell mass begins to burrow into the spongy uterine wall about the sixth day after fertilization. The burrowing is made possible by enzymes, produced by the outer layer, which dissolve a pathway into the uterine tissue. The dissolved tissue also nourishes the developing embryo for the next few weeks until the placenta forms.

As the blastocyst becomes enveloped in uterine tissue, the inner cell mass separates into two distinct layers that become the *ectoderm* and *endoderm*. At a later time, still a third layer of cells, the *mesoderm*, appears. From these three layers of cells all parts of the embryo will grow and gradually specialize (differentiate) into various organs of the body. Hence these are often called *germ layers* (from the word "germinal"). Their formation from the embryonic disc is called *gastrulation*. Gastrulation is a key event in the development of all vertebrate embryos although the events leading to it may differ from group to group.

Careful study of this differentiation procedure has established precisely which body parts of the individual derive from each germ layer. Ectoderm is known to form the epithelium of nervous system; endoderm forms the epithelium of digestive tube and its glands; and mesoderm differentiates into the skeleton and blood system. Table 14-1 lists additional derivatives of each germ layer. Many organs contain tissues from more than one germ layer. The lining of the intestine, for example, is derived from endoderm, but its muscle layers, connective tissues, and blood vessels all develop from mesoderm.

Basic Concept

Each part of the embryo develops from one of the three germ layers; tissues and organs are often made of combinations of the three.

Differentiation

Because all germ cell tissues look identical for some time in the embryo, the question of what causes these tissues to specialize (differentiate) eventually into dif-

TABLE 14-1. Derivatives of Ectoderm, Endoderm, and Mesoderm

Ectoderm	Nervous system
	Epidermis of skin
	Tooth enamel
	Lens of the eye
	Lining of the mouth, anus, and nasal chamber
Endoderm	Inner lining of the digestive tract
	Middle ear chamber
	Eustachian tube
	Lining of pancreas
	Thyroid gland
Mesoderm	Inner layer of skin
	Skeleton
	Muscles and connective tissues
	Part of excretory and reproductive systems
	Lining of the coelom (body cavity)

ferent organs has intrigued biologists for years. Part of the answer concerns the position or location of the tissues in the gastrula. If a piece of mesoderm is transplanted to the ectoderm of an *early* embryo, it will undergo the same development as does the ectoderm in that region. But in just a slightly older embryo, the same experiment will produce dramatically different results. A tissue transplanted later will grow into one particular body organ despite its transplanted location. Hence a piece of neural ectoderm in an older embryo will grow into a neural tube wherever it is placed. Evidently, then, each tissue is influenced at some stage by its position in the embryo, and at that point it becomes committed to a certain anatomical fate.

What produces this change or commitment is not known. Because every cell in an embryo contains identical informational material in its DNA, it seems reasonable that differentiation must involve a selective use of a part of each cell's DNA. If changes, or mutations, are induced in the DNA of a developing embryo by experimental means such as X rays, abnormal growth often takes place. Tissues do not follow their normal developmental pathways, and the embryonic "landscape" becomes distorted.

In short, it appears that the relationship between a cell in the developing embryo and the other cells with which it is associated, functions in some manner during development to determine which part of that cell's

DNA will function. Even this knowledge, however, does not reveal how a portion of a cell's DNA is selectively turned on and off.

Basic Concept

Differentiation is caused by a combination of factors: the location of the tissue in the embryo, association with other cells, the age of the embryo, and a selective functioning of the cell's DNA.

FETAL MEMBRANES AND PLACENTA

As the embryo is going through its early formative stages about the second week, several specialized parts develop that aid in nourishing and protecting the embryo. These are the *amnion, yolk sac, allantois,* and *chorion.*

The Amnion

In the human embryo, as the inner cell mass is thickening to form the germ layers, a cavity appears above it. This is the *amniotic cavity.* The cells that roof the cavity soon differentiate into a distinct structure, the *amnion* which is filled with fluid. The amniotic cavity rapidly enlarges until it completely envelops the embryo. The amnion and fluid constitutes the *amniotic sac* (Figure 14-4), which persists until birth; the embryo-fetus spends its existence immersed in the watery amniotic fluid. This liquid environment cushions the embryo as its mother moves about. It probably also keeps the embryo's tissues from adhering to adjacent tissues that it would otherwise touch.

Yolk Sac and Primitive Gut

The primitive gut is fashioned from endoderm soon after the germinal tissues are established. It is formed as the endoderm grows beneath the embryonic disc until it lines the inner surface of the blastocoel cavity. At first it cannot be distinguished from the rest of the cavity, called the *yolk sac* (Figure 14-4). But as the embryo grows longer, the gut reveals three distinguishable areas termed the foregut, midgut, and hindgut. These areas become distinguishable because the

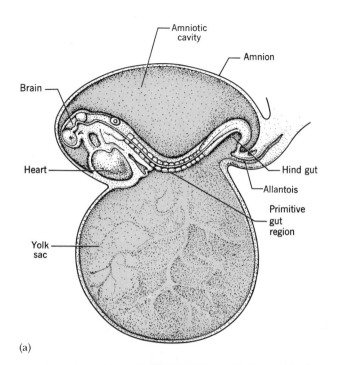

(a)

(b)

Figure 14-4   *Fetal membranes of the early embryo.* **(a)** *As shown, the embryo develops in the fluid of the amniotic sac. The yolk sac soon regresses and disappears in mammalian embryos as the placenta takes over the task of nourishing the embryo. The allantois is a small pouch that becomes incorporated into the umbilical cord.* **(b)** *Human embryo at 28 days. The amniotic sac has been opened to show the embryo and the large (at this stage) round yolk sac.*

embryo bends as it grows; also, the amniotic cavity increases in size and begins to push under the embryo, carrying endoderm with it. Thus a floor is formed at both the front and back ends of the gut. For a considerable period during development, the middle section of the primitive gut remains open, a broad space connecting it to the yolk sac beneath. Although in mammals the yolk sac does not actually contain yolk, in nearly all other vertebrates the yolk sac does enclose a supply of this rich nutrient that sustains the embryo through all of its growth. In mammals the yolk sac dwindles in size and disappears as the placenta takes over the nourishment of the fetus.

Allantois

At the same time that the yolk sac is developing, a portion of the hindgut pushes out to form a small pouch known as the *allantois* (Figure 14-4). This pouch grows outward until it makes contact with the developing placenta. The pouch disappears, but blood vessels develop in the area to become the umbilical cord somewhat later.

An allantoic sac is also found in the embryonic stages of reptiles and birds, even though their young develop inside eggs. In these vertebrates, the allantois grows into a prominent structure that lies adjacent to the inside of the covering (shell or membrane) that encases the egg. In this position, the allantois functions to exchange gases (O_2 and CO_2) between the embryo and the environment outside the egg. In some instances it also serves as a storage site for waste products from the embryo's body.

The Chorion and Formation of the Placenta

As implantation is proceeding, the outer cell layer of the blastocyst becomes the *chorion*. Small projections (villi) grow from the chorion into the uterine lining (Figure 14-5). The chorionic villi establish a vascular link between the uterus and the rapidly developing embryo. As the chorionic villi branch and penetrate the uterine tissues, the tissue cells around each villus disintegrate to leave a fluid-filled pool or space around each greatly branched villus. This complex of branched chorionic villi, in its intimate association with

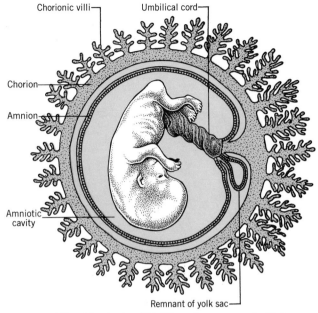

Figure 14-5 *Chorionic villi. These form an early link between the embryo and the uterine lining.*

the uterine lining, forms a distinctive structure called the *placenta* (Figure 14-6). The placenta is distinguishable by the third month of development. It continues to nourish and maintain the fetus until birth.

Although the placenta forms a very close association with tissues of the uterus, the mother's blood and fetal blood do not normally mix. In order for nutrients, hormones, antibodies, oxygen (and sometimes drugs) to get into the fetus from the mother's blood, these substances must first pass from the capillaries located in the uterine wall into the fluid-filled areas around the placental villi. Then they must diffuse through the villi walls and into the capillary beds inside the villi. Only at this point do these substances actually enter the blood supply of the fetus. The umbilical vessels then transport them into the fetus itself. In a similar manner, metabolic wastes from the fetus follow the same route in reverse in order to reach the mother's body. For the remainder of fetal life, the placenta will serve the diverse functions of a nutritive organ, a respiratory organ, and a kidney.

Basic Concepts

The fetal membranes—amnion, yolk sac, and allantois,

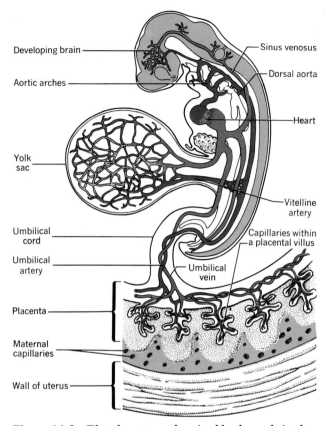

Figure 14-6 *The placenta and major blood vessels in the early fetus. Materials are exchanged between the mother and fetus via capillary beds in the placenta. All materials must diffuse through capillary walls, hence there is no direct blood connection normally.*

function in protecting and aiding the development of the embryo.

Following implantation, the chorion forms the placenta that establishes the respiratory, nutritive, and excretory links between mother and fetus.

DEVELOPMENT OF ORGAN SYSTEMS IN THE EMBRYO

With establishment of the germ layers, differentiation of the various organ systems begins. Rather than trying to describe all of these complex processes and sequences, we will look at some features of two important ones, the nervous system and the circulatory system.

Nervous System

The nervous system is the first organ system to form in the embryo, but it does not become functional until much later in development. The process begins with a portion of ectoderm at the surface of the embryo that thickens, early in development, to form the *neural plate*. It then begins to form a V-shaped *neural groove*. The neural groove becomes deeper and larger as it extends the length of the embryo. Finally the walls of the groove meet to form a *neural tube* (Figure 14-7).

The front end of the tube grows and enlarges rapidly to become eventually the five basic chambers of the brain. Concurrent with this growth, a marked flexion (bending) carries the first two brain chambers

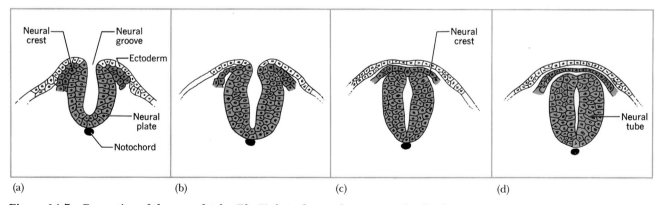

Figure 14-7 *Formation of the neural tube. The U-shaped neural groove gradually closes over to become a tubelike structure.*

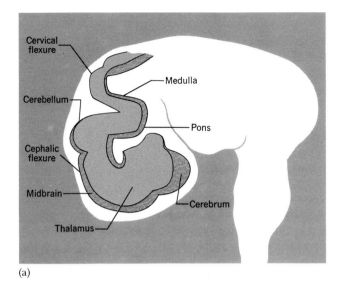

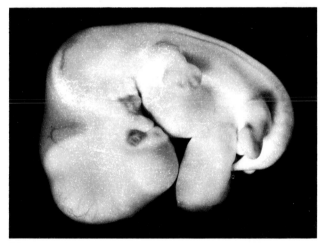

Figure 14-8 (a) *Formation of brain chambers and head flexures at about six weeks. The brain at this stage is a large, hollow body that develops rapidly. The flexures (bends) help fit the brain in a compact, rounded space.* (b) *Human embryo at 41 days. The head and brain are very large compared with the rest of the body and are flexed beneath the body.*

down and almost beneath the back chambers (Figure 14-8). This flexion is, at least partly, an evolutionary adaptation for fitting a large volume of nerve tissue into the compact chambers of a rounded cranium. During the subsequent months the hollow chambers of the embryonic brain gradually fill with nerve tissue, until only small canals called ventricles remain in the

fully developed brain. The brain will exceed all other organs in the embryo in both developmental rate and size for quite a long time. Not until late in childhood (6–7 years of age) does the remainder of the body finally catch up in proportionate size to the head.

As the brain is forming, 12 pairs of *cranial nerves* grow from it to innervate the many structures associated with the head and face. For example, one pair, the oculomotor nerves, serve the muscles that move the eyeballs. Another pair, the olfactory nerves, connect the nasal lining to the olfactory area of the brain, and so on.

The region of the neural tube not involved in brain formation becomes the spinal cord. It too is hollow and gradually fills with nerve cells. From its sides, paired *spinal nerves* grow into all parts of the body to serve muscles, skin, and organs.

As one of the two major control and coordinating systems for the body (the hormonal system being the other one), the nervous system undergoes an immense amount of complex differentiation and size increase, finally attaining the adult form. During its formation, the nervous system, especially the brain, forms a close relationship with the hormonal organs. Also, the pituitary gland, the "master gland" of the endocrine system of glands, forms in part from the floor of the brain. Hence control and coordination of the body eventually are the result of interactions between the two systems. We saw evidence of this relationship in previous chapters.

Basic Concepts

The nervous system develops from a tube that forms from ectoderm at the surface of the embryo.

It is the first organ system to develop in the embryo, exceeding all other systems in rate of development and in size for quite a long time.

Circulatory System

The circulatory system is the first *functioning* organ system to form in the embryo. Embryonic growth is so rapid that the embryo consists of an immense number of cells, each of which must constantly receive nutrients and oxygen and have its wastes carried away. Therefore, early formation of a transport system for these functions is vital. The system must have several basic parts in order to function, including the transport medium itself (blood), a muscular pumping device (heart), blood vessels, and capillaries. As you might expect, fetal circulatory development quickly becomes exceedingly complex (see Figure 14-6). Description of the development of the heart provides some idea of this complexity.

Early in development (about the twentieth day) a pair of tubes forms on the underside of the embryo near the head region. Eventually the two tubes fuse into one as the embryo bends or flexes on itself in its growth. This single *cardiac tube* soon begins to pulsate and thereby to function as a simple heart. It lies on the outside of the embryo's body for a period of time; later it is enveloped in a pericardial chamber inside the body.

The cardiac tube begins to grow in length. However, due to its confinement in the pericardial chamber, it can only do so by forming an S-shaped loop. At this time several chambers can be distinguished. The major veins of the embryo converge near the back end of the heart to empty into a chamber called the *sinus venosus*. The sinus venosus opens into the next portion, termed the *atrium*. From the atrium, blood flows into the loop of the primitive heart, the *ventricle* (Figures 14-6 and 14-9). When the ventricle pulsates, blood is forced out through the arterial trunk into various parts of the embryo. Valves between the chambers keep the blood flowing in one direction.

This is a simple but effective heart for the early embryo. Interestingly, it is not basically different from the type of heart seen in fishes. The mammalian heart does not long remain simple however. As time passes, the sinus venosus is incorporated into the atrium, and the atrium itself becomes divided into right and left chambers. However, a complete separation of these chambers does not occur until birth. Similarly, the ventricle develops into two compartments. What began as a simple, pulsating tube eventually becomes a complex multichambered muscular organ. By this time the rest of the embryo's body has also become markedly complex, and hence it requires such a specialized pump to serve its many tissues and organs. At birth the heart must make a dramatic change in its functioning. Oxygenated blood had previously been provided via the placenta and umbilical vessels, but it must now be provided via the lungs. This changeover to another source of oxygen is accomplished, remarkably, in a few minutes.

Basic Concepts

The circulatory system develops rapidly and becomes the first functioning organ system in the embryo.

The heart begins as a simple cardiac tube that eventually becomes a complex, four-chambered organ.

THE TRIMESTERS OF FETAL DEVELOPMENT

As the nervous and vascular systems take form and grow, the other body systems also develop. Muscles, bones, excretory organs, the reproductive organs, digestive organs, and the lungs all differentiate and become highly specialized. Most of them begin functioning long before the baby is born.

A traditional and convenient way to summarize all of these developmental events is to divide the nine months of intrauterine life into three equal trimesters.

Trimester One

The first trimester (months one through three) begins with the fertilized egg and carries through the early embryonic stages already described. Some of the major events of early development are summarized in Table 14-2. At the end of the first trimester (Figure 14-10), the fetus is about 3 inches (8 centimeters) long and weighs ½ ounce (about 15 grams).

As the early embryo develops, its huge head flexed almost beneath its tiny body, its large bulging liver, and its limb buds give it a grotesque appearance. Its eyelids

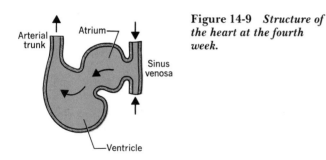

Figure 14-9 *Structure of the heart at the fourth week.*

TABLE 14-2.   Events in Early Fetal Development[a]

Day	Event
1	Fertilization of the egg in the Fallopian tube.
5	Blastocyst forms and enters the uterus.
6	Implantation begins; germ cell tissues forming.
14	Formation of primitive streak and embryonic membranes.
15	Gastrulation.
17	Notochord growing.
18	Neural plate and groove formation; heart primordium and blood formation.
21	Appearance of somites.
24	Neural tube forming.
28	Limb buds appear.
35	Brain with five chambers.
49	Heart with four chambers.
60	Embryo three centimeters long from top of head to rump. Beginning of fetal period: all basic organs established and body shape clearly defined.
61 to birth	Fetus increases from 3 cm to 50 cm; many organs functioning prior to birth. Birth occurs about 280 days, or 40 weeks, from the first day of the last menstrual period (267 days from fertilization).

[a] Days are counted from fertilization of the egg.

seal shut after the eyes develop and will not open again until the seventh month. After eight weeks the embryo is arbitrarily termed a *fetus*. About this time it begins to be more distinctly human in appearance. Also, at this time, the fetus is differentiating into a male or female. By the end of the first trimester all organ systems have been established, but the fetus cannot yet survive outside the womb. Like a parasite, it is utterly dependent on its host's body for nourishment and survival.

Because the organ systems are rapidly developing at this time, it is not surprising to find that the embryo-fetus stage is especially sensitive to events that take place in its environment, that is, in the mother's body. Detrimental events include diseases of the mother and various drugs or other substances that she may use. Maternal diseases such as German measles (Rubella), influenza, polio, syphilis, and infectious hepatitis are especially dangerous to the fetus. Radiation can also cause major disruptions of fetal growth; hence X rays should be avoided. A pregnant woman should also avoid using drugs, unless prescribed, because many are known to cause growth abnormalities in the fetus.

Included among such drugs are alcohol, tobacco, and heroin.

Obviously, then, a pregnant woman should be extremely cautious concerning her health, diet, and drug use while the fetus is developing.

Trimester Two

During the second trimester (months four through six), the fetus grows to about 1½ pounds (700 grams) and becomes 12 to 14 inches (30 to 36 cm) long (Figure 14-11). To accommodate this vigorous growth, the uterus must stretch appreciably. With all the organ systems already established, the second trimester is a period of continued differentiation of the fetal organs. In addition, there is a change in the proportionate size of some parts; in particular, many organs begin to catch up with the nervous system in development. And, of course, there is a general overall size increase.

During this trimester the fetus becomes immersed in a considerable pool of amniotic fluid. The volume of this fluid is great enough that by three and a half

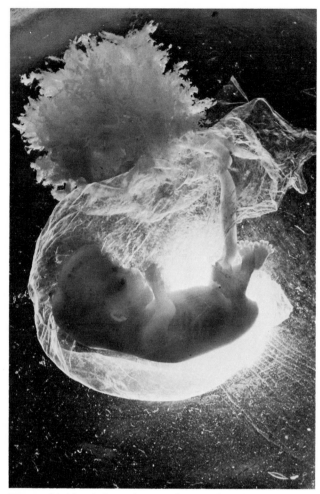

Figure 14-10 *A first trimester fetus with its large, flexed head and developing limb buds.*

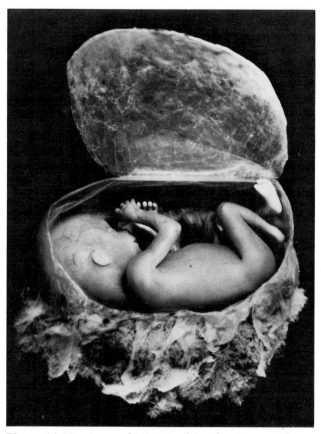

Figure 14-11 *A second trimester fetus showing considerably more development of its organs and larger body size.*

months a physician can carefully insert a needle through the abdominal wall and withdraw some of the fluid for examination. A few loose cells from the embryo can be found in the sample (Figure 14-12). By examining them, the physician can determine several important things about the embryo, including its sex (XX or XY), whether it has the normal number of chromosomes, and whether certain enzyme systems are functioning properly. This procedure, *amniocentesis,* is usually not done unless there are indications that the fetus risks a hereditary disease or is not developing normally. Over 40 genetic diseases can be detected in this manner, including Down's syndrome and

other conditions involving an abnormal number of chromosomes, a severe neurological disease called Lesch-Nyhan syndrome, and Tay-Sachs disease, which produces severe mental retardation and early death. With this valuable information available, many pregnant women choose to have abortions rather than give birth to severely defective infants.

The amniotic fluid is at first derived from the mother's blood serum, but by five months it is made up almost entirely of urine produced by the fetus. At this time the fetus becomes covered with a whitish, greasy material (vernix caseosa) which perhaps protects its skin from the constant immersion.

The fetus is quite active during the second trimester, making jerky, uncoordinated lumb movements. By the end of the fifth month, the mother begins to feel these movements, referred to by the rather quaint

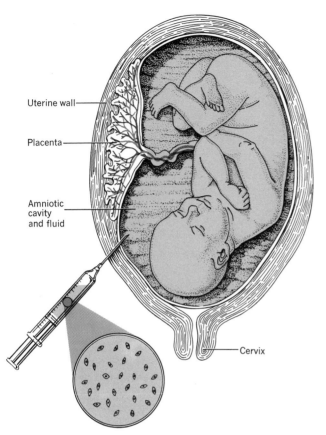

Figure 14-12 *Amniocentesis. A sample of amniotic fluid is removed and the embryo's cells, contained in it, are grown in a culture medium. A variety of tests can be performed on the cultured cells.*

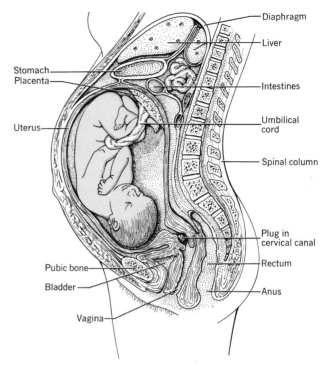

Figure 14-13 *A full-term fetus in the uterus. Note how the mother's organs are pushed out of place and compressed.*

term, *quickening.* These spasms signify to the mother, happily, that her baby is probably developing normally.

After six months, should an emergency demand it, the fetus has a chance of surviving if delivered by Caesarean section. If need be, the fetus is removed through an incision made in the abdomen and uterus. It must be placed in an incubator and carefully tended until it attains the development of a full-term baby.

Trimester Three

During the third trimester the fetus attains its birth size of 6 to 8 pounds (3 to 4 kilograms) and about 20 inches (50 cm) in length. Obviously the uterus must stretch considerably to accommodate it—to about *60*

times its normal size. By the time of birth, the mother's abdominal cavity becomes largely filled by the rapidly enlarging fetus (Figure 14-13). Her own abdominal organs are necessarily squeezed to one side. And the fetus itself finally is so tightly confined in its amniotic chamber that its movements are greatly restricted. During its seventh month, fine downy hair often covers the fetus, but this hair is usually lost by birth. The eyes open during the third trimester and remain so through birth; many other mammalian infants are born with their eyes closed and are "blind" for some time after birth.

During the final trimester the fetus makes its greatest demands on the mother's body, both for growing room and for nutrients. The fetus needs considerable amounts of calcium, iron, and nitrogen for its own growth. If the mother does not make an allowance for these nutrients in her diet, they will be removed from her tissues and provided to the fetus. This would obviously take a toll on the mother's health. The mother also provides the fetus with antibodies from her own body, protecting the fetus against a variety of diseases for some time after birth. A tremendous number of other changes also take place in the mother's body during pregnancy, because her body is providing many of the vital functions for two organisms. For this reason it is highly advisable for the woman to be under the care of an obstetrician, or to have some type of medical supervision throughout her pregnancy.

About two weeks before birth, the fetus moves down into the pelvic cavity, an event termed *lightening.* The mother is aware of this event, as her abdominal organs are less crowded and she can breathe with less effort. The fetus should be in a head-down position at this time, because this is the normal birthing position. If the baby is not in the correct position the physician may attempt to rotate it in the uterus just before birth. If this does not succeed, the baby may be delivered buttocks first, a *breech* birth. This complicates the birth and may endanger both the baby and the mother.

As the time of birth nears, occasional contractions of the uterus may take place, sometimes misleading the mother into thinking that labor has started. If the amniotic sac breaks and its fluid flows out, it is a fairly reliable signal of an impending birth. However, the mother should not wait for this sign if her contractions are close together.

Pregnancy normally lasts about 267 days in the human female from the time of fertilization or 280 days from the last menses. Because the time of fertilization is seldom known, and the duration of pregnancy is not precise, it is difficult to predict the date of an impending birth with any accuracy.

Basic Concepts

Fetal life may be described in terms of three trimesters:

Trimester One: **Major organ systems established; fetus completely dependent on its mother's body.**

Trimester Two: **Rapid growth, first movements; fetus may survive outside the mother's body with special care at the end of this trimester.**

Trimester Three: **Continued growth, eyes open; fetus makes strong nutrient demands on the mother's body; fetus is prepared to be born.**

LABOR AND BIRTH

Labor is the term applied to the strong and often painful contractions of the uterus that open the cervix and ultimately push the fetus through the birth canal into the outside world. It is thought that these contractions are triggered by prostaglandin hormones released by uterine cells. Prostaglandins also stimulate the posterior pituitary gland to release oxytocin hormone which evidently causes the strong contractions found during the second stage of labor.

The contractions, muscle spasms of the uterine wall, normally begin at the upper end of the uterus and pass rhythmically toward the cervix. At the start, these spasms occur at 10–15 minute intervals and last from 2 to 5 minutes each. The intervals lessen as labor proceeds, and the contractions become increasingly strong. Women who are under medical supervision during their pregnancy are usually instructed in exercises and breathing techniques that better enable them to help in the birthing process.

The *first stage* of labor begins with regularly spaced intervals of contraction and takes from 7–12 hours (but may last as long as 20) with the first child and 3–7 hours with subsequent births. The cervix gradually opens and the amniotic sac may break during this time, releasing its fluid. When the cervix is fully dilated (about 10 cm), the *second stage* commences. The uterine contractions are frequent, the amniotic sac breaks if it has not done so previously, and the baby's head begins to emerge through the cervix of the uterus, usually in a face down position. Contractions of the uterus then push the baby out through the birth canal (vagina) to the outside world, a dramatic event to say the least. This delivery stage lasts approximately 20 minutes but may take more than an hour in some cases.

The newborn infant must begin breathing promptly because its oxygen supply via the umbilical cord has lessened during labor and soon stops altogether as the umbilical blood vessels collapse. The physician clears the mucus from the baby's nose and throat and it inflates its lungs for the first time. Once breathing has begun the physician ties and cuts the umbilical cord and the infant is dried and examined. About 15 minutes after birth the *third stage* of labor occurs as the placenta and the remainder of the attached umbilical cord are expelled from the womb; hence the term *afterbirth.*

MULTIPLE BIRTHS

Women typically bear one child at a time, but twinning is not a rare event. About one birth in 85 to 100, depending on the race of the individual, produces twins. The odds of having triplets or quadruplets are one in many thousands.

The most common occurrence of twins (70 percent) results when two eggs are ovulated about the same time and both are fertilized by different sperm cells. Both zygotes implant and develop in the uterus (Figure 14-14a,b). The resulting offspring are like siblings conceived separately: They may be of different sexes, they will not be identical physically even if of the same sex, they have different blood characteristics, and their immune systems will reject transplanted parts from each other's bodies. These twins are termed *dizygotic* (two zygotes), or fraternal twins.

Less common (30 percent) are *monozygotic,* or identical twins. As the name implies, monozygotic twins develop from the same egg. Exactly how this occurs in humans is uncertain. Most often it is probably the result of the inner cell mass or embryonic disc separating into two cell masses, each then becoming a separate embryo (Figure 14-14c). These embryos usually share a single placenta but develop in separate amniotic sacs. Rarely, identical twins form in the same amniotic sac; they may even be joined by various parts of their bodies if the embryonic disc does not separate completely. In any event, monozygotic twins are always of the same sex and identical in physical appearance and blood characteristics, and they will tolerate transplanted parts from each other's bodies.

Other types of multiple births occur, but their probability is very low under natural conditions. However, gonadotropic hormones are now used in treating fertility problems. The drugs succeed in bringing about ovulation, but frequently more than one egg is ovulated at a time. The result is a higher incidence of multiple births. Infant mortality is high in these instances, because the newborn are small and usually premature.

Basic Concept

Multiple births are the exception in humans. Twins may be dizygotic or monozygotic.

THE NEWBORN INFANT

A newborn infant cannot be considered just a miniature adult, because many of its organs differ in proportionate size as well as in absolute size from those in an adult. For example, the central nervous system makes up a surprising 15 percent of the newborn's total body weight although its nervous system has not yet finished developing. (Only about 2 percent of an adult's weight is nerve tissue.) The newborn's muscular system comprises only one-fourth of its weight, but almost half of an adult's weight. The skeletal system is about 15 to 20 percent of body weight in both the infant and the adult. Compared with adult proportions the infant's head is very large and the legs are quite short (Figure 14-15).

An infant's weight normally doubles in about six months after birth and triples by the end of a year. But the child's growth from infancy to adulthood is not steady; rather it takes place in spurts. There are lag periods during early childhood, and thus children of the same chronological age may differ considerably in height. Even after puberty there are periods of rapid growth for some individuals. By the early twenties, however, the cartilage growth plates near the ends of the long bones turn into bone. After this event an individual ceases to grow in height.

Nutrition and Infant Development

As indicated previously, normal growth of the fetus requires a variety of nutrients, which must be provided by the mother's blood. If the mother's diet is deficient in a substance such as calcium, the fetal demand may be met at the expense of the mother's tissues — in this case, her bones. In some instances the mother cannot supply a necessary nutrient and the development of the fetus is retarded.

The effects of malnutrition on infants and children have been documented for some years owing to the tragedy of starvation in many parts of the world.

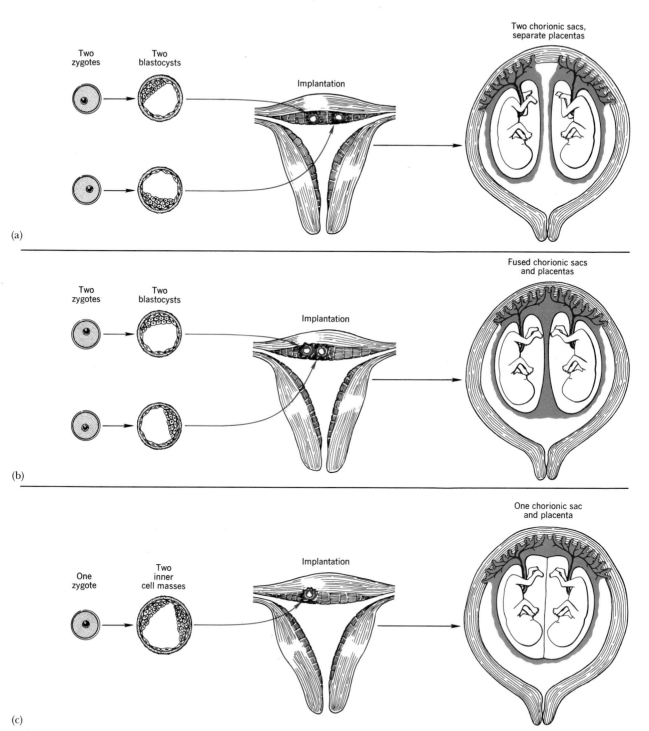

Figure 14-14 *Formation of twins. (a) and (b) are examples of dizygotic twins. In (a) the blastocysts implant separately. In (b) the blastocysts implant near each other, sharing the same placenta and chorionic sac. (c) Monozygotic twins develop from one zygote in which the inner cell mass separates into two masses and thus forms two embryos. As shown here, the embryos are in separate amniotic sacs; rarely do they share the same sac.*

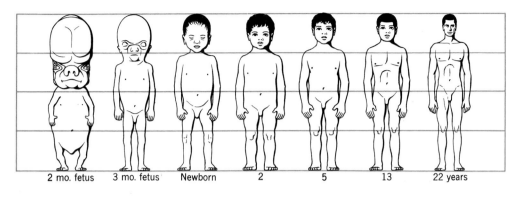

2 mo. fetus   3 mo. fetus   Newborn   2   5   13   22 years

Figure 14-15 *Comparison of the proportions of a fetus and infant's body to those of an adult human.*

Starved and malnourished children do not develop normally and often succumb to a variety of diseases.

In Chapter 4 we discussed another insidious consequence of starvation: PCM, *protein-calorie malnutrition.* The early stages of this gross undernutrition show as a wasted body and severe diarrhea. This condition is termed *nutritional marasmus;* later stages that affect the nervous system are called *kwashiorkor.* Although marasmus and kwashiorkor are unfamiliar terms to most of us, they constitute the most common childhood diseases found in many parts of the world. As an indication of the extent and seriousness of these diseases, it is estimated conservatively that 10 million infants and children in the world suffer from PCM and millions more hover on the verge of it.

Kwashiorkor is directly linked to development. If a fetus in its last trimester, or an infant during the first two years after birth, is forced to live on a high-carbohydrate, low-protein diet, its rapidly growing nervous system cannot develop properly. This damage appears to be permanent and dooms the child to a life of subnormal mental activity. Symptoms of this malnutrition effect, which usually appear in the second or third year of life, include a smaller body and head size as compared with children of the same age, and swollen body tissues (edema). These children are also socially retarded.

The remedy for PCM seems obvious, if not possible: Provide adequate nutrition for the world's hungry children, especially during the first two years of their lives. However, a number of recent studies indicate that there are additional factors that can produce symptoms similar to PCM, that is, retarded mental, behavioral, and physical growth. These factors include infectious diseases, inferior living conditions, and the absence of culturally stimulating experiences. A number of studies on humans and laboratory animals indicate that a stimulation-deprived infant may show retarded development even though its diet is adequate. All of this indicates pretty strongly that normal development of the nervous system at a particularly crucial time period of life requires a stimulating, hospitable environment in conjunction with an adequate diet. Recognizing the importance of this concept, the United States provided the Head Start Program some years ago for young children from culturally deprived homes. The success of this program in helping these children is just now becoming evident.

Basic Concept

The newborn infant's body proportions are different from those of an adult. Adult body size and shape are attained in about 21 years.

THE HUMAN LIFE CYCLE

Human beings follow a life-cycle pattern essentially like that of all sexually reproducing animals: fertilization, development of the zygote, attainment of sexual maturity, senescence, and death. We strive vigorously to defeat or at least deter the latter two stages, but most of our efforts to date are in vain. If we plot on a graph the percent of survivors in a human population against their age in years, the result is a *survivorship curve* (or a dying off curve, if you prefer) (Figure 14-16). The graph varies somewhat with different groups of people and at different times in the past, but overall the graph

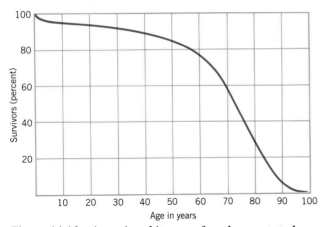

Figure 14-16 *A survivorship curve for a human population. Can you interpret the meaning of this graph in your own words?*

says the same thing. People are born, most of them survive to an age of 60 or 70 years and then they die rather rapidly. Many mammals have similar survival curves but it is by no means universal for the animal kingdom. Birds, for example, show an almost diagonal curve, and marine invertebrates such as oysters have an L-shaped curve, at least under natural conditions. One conclusion from a study of survivorship curves is that different groups of organisms have typical age-mortality patterns, suggesting that their life spans are genetically determined.

The human life cycle has a number of fairly distinct stages (Figure 14-17). One of these embraces the period from conception to birth. Many of its events are described in this chapter. Another stage involves infancy, the first two years after birth when dramatic changes occur in body proportions (briefly described earlier) and in behavioral development.

Following infancy is a long period, compared with other organisms, of childhood development that lasts until the onset of puberty or adolescence (about 10 years). Accompanying the overall growth in size and changes in bodily functions (physiology) are changes in the child's psychosocial development. The experiences of childhood are considered to have a profound influence on the personality development of the adult.

Adolescence is a period in life that involves numerous changes in the body associated with attainment of sexual maturity and ability to reproduce. Among many

alterations are development of the secondary sex characteristics and thus the sexual dimorphism characteristic of our species. For each individual it is often a somewhat stressful time of making numerous adjustments to social and emotional needs, achieving a stable self image, and making vocational plans for the future.

The next stage, so-called adulthood, is not as clearly defined biologically as some of the earlier stages. Generally it embraces a period of the midtwenties to the midsixties, but the dividing lines are indistinct. Aging, the slow decline of many body systems occurs through most of the stage, accelerating from the fifties on. An almost overwhelming number of different experiences occur: vocational achievement, marriage, children, seeking fulfillment, coping with disappointments and tragedies, dealing with aging and dying parents; also, realizing vocational success, achieving mature relationships with individuals around you, experiencing love and fulfillment, feeling a sense of awareness for the present, and sometimes just being glad that you are alive. The forties and fifties are decades when neglect of health care in earlier years may begin to take its toll in terms of heart disease, lung cancer, consequences of alcoholism, and severe dental problems. This is reflected in the most common causes of death in middle age: cardiovascular diseases, cancer, and hypertension.

The final stage of life involving the many processes we associate with advanced aging is not clearly marked by any sudden biological change in the body, in fact, aging is an ongoing process throughout our entire lives. We eventually die not of old age but of some disease or disorder that our bodies can no longer cope with, usually between the ages of 65 and 80. The study of aging, (gerontology) has documented the changes that occur in the body with aging. Especially important are changes in the skeleton (increase in brittleness of bones and a decrease in bone mass), decrease in muscle mass and consequent decline in muscle strength, and a deterioration of the nervous system, resulting in reflexes that are slowed and sensory perceptions that are impaired. Table 14-3 lists some of the organ functions that are diminished. In addition to changes in organs, the body's homeostatic systems become less able to handle stress or emergency situations. This is reflected by impaired regulation of body temperature, blood sugar, pulse rate, blood pressure, and kidney function to name a few. As the body of an aging person notably

Figure 14-17   *Three of the stages of the human life cycle: childhood, young adulthood, and old age.*

TABLE 14-3.   Changes in the Body Associated With Aging

Type of Change	Remaining Functions of Tissues in percent*
Brain weight	56
Cardiac output at rest	70
Number of glomeruli in kidney	56
Nerve conduction velocity	90
Number of taste buds	36
Maximum oxygen uptake during exercise	40
Vital capacity of lungs	56
Hand grip	55
Basal metabolic rate	84
Body weight	88

Source: Data from *The Physiology of Aging* by Nathan W. Shock, *Scientific American,* January 1962.

* Approximate percentages of functions or tissues remaining in an average 75-year-old male, taking the values found in an average 30-year-old male as 100 percent.

shrinks in size (muscle mass), we note the increasing graying of hair as melanocytes fail to produce pigment for the hair, and that the skin loses its elasticity and thus becomes wrinkled and creased.

There are many theories of aging, but virtually all fall into the category of being hypotheses with a minimum of supporting evidence. One viewpoint is that aging occurs as the body's organ systems become less efficient. Thus failures in the immune system, hormonal system, and nervous system could all produce characteristics that we associate with aging. Many current researchers are looking for evidence at the cellular and subcellular level. It has been shown that cells such as human fibroblasts (generalized tissue cells) grown in culture divide only a limited number of times and then die. (Only cancer cells seem immortal in this respect.) Fibroblast cells from an embryo divide more times than those taken from an adult. Thus some researchers believe that aging occurs at the cellular level and is part of a cell's genetic makeup. Any event that disturbs the cells genetic machinery such as mutation, damaging chemicals in the cell's environment, or loss of genetic material could cause cells to lose their ability to divide and thus bring on aging.

It should be emphasized that these and many additional changes which occur with aging should be looked upon by society as naturally occurring events and not as detrimental features or ailments that detract from an individual's worth to society. An impressive number of the eminent people have reached their peak of productivity in their supposedly "declining" years. In a youth-oriented society like ours, the elderly are all too often pushed into the background and ignored even though we could often utilize their depth of experience and wisdom. Medical science has not increased our longevity significantly for many decades, but it has enabled many more people to attain their full life span. The number of elderly is increasing demonstrably and in time they will outnumber the youthful segment of our population. The social and economic effects of this population change are already being felt and will undoubtedly bring about considerable social readjustments in the future. Perhaps the time is near at hand for us to redirect our energies from the elusive search for the fountain of youth and strive instead to find ways to improve on the quality of life for our citizens.

Basic Concept
The human life cycle consists of the stages of fetal development, infancy, childhood, adulthood, and old age.

SOME RECENT ADVANCES

In vitro Fertilization

It is a simple procedure to obtain eggs and sperm from a starfish, place them together on a microscope slide, and watch fertilization take place. Subsequent stages of development can also be observed. Such external fertilization and development is the normal course of events for many marine invertebrates and for vertebrates such as fishes and amphibians.

To observe these same events in a mammalian egg outside the body is not simple. Mammalian eggs are small, fragile, and highly sensitive to the chemical makeup of their surroundings. Special techniques must be used for obtaining the eggs and keeping them alive in vitro (i.e., in glassware or other containers) in a special culture medium. In addition, the sperm for use in in vitro fertilization must undergo capacitation (see page 223) before they can penetrate the eggs. This is accomplished in the laboratory by exposing the sperm to uterine fluids or blood serum. In the usual procedure, the female mouse, hamster, rabbit, or rat is

injected with gonadotropic hormones to cause multiple ovulation. The reproductive tract is removed from the female and the eggs are then washed out and placed in a specially prepared culture medium. Sperm are obtained, capacitated, and placed with the eggs. Fertilization can then be observed and studied in vitro.

In these cases, development usually goes as far as the blastocyst stage and then stops. However, some mouse embryos have been carried through almost half of their 21-day gestation period. It is also possible to transfer the externally fertilized, developing eggs of a variety of animals back into the mother's womb to complete their development. This procedure is called *artificial inovulation.*

The point of all of these experiments is to learn more about the processes of fertilization and development for their application to human biology and animal husbandry.

In 1978 the birth of Louise Brown provided evidence for the first successful in vitro fertilization and artificial inovulation in a human being. This notable biological event took place in England under the direction of Dr. Patrick Steptoe, a gynecologist, and Dr. Robert Edwards, a physiologist. The procedure involved using HCG (human chorionic gonadotropin) hormone to induce ovulation in the mother. A laparoscope was used to remove eggs from the ovary just prior to ovulation. The eggs were placed with capacitated sperm from the husband. A few days later a blastocyst was transferred into the uterus where it implanted and grew. The press called this event the first successful "test tube baby" a misleading designation, because only fertilization and cleavage took place in glassware. Since 1978, 19 additional artificial inovulations have been achieved (mostly in Australia) although the failure rate for achieving a successful pregnancy by this method remains high. Even so, it does offer a possible solution for women who are sterile because of defective Fallopian tubes but who have normal ovaries and uteri and wish to bear children.

The procedure has also been questioned on bioethical grounds because it involves manipulation of human reproductive cells. Perhaps a reasonable view is that, in reality, the technique involves cells (the blastocyst) and not persons or even pre-embryos. Therefore the rights of personhood are not endangered. What do you think?

Embryo Transfer

The technique of transferring embryos from one female animal to another is beginning to have wide application in animal husbandry, especially cattle breeding. In this technique, a cow is treated with gonadotropic hormones to make her ovulate numerous eggs (a process called superovulation). These are fertilized in the uterus, often by artificial insemination with sperm from prize bulls. The early stage embryos are flushed out of the uterus with saline and placed in surrogate cow mothers. The embryos can also be frozen and stored for later use or for shipping to cattle ranches in other parts of the country. More than 17,000 cow pregnancies per year are brought about in North America by the embryo transfer technique. It has also been accomplished in horses, sheep, dogs, cats, and one primate, a baboon. A unique application was the use of a Holstein cow in the Bronx Zoo to deliver a rare Indian quar (a type of wild ox).

Cloning

In the early 1960s, a plant physiologist, F. C. Steward of Cornell University, performed an experiment with plant cells that had rather profound implications for developmental biology. He was studying the growth of carrot root cells in special culture media. Normally the cells would multiply and grow into a mass or clump of unspecialized cells. But in one particular experiment, coconut milk was added as a nutrient to test its effects on cell growth. The tested cells did indeed grow rapidly but, more surprisingly, some of them subsequently developed into tiny carrot plants. The coconut milk had somehow allowed the DNA in the carrot cell nucleus to express its full potential of growing into an entire carrot plant. This experiment reinforced a general biological principle: Every cell of an organism has the same genetic message in it and therefore, theoretically, has the capability of growing into another organism. All the individuals developed from the body cells of one parent, in this case a carrot, are termed a *clone*. All members of a clone are identical to the parent and, of course, to each other.

Biologists working with animal cells attempted to find the same kind of "magic" agent for animal cells but were unsuccessful. However, a developmental biologist, J. B. Gurdon, and his associates at Oxford University, accomplished a type of cloning by a different method. Using special microtechniques, nuclei were removed from intestinal cells of tadpoles and placed in unfertilized frog eggs whose own nuclei has been destroyed with ultraviolet light (Figure 14-18). The number of these eggs that successfully developed into tadpoles and frogs was very small (1.5 percent); nevertheless, the method worked. Gurdon's experiments have been repeated using other kinds of body cells and with similar results: that is, a set of frogs identical to the parent frog, a clone of frogs. This again supports the concept that body cells of an organism contain the information to form a complete organism if the genes can, in effect, be turned on.

Two conclusions can be drawn from Gurdon's work. The first is that even after a cell has differentiated, its genes are not permanently inactivated. The second is that development takes place only if egg cytoplasm is present. A logical hypothesis is that something in the egg cytoplasm activates the genes that in turn control development of the organism. This substance is still being sought.

Although there is a huge biological difference between a frog and a human being, Gurdon's work inspired considerable speculation in the news media about the possibility of cloning humans. If this could be done, identical genetic copies could be obtained of individuals deemed outstanding by society. Conversely, an unscrupulous leader might have clones made of himself for posterity.

The prospect of cloning human beings or any other mammal seems enormously difficult because of the small size of their eggs and their complex growth requirements. Nevertheless, significant progress was achieved in 1981 at the University of Geneva. Researchers succeeded in transplanting nuclei from embryonic mouse cells into fertilized mouse eggs whose own nuclei had been removed. The eggs were then transplanted into an unrelated surrogate mouse mother. Only three "cloned" mice were obtained in this manner, and each of these came from a different embryo, but at least this experiment and subsequent ones show that multiple copy cloning is possible in mammals. The bioethical implications of performing this procedure with human cells are numerous to say the least.

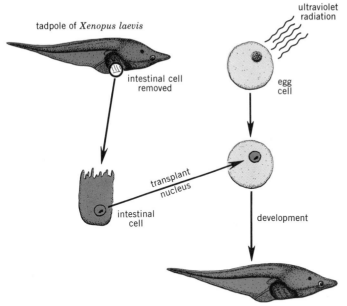

tadpole of *Xenopus laevis*

intestinal cell removed

ultraviolet radiation

egg cell

transplant nucleus

intestinal cell

development

Figure 14-18 *Cloning of the frog,* Xenopus laevis.

KEY TERMS

afterbirth	cardiac tube	fetus	neural groove
allantois	chorion	gastrulation	neural plate
amniocentesis	cleavage	germ layers	neural tube
amnion	cloning	in vitro fertilization	PCM
artificial inovulation	differentiation	inner cell mass	placenta
atrium	dizygotic twins	kwashiorkor	sinus venosus
blastocyst	ectoderm	labor	survivorship curve
blastula	endoderm	mesoderm	yolk sac
breech birth	fertilization	monozygotic twins	

SELF-TEST QUESTIONS

1. Define fertilization and discuss its significance.
2. (a) Describe the events of cleavage.
   (b) What determines the kind of cleavage seen in different types of eggs?
3. Define gastrulation and tell how it takes place in the blastocyst.
4. List several derivatives of each germinal layer.

5. What are some important factors that cause differentiation in the embryo?

6. List the fetal membranes and give a function of each.

7. Describe how the placenta is formed and its role in fetal life.

8. (a) What is the first organ system to develop?
   (b) Describe how it forms.

9. Describe how the circulatory system and heart form in the embryo.

10. Tell two major developmental events that occur in each of the three trimesters of fetal development.

11. What happens during each of the three stages of labor?

12. Describe the difference between dizygotic and monozygotic twins.

13. Give examples to show how an infant's body proportions are quite different from those of an adult.

14. Describe the cause and effects of kwashiorkor.

15. What is meant by a survivorship curve?

16. How does the human life span look on a survivorship curve?

17. List five general stages in the life cycle of a human being.

18. Describe some of the major effects of aging on the body.

19. Discuss some of the major theories of aging.

20. Define what is meant by in vitro fertilization, embryo transfer, and cloning.

chapter 15

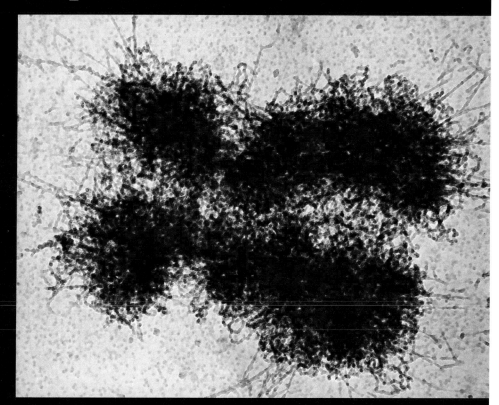

Two human chromosomes in metaphase (37,500×).

GENETICS PART I: WHAT ARE GENES?

chapter 15

questions to think about

How do we know that genes are made of DNA?

How is information coded into DNA?

How does DNA direct cellular activities?

What is the relationship of DNA to chromosomes?

How is gene splicing accomplished?

The idea of the genes' being immortal smelled right, and so on the wall above my desk I taped up a paper sheet saying DNA→RNA→protein.

James D. Watson

Genetics, the study of heredity, is a relatively new scientific field. Most of our knowledge about this area of biology has accumulated since 1900, and knowledge of the gene itself is quite recent. The "father of genetics" is an Austrian monk, Gregor Mendel, who lived in the mid 1800s. As a result of experimenting with garden peas, he derived some of the basic concepts about how traits are passed from parents to offspring. The significance of these concepts was not realized until long after his demise. The discovery of his publications by geneticists in 1900 marked the beginning of the field of modern genetics.

This chapter begins with a consideration of the gene because genes are the basic units of inheritance. With a knowledge of what genes are and how they work, we can better understand the concepts presented in the next chapter on inheritance in human beings.

THE GENETIC BLUEPRINT—DNA

In Chapter 14 we described how the development of an organism proceeds from a simple undifferentiated

zygote to a complex organism possibly consisting of trillions of cells. Utilizing just a few basic building materials such as organic molecules and minerals, the developmental events unfold as though following a blueprint or plan of some sort. And, indeed, there is a molecular blueprint, in the form of a compound called *deoxyribose nucleic acid* (*DNA*). Later in this chapter we will learn how the structure of DNA allows it to function as the blueprint of the organism as well as the controller of all cellular activities.

DNA has yet another indispensible function: It is the hereditary material passed from one generation to the next—the bridge of biological immortality. This is a truly remarkable achievement when you consider that the entire genetic heritage passed from a parent to its offspring consists of a microscopic dab of DNA packaged into either an egg or a sperm.

THE GENE CONCEPT

Prior to our knowledge about the vital roles of DNA, geneticists referred to units of heredity as *genes*. From the results of their many experiments with laboratory organisms such as fruit flies and plants, biologists learned a lot about the transmission of traits (characteristics) from an organism to its offspring. They were sure that genes were distinct units of some sort. They also had evidence that genes usually occurred in pairs, that they were distributed along *chromosomes* somewhat like beads along a string, and that their inheritance followed predictable patterns and ratios in most cases.

Because hereditary factors (genes) were known to be located on chromosomes, it was obvious that genes had to do whatever chromosomes did—as in mitosis, for example. Also, the duplication of chromosomes implied that the genes found on them must also be duplicated and parceled out into the daughter cells. Moreover, all of the genes located on the same chromosome would necessarily move with it as a unit. In meiosis, chromosomes exchange portions of their chromatids and accompanying genes with each other (crossing over). Knowing about this event made it possible to construct chromosome maps showing the approximate locations of many genes, as we shall see in the next chapter. It also became known that exposure of organisms to radiation, such as X ray or ultraviolet light, caused inheritable changes in some of their genes. These genetic changes were termed *mutations*. In short, a great many events that are important in heredity became known. However, the exact nature of the hereditary material itself remained in doubt for many years.

Basic Concepts

Genes, as hereditary units, have the following features:

They are discrete units.

They usually occur in pairs.

They are located on chromosomes.

Their locations on chromosomes can be determined by crossing-over experiments.

Their transmission follows predictable patterns.

Identifying the Gene

After 1920, many researchers attempted to identify the chemical nature of the elusive gene. From what was already known about heredity, geneticists reasoned that genes had to meet the following qualifications:

1. A gene must contain a chemical code of some sort, because organisms are extremely complex and a code is the simplest way to pack a lot of information into a limited space;

2. A gene must be able to duplicate itself precisely so that it can be passed along to daughter cells and still carry out the same functions;

3. A gene must be able to control activities out in the cytoplasm of a cell, even though the gene itself is probably confined to the nucleus.

When researchers analyzed nuclear material, they found two substances that might meet all of these criteria: proteins and DNA. These two substances together form the chromosomes. Of these two, proteins seemed at first to be the most likely candidate. For one thing it was known that they were made of numerous amino acid subunits, and the order of the subunits might function as some kind of code. Proteins are made throughout the cell, solving the problem of how they are duplicated. Finally, the enzymes that control all major chemical actions in cells are made of proteins. DNA seemed less likely at the time to be the genetic material, because what little was known about its chemical structure made it seem too simple to carry out the complex actions required of a gene. Consequently, much more of the early research effort was devoted to the study of proteins than to the study of nucleic acids.

Not until the early 1950s did enough evidence accumulate to persuade biologists to the opposite viewpoint, and even then there were many doubters. The doubt was ended in 1953 when an Englishman, Francis Crick, and an American, James Watson, pooled their research efforts and inspiration to work out the chemical structure of DNA. The model they proposed—the DNA molecule—immediately showed how information could be coded into it. This discovery was a tremendous breakthrough, for which they received a Nobel Prize in 1962.

Basic Concepts

A gene must meet three qualifications:

1. **It must contain a chemical code.**

2. **It must be capable of precise self-duplication.**

3. **It must be capable of controlling all cellular activities.**

DNA as the Hereditary Material

The following list is only a sampling of the many kinds of evidence supporting DNA as hereditary material. Some of the evidence, such as described in the first item, sounds simple, and you may wonder why it was not discovered long ago. The answer is that most of these findings are based on sophisticated laboratory techniques that were only recently developed. For example, how does one determine how much DNA is contained in the nucleus of a cell or in a gamete? The answer was not possible until an unusual optical measuring technique was developed. Most advances in biology, and in other sciences, depend heavily on finding improved ways to answer such basic questions.

1. All cells of an organism contain the same amount of DNA except the gametes, which contain one-half as much. This consistency is what one would expect of hereditary material that is passed from cell to cell via mitosis or meiosis. By contrast, the amount of protein in cells is *not* constant within an organism, and the amount of protein in gametes is not one-half that found in the body cells.

This does not mean to imply that all organisms contain the same amounts of DNA in their cells. Different kinds of organisms often have different numbers of chromosomes, and therefore different amounts of nuclear material. Nevertheless, the statement evidently holds true for virtually all the cells in a particular animal or plant, and for all the members of the species to which it belongs.

2. When DNA is tagged with a radioactive isotope for recognition purposes, it is found that this radioactive label stays with the DNA throughout the life of the cell. This shows that DNA is a highly stable substance, an important quality of genetic material. In contrast, other large molecules such as proteins, tagged in a similar manner, soon lose their labels, indicating that they are being continually broken down and resynthesized.

3. It has long been known that ultraviolet light can cause mutations—hereditary changes—in cells, presumably because it specifically affects the genetic material in the cell. When techniques became available for extracting DNA from cells, and purifying it, investigators found that DNA strongly absorbed light in the wavelength band of ultraviolet light (Figure 15-1). Moreover, exposure of cells such as bacteria to this same wavelength greatly increased their mutation rate. Other wavelengths did not produce this effect. Proteins have absorption spectra that are different

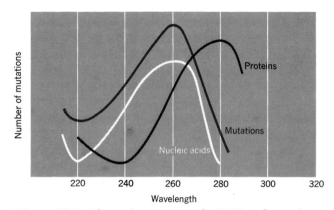

Figure 15-1 *Absorption spectrum for DNA and proteins. Mutation-causing ultraviolet light is absorbed by DNA but not nearly so much by proteins.*

from that of DNA. Furthermore, the light waves that proteins absorb the most do not increase the mutation rate in cells significantly.

Additional experiments have shown that ultraviolet radiation influences many processes of development, presumably because it alters the DNA in the developing zygote.

4. In the late 1920s, a bacteriologist named Fred Griffith found that freshly killed bacteria of one strain could cause a major change in the genetic characteristics of a related living strain merely by being mixed with it in the same culture medium. It strongly appeared that a substance from the bodies of the dead bacteria was being absorbed by the living forms; in some manner this substance changed the genes in the hosts (Figure 15-2). Griffith thought that the substance, which he called a "transforming factor," might be a protein, but he was unsuccessful in analyzing it. In the early 1940s a team of researchers consisting of A. T. Avery, C. M. MacLeod, and M. McCarty realized the significance of Griffith's work and set out to discover the nature of the transforming substance. One of their major experiments involved purifying the transforming substance from the killed bacteria until they had an extract that had the same transforming qualities as the killed bacteria themselves. When this purified transforming substance was mixed with an enzyme that destroys DNA (a DNAase), it lost its activity. On the other hand, when it was treated with a *proteinase,* its ability to transform was not affected. Therefore, the active agent in their purified extract

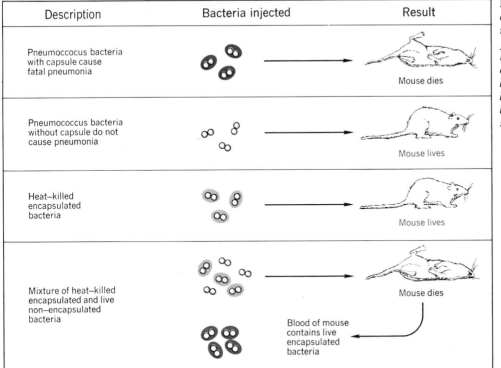

Description	Bacteria injected	Result
Pneumoccocus bacteria with capsule cause fatal pneumonia		Mouse dies
Pneumococcus bacteria without capsule do not cause pneumonia		Mouse lives
Heat–killed encapsulated bacteria		Mouse lives
Mixture of heat–killed encapsulated and live non–encapsulated bacteria	Blood of mouse contains live encapsulated bacteria	Mouse dies

Figure 15-2 *Fred Griffith's experiments with pneumonia-causing bacteria. He found that a substance from killed virulent bacteria was capable of converting harmless bacteria into the virulent form. About 20 years later this substance was shown to be DNA.*

was DNA. Additional experiments showed that purified protein extracts had no transforming qualities. Thus the experiments of Avery, MacLeod, and McCarty along with those of a number of subsequent researchers added further proof that DNA and genes are synonymous.

It is ironic—but in some ways so typical of scientific advances—that the actual discovery of DNA in the nucleus of cells came long before its importance was appreciated. A German physiologist, Friedrich Miescher, isolated a substance from the nuclei of pus cells in 1869, calling it "nuclein." He thought that nuclein must be important to the cell, but many years of study failed to reveal its function. In the later 1800s a group of cell biologists even suggested that nuclein was associated with the transmission of hereditary traits, but no one took the idea seriously. This significant potential advance in biological knowledge was destined to lie dormant for more than 50 years.

Basic Concepts

A large body of evidence indicates that hereditary material

is composed of DNA. Among this evidence are the following:

1. **The same amount of DNA occurs in all of the cells of an organism; the gametes contain half this quantity.**

2. **DNA is a highly stable molecule.**

3. **DNA strongly absorbs ultraviolet light, which is known to cause mutations.**

4. **Purified DNA extracts from one strain of bacteria will cause a major change in the genetic characteristics of a related strain.**

STRUCTURE AND FUNCTION OF DNA

It is necessary to understand the structure of DNA, at least very generally, in order to appreciate how it functions. To begin with, DNA is an extremely large, long molecule—the largest molecule in living matter, in fact. Like proteins and large carbohydrates, it is a *polymer*—that is, it is made up of a relatively few kinds of chemical units occurring over and over.

(a)

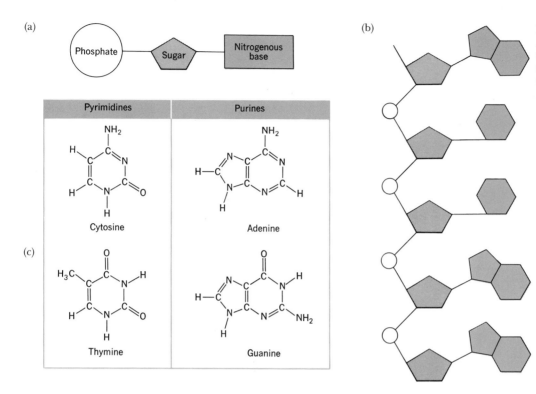

(c)

(b)

Figure 15-3 *The general structure of a nucleotide (a), a chain of nucleotides (b), and four organic bases (c).*

In DNA the repeating units are *nucleotides* arranged in two long intertwined chains. Each nucleotide consists of three parts: an organic base, a sugar (deoxyribose), and a phosphate group (Figure 15-3*a*). A bond forms between the phosphate group of one nucleotide and the sugar of the next nucleotide. The phosphate of this second nucleotide is bonded to the sugar of the third, and so on, to form a chain (Figure 15-3*b*).

There are four different organic bases that may occur in a nucleotide: *guanine, thymine, adenine,* and *cytosine* (Figure 15-3*c*). These bases extending out from one nucleotide chain are bonded by weak bonds (*hydrogen bonds*) to the bases of the other chain. The bonds form only between specific pairs of bases: adenine in one chain always bonds with thymine in the other chain, and cytosine is always bonded to guanine. Because of the specific pairing of bases, the sequence of bases in one chain defines the sequence of bases in the other. Therefore, the two chains are *complements* of each other.

It is the sequence of bases in the nucleotide chain

that constitutes the chemical code of life. It has been found to be a *triplet code* in that a sequence of three consecutive bases functions like an individual letter in an alphabet. In this chemical code, the letters represent amino acids. Because there are four different bases, there are $4 \times 4 \times 4$, or 64 different combinations (triplets) that can be formed. These combinations are more than sufficient to code a huge amount of information into a DNA molecule. It is thought that a sequence of 900–1500 base pairs constitutes the information unit we call a gene. These are the units that are passed from one generation to another and also function in directing activities within cells.

The two nucleotide chains that make up a DNA molecule are twisted around each other in precise symmetrical coils, forming a double helix. To help visualize this, DNA can be compared with a long, twisted ladder. The rungs of the ladder are the nitrogeneous base pairs, adenine—thymine and guanine—cytosine. The twisted legs or uprights of the ladder are the two sugar-phosphate chains (Figure 15-4).

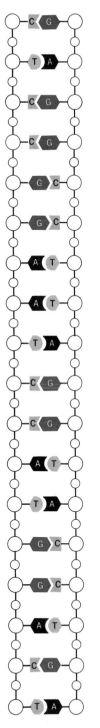

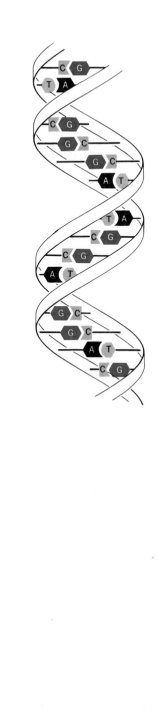

Figure 15-4 *Two chains of nucleotides on the left shown twisted into a double helix on the right. Note that cytosine (C) always bonds with guanine (G) and that thymine (T) bonds only with adenine (A).*

Basic Concepts

DNA is a long polymer made of two chains of nucleotides wrapped about each other in a double helix.

The sequence of bases in the nucleotide chain constitutes the chemical code of life. It is a triplet code.

A gene is a sequence of organic bases in a DNA molecule.

How New DNA Is Made: DNA → DNA

If DNA is to function as genetic material, it must have a mechanism for precisely duplicating itself. The *Watson-Crick model* for DNA structure indicates how this is possible.

At the time of duplication, the hydrogen bonds that hold the base pairs together must be broken so that the two intertwined nucleotide chains can separate. This is accomplished with an appropriate enzyme. Its action is somewhat like the slide on a zipper which can either interlock the two rows of teeth of the zipper or else separate them. As a result, the two chains of the helix are separated but each retains its sequence of attached nitrogen bases. Each chain is now available to act as a pattern or template for making a new and complementary chain.

To do this the nucleotide units of each chain begin to link up with complementary nucleotide units, from a "pool" that exists in the nucleus. As these nucleotides become associated with the ones in the preexisting chain, they are joined to each other by sugar-to-phosphate bonds, thereby forming a new chain with a sugar-phosphate backbone (Figure 15-5). It is important to note that *each* of the old nucleotide chains serves as a pattern for the production of a new complementary chain. Each of the old strands becomes one-half of the new double helix. This process proceeds along the entire length of the molecule until there are ultimately two copies of the original double helix (Figure 15-5).

As you undoubtedly suspect, this explanation is greatly simplified. In reality, the actual chemical events involve a series of reactions and at least seven different enzymes. Some of the steps in the reaction are still being studied. However, the major events in DNA duplication seem to be worked out, because a DNA strand can now be made in a test tube by mixing the precursor units and proper enzymes. The process by

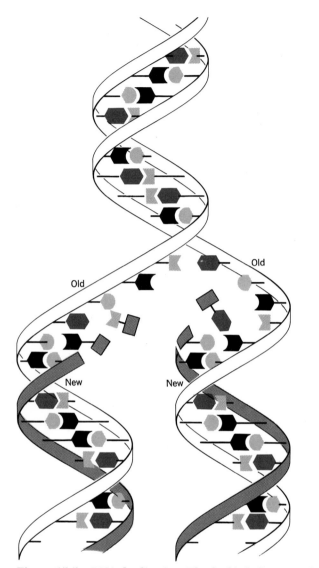

Figure 15-5 *DNA duplication. The double helix unwinds and each of the old nucleotide chains serves as a pattern for the formation of a new and complementary chain as shown.*

which DNA is duplicated is customarily referred to as *replication* (because an exact replica is made) rather than duplication, although the two terms are synonymous.

Following replication, the copies of DNA may be parceled out into two daughter cells through mitosis. In ovaries and testes, the similar but more complex events of meiosis also assure that each of the gametes receives identical amounts of the replicated DNA. The precision of the DNA copying mechanism and the exactness of its distribution into daughter cells are remarkable.

At this point we have seen how DNA meets two of the requirements of a gene: It contains a chemical code, and it can be duplicated precisely. In the next section we will see how DNA controls activities throughout the cell even while remaining inside the nucleus.

Basic Concept

In DNA replication, each nucleotide chain of the double helix serves as a pattern for the formation of a new complementary chain. Each of the original chains becomes one-half of the new DNA molecule.

How DNA Directs Cellular Activities

Although DNA is confined inside the nucleus, it nevertheless controls the making (synthesis) of proteins outside the nucleus. One vital category of proteins, the enzymes, directs all chemical reactions in the cytoplasm. Thus, indirectly, DNA controls practically every action undertaken by a cell. Numerous studies indicate how this control occurs.

Transcription: DNA → RNA. The first step is the transcription (copying) of a portion of the chemical code in a DNA molecule into ribonucleic acid, RNA. In order for this to occur, the two nucleotide chains of the DNA helix are separated along part of their lengths by enzyme action. A segment of one of the chains then serves as a pattern for making an RNA strand (Figure 15-6). The strand is made by matching organic bases (nucleotides) from a supply in the nucleus with complementary bases on the DNA chain. In this process, cytosine pairs with guanine, and thymine (in the DNA) pairs with adenine. Thus, if the DNA chain "reads" cytosine—thymine—guanine, the RNA bases line up to read guanine—adenine—cytosine. (For some reason, however, thymine does not occur in RNA but is replaced with a very similar base, *uracil.* Thus adenine on the DNA chain is matched by uracil on the RNA.) To complete this single-stranded RNA molecule, the nitrogen bases are linked by phos-

Figure 15-6 *Transcription of DNA into RNA. A portion of the double helix separates as shown and one segment serves as a template for making a strand of messenger RNA. The mRNA eventually moves out into the cytoplasm.*

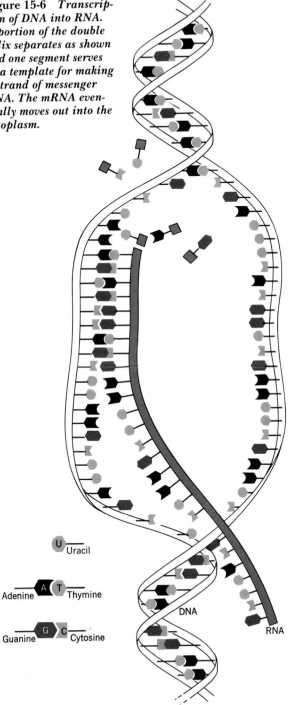

Uracil

Adenine — Thymine

Guanine — Cytosine

DNA

RNA

phate-to-sugar bonds. The result is a product called the primary RNA transcript. For reasons not yet known, this RNA chain of nucleotides contains sequences of triplet codes that appear to be nonfunctional. Some geneticists call them genetic gibberish or "junk" sequences for this reason. These nonsense triplets are removed (edited out) by enzymes to leave a shorter strand of RNA called messenger RNA (mRNA). The mRNA then moves out through a pore in the nuclear membrane into the cytoplasm.

Some geneticists speculate that the junk sequences may be a form of chemical communication between genes. Others believe that the editing process is necessary for stabilizing the RNA before it moves from the nucleus into the cytoplasm. All agree that further studies are needed to understand the function of this interesting step in the transcription of DNA into mRNA.

Basic Concept

DNA is transcribed into mRNA, which in turn exits from the nucleus and carries its copy of the DNA triplet code into the cytoplasm.

Translation: RNA → Protein. In the cytoplasm, the mRNA attaches to one or more ribosomes, where it serves as a pattern on which a protein can be put together from amino acid building blocks. This step is referred to as the *translation* of the mRNA code into protein (Figure 15-7). Each triplet (a *codon*) in the mRNA specifies a particular amino acid. The job of obtaining these amino acids requires another type of RNA called *transfer RNA* (tRNA). Its function is to transport, or transfer, amino acids to the ribosomal sites where mRNA strands are located. Each tRNA combines at one of its free ends with a particular type of amino acid. Another portion of the tRNA has three consecutive organic bases (an *anticodon*) that determines the site on the mRNA at which the tRNA may attach (Figure 15-8). For example, a tRNA that transports the amino acid glycine has an anticodon of GGU (guanine—guanine—uracil). The GGU anticodon will match only with a *codon* of CCA (cytosine—cytosine—adenine) on the mRNA. In this way the base sequence on the original DNA molecule, by dictating the base sequence of the mRNA, indirectly determines the precise structure of the protein being built.

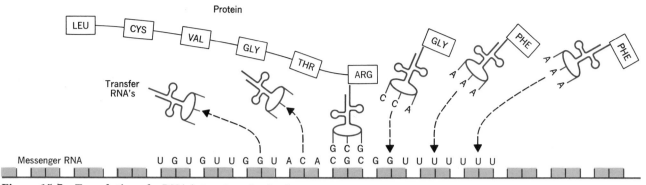

Protein

Figure 15-7 *Translation of mRNA into a protein. In the cytoplasm, mRNA attaches to one or more ribosomes and provides a pattern for building a protein. Transfer RNAs transport amino acids to the mRNA and match with an* *appropriate triplet code as shown. The amino acids gradually form a chain. The amino acid chain eventually becomes a protein.*

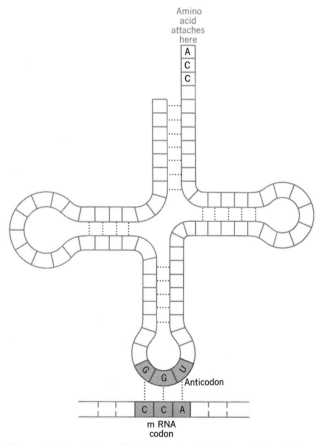

Figure 15-8 *A transfer RNA molecule. One end is coded for a specific amino acid as shown. The opposite end has another triplet code, an anticodon, that functions as a binding site with a complementary triplet, a codon, on a messenger RNA.*

Ribosomes also play a part in this translation process. The ribosomes seem to function in small groups that move along the mRNA, attaching the incoming tRNAs to appropriate sites on the mRNA. And as each tRNA pairs with its triplet on the mRNA, the amino acid it carries is attached to the amino acids of the forming protein. In this manner, the amino acid chain continues to grow longer.

When the amino acid chain (a *polypeptide chain*) is completed, the ribosomes leave the mRNA and free the chain, which in turn becomes a protein (Figure 15-9).

There are about 20 amino acids used in making proteins but over 64 possible triplet codons. Consequently, some of the amino acids are represented by more than one codon, and some codes serve as "punctuation marks" in the process. The deciphering of this genetic code, matching amino acids with their triplet codes, was an immense job. It was finally accomplished during the 1960s through the efforts of many biologists using numerous ingenious experimental techniques. Eventually the complete genetic code was worked out.

The events of transcription and translation occur in all cells, and they constitute the only known mechanism by which proteins are made. Hence, biologists often refer to the mechanism as the *DNA → RNA → protein synthesis dogma*. It seems strange to see the term *dogma* used in a field of knowledge such as science, which claims not to be bound by any unquestionable

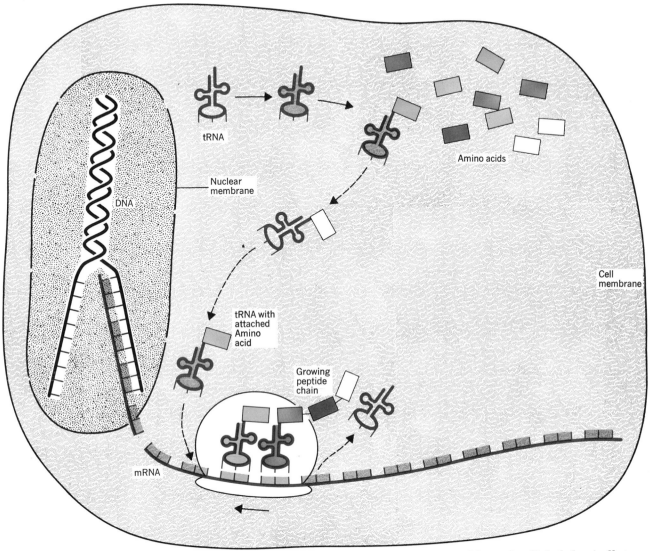

Figure 15-9   *Summary of protein synthesis as directed by DNA. Messenger RNA leaves the nucleus and attaches to ribosomes. tRNA molecules, carrying amino acids, match their anticodons with codons on the mRNA with the aid of ribosomes. The amino acids are then linked chemically to form peptide chains. Eventually the chain becomes long enough to become a protein.*

beliefs. Like "law," the term is chosen to emphasize the universality of the protein-making process.

Verification of the DNA → RNA → protein synthesis mechanism also confirms another hypothesis, which can be abbreviated as *one gene–one polypeptide*. This hypothesis means that each kind of polypeptide chain in a cell is made under the direction of a different gene (a different segment of the DNA code). This is a useful concept, because it provides a way of estimating the number of genes in a cell.

Now that we understand the events of replication, transcription, and translation, we can see that DNA fulfills all the requirements of the genetic material. The clarification of the structure of DNA and how it functions is considered by some scientists to be the major biological advance in this century.

Basic Concepts

Messenger RNA is translated into amino acid chains with the aid of transfer RNAs, which transport amino acids to the ribosomes attached to the mRNA strands.

The end product of the translation process becomes one of a variety of proteins used in the cell.

The events of transcription and translation are summarized as the DNA → RNA → protein synthesis dogma.

DNA AND CHROMOSOMES

The DNA in all eukaryotic cells is part of the chromatin complex, a tightly packed mass of threads in the nucleus. During cell division these threads coil to form the short rodlike bodies called chromosomes that are distributed, in turn, into daughter cells during cell division.

How DNA and the proteins fit together in a chromatin complex has been extensively studied, but it has proved to be a difficult problem to solve. The traditional theory, the *supercoil theory,* depicts the DNA molecules as being coated with proteins. It proposes that this complex is twisted into larger coils (supercoils) which make up what we call a chromatin thread (Figure 15-10a). A newer concept, the *beads-on-a-string theory,* proposes that instead of a supercoil, certain regions of the long DNA molecule associate with globular clusters of proteins (of a special class called histones), rather like beads along a string (Figure 15-10b). (The regions between the globules, or beads, contain the remainder of the DNA molecule coated with other kinds of proteins.)

Evidence for the beads-on-a-string theory is based on electron microscope studies that show chromatin in the form of thin fibers connecting successive globular particles (Figure 15-10c). The globules are resistant to the action of nucleases, enzymes that destroy naked DNA strands. In addition to protecting the DNA, the globular proteins (histones) are thought to play a role in the functioning of its contained DNA.

Basic Concept

DNA in eukaryotic cells forms a chromatin complex with nuclear proteins.

Chromosomes

The number of chromosomes in an organism is significant in the sense that each chromosome carries different sequences of nucleotides. An organism that lacks one of its usual chromosomes, or that has too many, shows gross abnormalities. One human example is the genetic disorder called mongolism or Down's syndrome, whose victims have an extra chromosome in all of their cells. They are mentally retarded and have various kinds of physical abnormalities.

Closely related species have similar numbers of chromosomes, and sometimes even the same number. Thus a horse and a jackass both have 33 pairs of chromosomes. Man's closest relative, the chimpanzee, has 24 pairs of chromosomes. These chromosomes are quite similar to the human's 23 pairs (Figure 15-11). Chimps and human beings have many similar proteins and undoubtedly share many genes. However, the total number of chromosomes or the total amount of DNA in organisms is not particularly meaningful on an evolutionary scale. Complex organisms may have fewer chromosomes than simpler ones. A salmon, for example, has 48 pairs of chromosomes, and an opossum has only 11 pairs.

Human beings have a characteristic number of 46 chromosomes. Twenty-two of these occur in pairs and are termed body chromosomes, or *autosomes.* The additional two are known as the *sex chromosomes.* The two sex chromosomes in females are paired and designated as X chromosomes. Males have one X chromosome paired with a slightly smaller one called the Y chromosome. (The designations *X* and *Y* do not refer to the shapes of these chromosomes but rather are simply convenient labels.)

Determining the number of chromosomes in a cell is not a simple matter because they are so small and often appear jumbled together. In fact, until about 20 years ago, it was thought that human cells had 48 chromosomes. Improved viewing and counting techniques finally established the correct number as 46.

A common method today for studying human chromosomes involves the use of lymphocyte white blood cells. These cells are grown in a special culture medium and induced to go into mitosis. Then, with the aid of colchicine (a mitotic poison), many of the cells are stopped at metaphase when the chromosomes are most spread out and most easily visible. Additional treat-

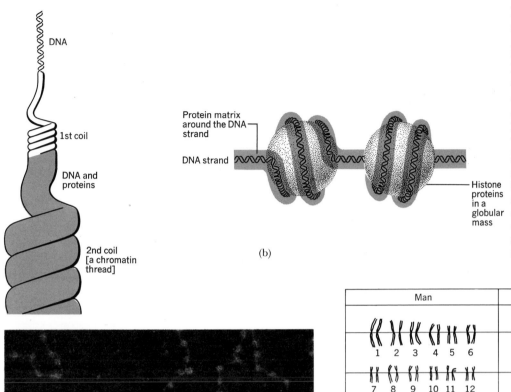

DNA

1st coil

DNA and
proteins

2nd coil
[a chromatin
thread]

(a)

Protein matrix
around the DNA
strand

DNA strand

Histone
proteins
in a
globular
mass

(b)

Figure 15-10 *Two models to show how DNA and proteins fit together in a chromosome.* (a) *The* supercoil hypothesis *suggests that the DNA and protein complex is wrapped into tight coils.* (b) *The* beads-on-a-string hypothesis *depicts DNA in a protein matrix wrapped around globules of another type of protein (histones).* (c) *Highly magnified chromatin fibers from the nucleus of a chicken red blood cell (×156,000). Globular particles show distinctly along the fibers.*

(c)

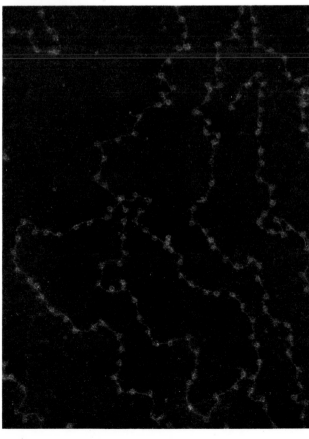

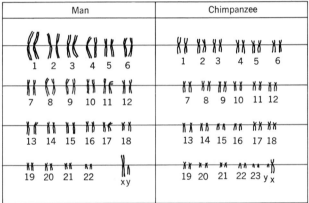

Man						Chimpanzee					
1	2	3	4	5	6	1	2	3	4	5	6
7	8	9	10	11	12	7	8	9	10	11	12
13	14	15	16	17	18	13	14	15	16	17	18
19	20	21	22		xy	19	20	21	22	23	y x

Figure 15-11 *Karyotypes from a male human and a chimpanzee to show their similarities. Do you see any differences in their chromosomes?*

ment makes the chromosomes swell and spread out even more. Then they are photographed, and the photo is enlarged so that the special features of each chromosome can be seen clearly. The chromosomes can then be cut out of each photo and matched up to form the human karyotype seen in Figure 15-12. This technique has been immensely useful in studying various chromosomal abnormalities and genetic diseases such as Down's syndrome (mongolism). The karyotype of a Down's syndrome victim always shows three chromosome 21s, the result of an abnormal meoisis in the gametes.

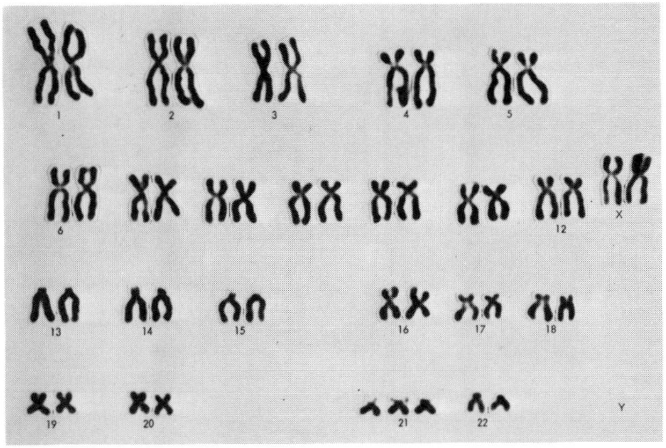

Figure 15-12 *A human karyotype from an individual who exhibits Down's syndrome. Note that there are three members of chromosome 21 instead of the usual two.*

Diploidy and Haploidy

It is of considerable genetic significance that chromosomes in eukaryotes occur in pairs (i.e., in the *diploid* condition). This means that all of the genes on the autosomes are present in duplicate. One significance of this arrangement is that if one of the genes in a pair changes (mutates), and becomes harmful, its effects are usually masked by the actions of its normal mate. For example, a number of people carry an inherited gene for albinism but are not albinos because they also carry the gene for normal skin pigmentation, and this latter gene is dominant in its influence over the one for albinism (Figure 15-13). In like manner, it is thought that everyone carries a small "load" of lethal (death-causing) genes, but we survive because each lethal gene is masked by its normal gene mate. Therefore the diploid condition has considerable survival value. Being diploid, however, requires special types of cell division such as mitosis and meiosis in order that the chromosomes can be distributed equally into daughter cells and gametes. Gametes of course are *haploid,* that is, have one of each kind of chromosome.

Prokaryotes (bacteria and blue-green algae) are not diploid and, in fact, do not even have eukaryotic-type chromosomes. Their DNA lies "naked" in the cell, and often consists of a single long DNA molecule in the form of a ring. Inheritance patterns are different and simpler than in eukaryotes, and genetic changes such as mutations show up immediately because their genes are not in pairs. For these reasons, geneticists have

Figure 15-13 *Examples of the trait of albinism. Albinism is recessive to normal pigmentation, hence an individual must inherit two albino genes in order to be an albino.*

focused a great deal of research in recent years on the heredity of prokaryotic organisms and on their genetic material. From this type of research came an understanding of the DNA code, a lot of knowledge about how viruses function, and techniques for making new kinds of bacteria that never existed before. All of this may seem pretty far removed from the genetics of diploid organisms, humans, for example. In reality this is not so. The DNA of both bacteria and mammals (or any form of life) uses the same nucleotides and the same triplet codes, and it codes for many of the same kinds of amino acids. It is difficult to imagine any stronger evidence of the relatedness of all life forms.

Basic Concepts

Related organisms usually have the same or similar numbers of chromosomes.

The human karyotype consists of 22 pairs of autosomes and two sex chromosomes.

Diploidy helps protect an organism from the effects of its mutant genes.

RECOMBINANT DNA—A BIOETHICAL PROBLEM

The considerable knowledge gained from studies on the genetics of prokaryotes led to an interesting and far-reaching biological controversy involving *gene splicing*. It is now possible to remove some DNA from an *Escherichia coli* bacterium (the common colon bacillus widely used in prokaryote studies), add genes to it from another organism, and to reinsert the recombined (*recombinant*) DNA into an *E. coli* host where it becomes functional. By this technique, microbial geneticists have the power to create new organisms. Such bacteria can serve as living "factories" for the synthesis of enzymes, hormones, and other chemicals of immense medical usefulness.

The technique for producing recombinant DNA is very generally as follows. *E. coli* contains tiny rings of DNA called plasmids in addition to its regular DNA (Figure 15-14). These plasmids can be removed from *E. coli* and exposed to an enzyme that cleaves the DNA ring at a particular place. In effect, a small segment of it can be removed. This segment can then be replaced with a gene segment from another source, and the ring is reconstructed. The plasmid with its new piece of DNA can then be placed back into an *E. coli* host (Figure 15-15). This is possible because bacteria tend to absorb DNA from their surroundings and incorporate it into their own genotypes—a phenomenon called *transformation*.

In practice, when a large population of *E. coli* bacteria is exposed to a set of recombinant plasmids, only a few *E. coli* will be transformed. Hence these bacteria must somehow be separated from the remainder of the population. This is done by using genes that give the transformed bacteria some advantage over the others. For example, if one of the things added to the plasmid is a gene that makes an organism resistant to a particular antibiotic, then the transformed bacteria can survive in a culture medium containing that antibiotic, whereas the nontransformed bacteria all die. In this

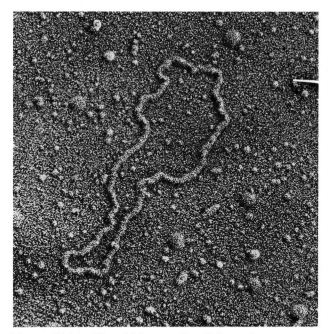

Figure 15-14 *Plasmid from an* E. coli *bacterium* (×100,000). *This tiny ring of DNA is used in gene splicing experiments.*

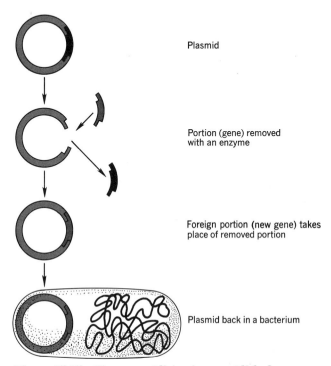

Plasmid

Portion (gene) removed with an enzyme

Foreign portion (new gene) takes place of removed portion

Plasmid back in a bacterium

Figure 15-15 *How gene splicing is accomplished.*

way, the experimenter can obtain a pure culture—a clone—of the transformed bacteria.

Experimenters have succeeded in adding DNA fragments from a variety of sources to *E. coli* plasmids, in fact, virtually any gene from any organism, including human genes, can be added.

As this technology developed in the early 1970s, a number of biologists and public officials became uneasy about the risks of this kind of experimentation. For example, suppose a geneticist created a virulent bacterium or a new kind of cancer-causing organism for which there was no treatment—a "doomsday bug" so to speak? Many scientists and nonscientists questioned the right of geneticists to experiment with such potentially dangerous materials without outside restraints. So great was the concern that there was a brief moratorium on this type of research in the United States. In 1975, agencies such as the National Institutes of Health and the World Health Organization published guidelines for safely conducting studies on gene splicing.

Recombinant DNA technology is now advancing so rapidly that "genetic engineering" is a reality, at least for designing bacteria that can make useful products such as insulin. An even greater goal is that of inserting genes into the cells of a higher organism in order to correct genetic defects. Progress toward this extremely important goal is proceeding rapidly as techniques are already available for inserting genes into mouse cells (in culture, at least) and thereby correcting certain defective genes. Whole organism gene therapy will likely be achieved in the near future. The ability to cure genetic diseases such as cystic fibrosis, diabetes, and sickle-cell anemia is awesome in its potential benefits for humanity.

KEY TERMS

adenine	gene	one gene–one polypeptide hypothesis	thymine
anticodon	gene splicing		transcription
autosome	genetics	polymer	transfer RNA
beads-on-a-string theory	guanine	proteinase	translation
chromosome	haploid	recombinant DNA	triplet code
codon	hydrogen bond	replication	uracil
cytosine	Mendel	RNA	Watson-Crick model
diploid	messenger RNA	sex chromosome	X chromosome
DNA	mutation	supercoil theory	Y chromosome
Down's syndrome	nucleotide		

SELF-TEST QUESTIONS

1. Who is called the "father of genetics"?
2. When did the field of modern genetics actually begin?
3. What are the two major functions of DNA?
4. Name four basic characteristics of genes.
5. Why were proteins considered at one time to be genetic material?

6. What information about genes did Watson and Crick provide?

7. Describe how each of the following indicates that DNA is hereditary material:
   (a) amount of DNA in body cells and in gametes
   (b) molecular stability
   (c) absorption of ultraviolet light
   (d) the transforming qualities of DNA

8. Why is DNA called a polymer?

9. Define a nucleotide including its basic units.

10. (a) Name the four organic bases in a nucleotide.
    (b) Explain what is meant by "base pairing."

11. How is information coded in a nucleotide?

12. Explain how a DNA molecule is somewhat comparable to a twisted ladder.

13. In modern terminology, what is a gene?

14. Explain how DNA is replicated.

15. Describe how DNA is transcribed into RNA.

16. What is the difference between primary RNA and messenger RNA?

17. Explain how mRNA is translated into a polypeptide chain.

18. What is meant by the one gene–one polypeptide hypothesis?

19. Compare the supercoil theory with the beads-on-a-string theory of chromosome structure.

20. How can two organisms have the same number of chromosomes and yet be different?

21. (a) What is the diploid number of human chromosomes?
    (b) What is the haploid number?
    (c) What is the advantage of being diploid rather than haploid?

22. Explain what is meant by gene splicing.

GENETICS PART II: GENES AND HUMAN HEREDITY

chapter 16

major topics

questions to think about

What are the sources of genetic variation?

What are some patterns of human heredity?

Where do new genes come from?

Is the human race becoming overburdened with mutant genes?

Nature never rhymes her children, nor makes two men alike.

Emerson

If you had to make a guess, how many genes would you think are necessary to specify a human being? One thousand? Ten thousand? A hundred thousand? In reality, no one knows, so your "guesstimate" might be about as valid as another person's. Some geneticists estimate that there are about 50,000 genes in a human cell. However, somewhat fewer than 1500 are actually known. The numbers of individual genes are not really very meaningful anyway, because geneticists have found that sequences of genes are repeated in the DNA of many organisms, undoubtedly in human cells also. The reason for this redundancy is not known.

Whatever the number, there are obviously enough genes not only to program the formation of the whole human animal from a fertilized egg but also to allow almost infinite variations among human beings, as the quotation from Emerson that opens this chapter implies. Emerson was a poet, not a geneticist, but he observed what you too can observe: Human beings show tremendous variation in their shapes, sizes, and colors (Figure 16-1). This fact raises a fundamental question: If all of us have 46 chromosomes and the same amount of hereditary material, how can we differ so much from one another? The answer is not simple, because hereditary variation is the consequence of many interacting factors. These begin with the events of meiosis as the sperm or egg cells are forming, especially crossing over and *independent assortment* (see Chapter 13).

Independent assortment has been compared with dealing cards from a well-shuffled deck: each sperm or egg gets the same number of "cards" (chromosomes) but the spots (genes) on the cards are not the same. This variation in genes results from mutations—occasional changes in the genetic material—a topic discussed later in this chapter.

An additional source of variation arises when a sperm and an egg unite. This recombination of genetic material adds to the pool of variations because no two eggs or sperm cells carry precisely the same genetic material. Hence, each recombination creates a unique individual. This is the reason that children of the same parents differ from each other. And finally, the expression of the genes in a developing zygote is always guided and influenced to varying degrees by the environment. "Nature" and "nurture" always function together in producing the mature organism (Figure 16-2).

A large part of this chapter is concerned with how genetic variations in parents are transmitted in their

Figure 16-1 *Human beings, like nearly all organisms, vary greatly in their body characteristics. Quite a bit of this variation is hereditary.*

offspring, a topic that is termed *transmission genetics.* This is the aspect of heredity that many people find so interesting, especially expectant parents: "Will our baby be a boy or a girl?" "Which of us will our baby most resemble?"; "Will she have my red hair and your blue eyes?"; and then, of course, the most important question of all, "Will our baby be normal?" No one can answer all of these questions, but a genetics counselor can often tell parents the approximate odds of particular traits appearing in their child. You, too, can practice being an amateur genetics counselor if you understand the concepts presented in this chapter.

SOME METHODS OF STUDY

Humans have had at least a crude understanding of some aspects of transmission genetics for thousands of years. All of our present-day domesticated plants and animals came from a wild ancestry. To the extent that they differ from the wild stock, it means that people have selected and bred them for certain characteristics. Much of this breeding was simple trial-and-error at first. But as time passed, a fund of knowledge about livestock and food plants accumulated (Figure 16-3). Along the way a lot of misinformation also became part of the folklore of many peoples, especially as related to

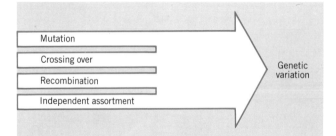

Figure 16-2 *Sources of genetic variation.*

Figure 16-3 *Animal domestication. The ram on the left is the result of a breeding program that began 160 years ago using sheep like the one shown at the right for breeding stock.*

human heredity. Particularly common were beliefs about the passing on of traits acquired by an individual during his lifetime. Thus, the bright red birthmark of a newborn infant's body was often thought to result from the mother's eating strawberries (or some other red-colored food). The "prodigal son" of a family was thought capable of passing on his "bad blood" to his children. Even today this inheritance of acquired characteristics is a widely held belief.

As stated in Chapter 15, the basic concepts of transmission genetics were discovered in the mid-1880s by Gregor Mendel. Unfortunately his published results were ignored until 1900. Nevertheless, many biologists honor him today by referring to Mendel's Laws of Genetics. As we go through the chapter we will see some of the applications of Mendelian concepts to human heredity.

A traditional way to study inheritance is to do experimental crosses with plants or laboratory animals, not unlike those that Mendel did with pea plants over a hundred years ago. The experimenter selects parents showing a particular trait, breeds them, and then observes how the trait appears in the offspring. The experiment can be repeated many times to give substantial quantities of data. The most commonly used organism until recently has been the ordinary fruit fly, *Drosophila*. This is the tiny insect that hovers around whenever fresh fruit is left out for a day or so. Geneticists began using *Drosophila* in the early 1900s because it lives and breeds rapidly in lab glassware, it has large numbers of offspring, and it shows a variety of distinct traits such as different wing shapes, eye colors, and body colors (Figure 16-4). Its small number of chromosomes (four pairs) also proved helpful in chromosome studies. A lot of our knowledge of transmission genetics came from experiments with this tiny insect.

It is considerably more difficult to study human inheritance, as you might imagine. Experimental crosses cannot be done. They would not be very useful anyway, because we bear so few young and our lifespan is so long. Consequently, other techniques for studying human heredity are used. These include pedigree analysis, statistical studies of hereditary diseases, and experiments done with cell hybrids. We will discuss these techniques later in the chapter.

Basic Concept

The traditional method of studying transmission of genetic traits is to perform experimental matings and observe the types of offspring that are produced.

SOME PATTERNS OF INHERITANCE

Monohybrid Inheritance

The simplest pattern of inheritance involves the transmission and expression of one particular trait, hence the term *monohybrid* inheritance. As an example, let us consider the common trait of freckles. The formation of freckles in a person's skin is controlled by a gene (noted by the letter *F*) located on an autosomal (non-sex) chromosome. Another form of this gene, called its *allele, f*, does not produce freckling. In a diploid organism, every cell has two genes related to this trait because its chromosomes occur in pairs; that is, each chromosome has a corresponding one (its homologue) carrying genes that affect the same trait. The two genes may occur in the combinations *FF, Ff,* or *ff*. If only one of the genes in a combination is an *F*, the person will have freckles. Therefore *FF* or *Ff* both produce freckling. A nonfreckled individual results only from the combination of the alleles, *ff*. For this reason, freckling is said to be dominant to nonfreckling. (The concept of dominant and recessive traits was first proposed by Gregor Mendel, and is often referred to as *Mendel's Law of Dominance*.) It is customary to let a capital letter symbolize a dominant gene and a lowercase letter its allele.

If two nonfreckled individuals mate and produce children, obviously all of their children would be nonfreckled because they would inherit *only* the genes for nonfreckling. This mating and its results can be shown symbolically as follows:

Parents: (father) *ff* crossed with *ff* (mother)
Gametes: (sperm) *f* × *f* (egg)
Offspring: *ff*

Note that each of the gametes contains only one of the two genes from a parent, because the gametes are haploid.

Next, let us see the results of a mating between a

(a)   (b)

Figure 16-4 *Wild type Drosophila fruit fly* (a) *with some of its mutant traits including white eye color* (b), *bar eye* (c), *and vestigial wing* (d).

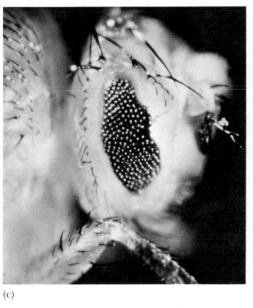

(c)   (d)

nonfreckled individual and a freckled person whose cells have the two dominant genes:

> Parents: *ff* crossed with *FF*
> Gametes: *f* × *F*
> Offspring: *Ff*

The children from this mating would be freckled, but would also be said to be *carriers* of the nonfreckled allele.

Now suppose a carrier individual and a nonfreckled individual marry and have a family. Would their children be freckled or not?

Parents: *Ff* × *ff*
Gametes: (*F* or *f*) × *f*
Offspring: *Ff* or *ff*

1 *FF*—freckled individual
2 *Ff*— freckled, carrier individuals
1 *ff*— nonfreckled individual

Note that the carrier individual produces two kinds of gametes, presumably in equal numbers. This is the consequence of the segregation (separation) of homologous chromosomes in the first part of meiosis. This is often termed *Mendel's Law of Segregation* because he proposed long ago that this type of mechanism was necessary to explain the results of genetic crosses. The offspring may be either freckled or nonfreckled, depending on the combination of gametes they receive. The chance or probability of these parents having a freckled child is one out of two (50 percent), and the same odds prevail for having a nonfreckled child.

Finally, we might examine the situation where both parents are carriers:

Parents: *Ff* × *Ff*
Gametes: (*F* or *f*) × (*F* or *f*)
Offspring: *FF* + *Ff* + *Ff* + *ff*

If you got lost in trying to put the gametes together to find the offspring, try putting the gametes in a *Punnett square,* as follows. List the possible alleles and genes that the parents can contribute:

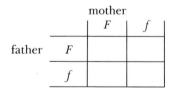

	mother	
	F	*f*
father *F*		
f		

Then go through the square and fill in the different combinations of gametes that can result from a union of sperm and egg:

	F	*f*
F	*FF*	*Ff*
f	*Ff*	*ff*

Now simply count the numbers and the kinds of offspring produced there:

Note that three of the four possible combinations are freckled and only one of the four is nonfreckled. This 3-to-1 ratio is typical of monohybrid crosses where both parents are carriers of the trait. Also typical is the fact that 50 percent of the offspring of such a mating will themselves be carriers.

A number of additional traits in human beings are known to be inherited in this simple, monohybrid pattern. These include the Rh blood factor, polydactyly (extra fingers or toes), albinism, and a number of diseases including sickle-cell anemia and cystic fibrosis. Most of these traits are controlled by dominant genes, although albinism and cystic fibrosis are caused by recessive alleles. Because a trait results from recessive alleles does not mean that it is necessarily rarer in a population than one produced by dominant genes. Cystic fibrosis, for example, is the most common hereditary disease found in the Caucasian (white) race, occurring about once in every 1000 to 1500 births.

Cystic fibrosis (CF), a serious hereditary disease, demonstrates how recessive traits are transmitted. CF is an abnormal metabolic condition produced by a pair of recessive alleles. The action of these genes causes the mucus produced in the body to be thick and viscous instead of thin and watery. Mucus is the body's lubricant, so to speak, and has widespread uses. It moistens the lining of the lungs, the windpipe, and nose: it helps transport digestive enzymes and secretions from the pancreas and liver into the intestines; and it is secreted by many cells in the wall of the intestine and by sweat glands. When the mucus is too thick to flow properly, the organs that depend on it cannot function properly. The lungs become congested with mucus, encouraging infections. The heart is eventually damaged in attempting to force blood through the congested lungs. Secretions from the pancreas and liver cannot empty into the small intestine, and this deficiency in turn interferes with digestion, causing malnutrition. So many organ systems are affected that CF victims show a broad range of symptoms requiring a variety of specialized and expensive treatments. And even with the best medical help, the average age at death of a CF patient is

18 years. Like all hereditary diseases, there is no simple cure; the basic cause is genetic, and there is no technique for replacing defective genes in diploid organisms at the present time.

Because CF results from recessive alleles, two apparently normal parents can produce a child with CF, provided both parents are carriers. Let c represent the allele for cystic fibrosis and let C symbolize the normal condition:

Parents: Cc $\times$ Cc
Gametes: $(C$ or $c) \times (C$ or $c)$

	C	c
C	CC	Cc
c	Cc	cc

According to these results, the probability that two carrier parents will have a child with CF is one out of four. However, another half of their children will likely be carriers who will be capable of transmitting the allele to future generations. Many parents would not have children if they knew that they were carriers of this tragic disease. On the other hand, some parents would consider that three chances out of four of having a normal child are pretty good odds, even though two-thirds of their normal children would be carriers.

It is this hidden nature of recessive alleles that keeps their occurrence (frequency) about the same in a population from generation to generation. The frequencies of defective genes can be decreased in populations only if there is an effective organized effort to convince carriers *not* to have children, and even then, progress would be slow. First of all, however, a screening test must be made available that detects carrier individuals. In the case of cystic fibrosis, there is none at present.

Interacting Genes

Although we usually refer to genes as being recessive or dominant, a number of genes are known that interact in such a way as to produce some type of intermediate condition. For example, when the gene for curly hair and the gene for straight hair occur together, they

Figure 16-5 *Normal red blood corpuscle (top) and sickled red blood corpuscle (bottom). Sickled cells having difficulty passing through capillary beds.*

produce wavy hair in an individual. If two wavy-haired people have children, their children could have straight hair, wavy hair, or curly hair. See if you can set up this mating in a Punnet square and obtain these results.

Sickle-Cell Anemia. Sickle-cell anemia is another example of interacting or *codominant* genes. This disease, found predominantly in black people, results from defective hemoglobin molecules that cannot carry oxygen as they should. Red blood cells containing this defective hemoglobin are sickle-shaped and cannot efficiently pass through capillary beds (Figure 16-5). We can designate the sickle-cell gene as Hb^S, in contrast to the normal gene for hemoglobin, Hb^A. Individuals who inherit both sickle cell genes, $Hb^S Hb^S$, are severely anemic. They have jaundiced coloring and suffer severe pains in joints about the body. Most victims die young.

By contrast, individuals who have only one of the abnormal genes, Hb^{A}Hb^S, are not as severely affected, because many more of their red cells contain normal hemoglobin. These individuals may not, in fact, suffer any symptoms unless placed under severe physical stress that rapidly uses up oxygen, such as athletic contests. In one such incident, a football player at the University of Colorado who collapsed and died during a practice session was found to be a sickle-cell carrier. The combination of physical exertion and high altitude probably caused his death.

About one black person in 10 is a carrier, which means that there are over 2 million carriers of this serious trait in the United States alone. In addition, about one in 200 whites in the United States is also a carrier. The disease is also fairly common among Greeks and Sicilians, who have brought the gene into this country through immigration. All in all, this disease presents a serious health problem. It is especially severe among black people, and its incidence is expected to rise steadily among whites as more interracial marriages occur.

The incidence of sickle-cell anemia is higher in tropical Africa than elsewhere in the world. Studies showed that carriers of the genes are less affected by malaria (still a serious tropical disease) than are non-carriers because the malaria parasite does not live well in their bloodstream. In other words there is an adaptive advantage in being a sickle-cell gene carrier in the malaria infested tropics.

ABO Blood Types. A second example of codominance is shown by the ABO blood types. A person's blood type is determined by a pair of genes. Actually there are three alleles in the population which we call i^A, i^B, and i^o. However, a given individual can inherit only two of these. The gene for blood type A (i^A) is dominant to the gene for blood type O (i^o). Gene i^B is also dominant to i^o. Genes i^A and i^B are *codominant*. Thus the different blood types express themselves as follows:

$$i^A i^A = \text{blood type A}$$
$$i^A i^o = \text{blood type A}$$
$$i^A i^B = \text{blood type AB}$$

$$i^B i^B = \text{blood type B}$$
$$i^B i^o = \text{blood type B}$$
$$i^o i^o = \text{blood type O}$$

From this chart you can make a number of predictions. One is that two type-O parents can have only type-O children. Another is that mating between a type-A person (i^{A}i^o) and a type-B person (i^{B}i^o) could possibly produce children showing all four blood types (i^{A}i^o × i^{B}i^o). Two type-A (or B) parents could have a type-O child if both carry the type-O gene, and so on. Blood-type information is occasionally used in legal cases in some states. It can help identify a blood stain with a certain person, or help determine who could or could not have been the father in a paternity dispute. For example, a type-O male could *not* be the father of a type-A or B child if the mother is also type O.

This example of inheritance also demonstrates another concept, that of *multiple alleles*. The "gene pool" of the population contains three alleles for blood types, but an individual can inherit only two of the genes. Table 16-1 shows the frequencies of the different blood types in the U.S. population.

Sex Determination

In human beings and other mammals, the sex of the organism is determined by its complement of X and Y chromosomes. Males normally contain X and Y, and a female has two Xs. However, the Y chromosome is known to control maleness even when extra X chromosomes are present. Thus, perhaps it would be more accurate to state that the Y chromosome determines

TABLE 16-1. Frequencies of Different Blood Types in U.S. Populations

Blood Type	Distribution in percent	
	White	Black
A	41	27
B	10	21
AB	4	4
O	45	48

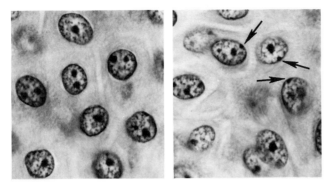

Figure 16-6 *Barr bodies, shown by arrows, in the nuclei of cells from a female (right). This sex indicator provides a convenient way to distinguish male (left) from female cells, and to diagnose cases where abnormal numbers of sex chromosomes are present.*

maleness and the *absence* of the Y leads to femaleness. The Y contains few genes. However, one that it does contain, for the formation of the testes, is evidently sufficient to control the formation of all other male features as well.

Mammalian sex chromosomes have been investigated thoroughly in order to better understand the sex-determining mechanism. In 1949 a Canadian biologist, Murray Barr, found that cells from a female cat could be distinguished from male cells by a darkly staining speck of material in the nucleus. Subsequently it was found that cells from female and male human beings also could be distinguished by this feature, now called a *Barr body.* The presence or absence of the Barr body offers a convenient way to separate female from male cells (Figure 16-6).

In the 1960s an English biologist, Mary Lyon, found evidence that only one of the two X chromosomes in a female cell is functional; the other is inactive. This means that in any woman, some of her cells have an active paternal X chromosome (derived from her father), and some have an active maternal X. In a female mouse, for example, this could result in a mottled fur coat pattern, as the paternal and maternal genes each express their specific coat colors. It appears now that the inactive X chromosome in a female's cell forms the Barr body in the nucleus. It is visible in nondividing female cells but does not appear in cells from a male.

This *sex indicator,* the Barr body, not only provides a convenient way to separate female from male cells, but it also helps diagnose unusual cases in which a person has an abnormal number of sex chromosomes. For example, occasionally a male may have an extra X chromosome (XXY) and exhibit a set of symptoms known as Klinefelter's syndrome: sparse body hair, enlarged breasts, and underdeveloped or abnormal male genitalia (Figure 16-7). Cells from such males show Barr bodies, an indication of an extra X chromosome. Other abnormal sex chromosome ratios are also known, including XXX in females. Such a woman's cells contain two Barr bodies. As the number of X chromosomes increases, so does the number of Barr bodies. Such chromosomal abnormalities probably occur because the sex chromosomes fail to segregate during meiosis.

Sex Inheritance. Because males contain the XY chromosome combination, these chromosomes segregate at meiosis to form equal numbers of X-carrying sperm and Y-carrying sperm. Ova carry only X chromosomes. Thus the mating between male and female in respect to sex chromosomes is a simple monohybrid cross:

$$
\begin{array}{ccc}
 & \text{Male} & \text{Female} \\
\text{Parents:} & XY & \times & XX \\
\text{Gametes:} & (X \text{ and } Y) & \times & X \\
\text{Offspring:} & XX \text{ and } XY &
\end{array}
$$

As you can see, the probability of having a boy or a girl is one out of two. These odds prevail at every birth, no matter how many boys or girls the parents had previously. Previous births have no influence over the odds at the next mating.

Sex-linked Genes. X chromosomes carry many genes including those for red-green color blindness, hemophilia, myopia, and muscular dystrophy. These genes are perhaps more accurately called *X-linked* than *sex-linked,* because they do not affect the sex of the individual in any manner. They do, however, appear more often in males than in females. The reason is that males have only one X chromosome, and all genes on it, recessive as well as dominant, will be expressed.

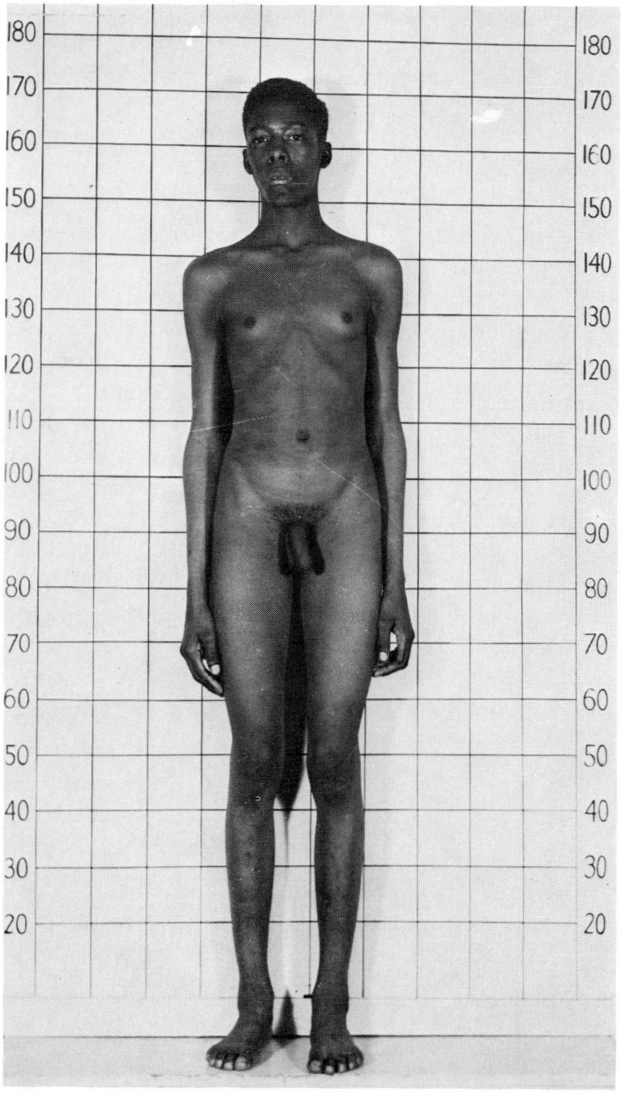

Figure 16-7 *An individual showing Kleinfelter's syndrome of sparse body hair, enlarged breasts, and abnormal male genitalia. The karyotype of such individuals contains an extra X chromosome (XXY). This is confirmed by the presence of a Barr body in the nuclei of his cells.*

For example, red-green color blindness is an X-linked recessive trait appearing in about 8 percent of American males but in only about 1 percent of females. Its inheritance pattern is as follows: Let X^c represent the gene for normal color vision, and X^c symbolize the gene for color blindness. Males are either X^cY or X^cY.

Females, with two X chromosomes, can be X^cX^c, X^cX^c, or X^cX^c. Note what happens in a mating between a color blind male and a normal female:

$$\text{Parents:} \quad X^cY \quad \times \quad X^cX^c$$
$$\text{Gametes:} \quad (X^c \text{ and } Y) \times X^c$$
$$\text{Offspring:} \quad X^cX^c \text{ and } X^cY$$

None of the children is color blind. The father passes the X^c chromosome only to his daughters, because the sons, to be male, must receive the Y chromosome. The daughters become carriers. If the daughter later has children by a normal male, note what happens:

$$\text{Parents:} \quad X^cX^c \quad \times \quad X^cY$$
$$\text{Gametes:} \quad (X^c \text{ and } X^c) \times (X^c \text{ and } Y)$$

		male	
		X^c	Y
female	X^c	X^cX^c	X^cY
	X^c	X^cX^c	X^cY

In summary:

X^cX^c	normal daughter
X^cX^c	normal but a carrier
X^cY	normal male
X^cY	color blind male

The trait has passed from father through daughter and now appears in half his grandsons. The "skip-a-generation" pattern is typical of sex-linked traits. More males than females exhibit the trait, but can inherit X-linked traits only from their mothers.

Quantitative Inheritance

The traits we have discussed to this point are mostly *either/or* — that is, the trait is either present or absent. Such characteristics could be called *discontinuous traits*. A number of characteristics, such as body height, weight, I.Q. score and skin color show numerous intermediate conditions and hence are termed *continuous traits* (Figure 16-8). Continuous traits are best explained by hypothesizing that they are controlled by a whole series of interacting genes rather than just one gene pair, that is, they are *polygenic traits*.

Figure 16-9   *A mulatto baby from the mating between a black father and a white mother. Interaction between the genes for dark skin color and light skin color results in intermediate color in the mulatto individual.*

Skin color, for example, can be explained as a polygenic trait. Skin color results mostly from the concentration of melanin skin pigments; dark-skinned (black) people have a greater concentration of melanin than do lighter-skinned (white) people. Let us postulate that the polygene complex that controls the amount of melanin in the skin consists of four gene pairs. Dominant gene pairs *AABBCCDD* produce the greatest amount of melanin and the blackest skin. The recessive alleles *aabbccdd* cause far less melanin production and

thus a light skin. The children from a mating between a black-skinned person a a light-skinned individual would inherit the genotype AaBbCcDd; they should then be intermediate in skin color. When this happens the intermediate skin color is referred to as mulatto (Figure 16-9). From this point on, however, there are a great many possible variations in skin color that could arise from different matings such as between two mulattos, or between a mulatto and a black-skinned individual, or between a mulatto and a white-skinned person. This variation arises from the numerous kinds of gametes that are possible from a gene combination such as AaBbCcDd: ABCD, abcd, AbCd, and so on — 16 different combinations to be exact.

This same polygene concept can be applied to explain the many gradations of height among people. It also explains how two parents of medium height could have a child taller than either parent, or a child who remains shorter than either parent (Figure 16-10).

The concept of quantitative inheritance by means of interacting polygenes is a very important one in modern genetics. The expression of many polygenic traits is greatly influenced by environmental conditions: Body weight and height are examples because both depend on whether organisms obtain adequate nutrients in their diets. In addition, many if not most sources of variation in human beings and other organisms are the consequence of quantitative rather than Mendelian inheritance.

Figure 16-10 *In this family photo the son is taller than either parent and the daughter is shorter than either parent. Can you explain this on the basis of the polygene concept?*

Our incomplete understanding of quantitative inheritance has led to controversy concerning the relative influences of heredity and environment on human intelligence. Intelligence is customarily measured as an intelligence quotient (IQ) score, but it is not yet known how many genes are involved and whether individual differences in IQ scores are due to different combinations of polygenes or to differences in the environment (culture) in which the individual was raised.

Other Modes of Inheritance

There are many more patterns of inheritance in addition to the brief sample presented here. For example, there are genes that prevent certain other genes from being expressed as well as genes that are lethal in certain combinations and thus kill the organism. Even this sample, however, gives you some basic ideas about what is involved in transmission genetics.

Basic Concepts

Some common patterns of heredity and examples of them include:

1. **Monohybrid inheritance (freckling, albinism, and cystic fibrosis)**
2. **Codominance with interacting genes (sickle-cell anemia and blood type)**

Figure 16-11 *Secretariat, a highly acclaimed racehorse. Horse breeders of fine animals such as these, keep careful records of the pedigrees of their animals. These records are sometimes useful in determining how certain traits are inherited.*

3. **Multiple alleles (blood types)**
4. **Sex determination by X and Y chromosomes**
5. **Sex-linked genes (hemophilia)**
6. **Quantitative inheritance (skin color)**

TECHNIQUES FOR STUDYING HUMAN INHERITANCE

Pedigree Analysis and Inherited Diseases

Just as dog fanciers or owners of thoroughbred horses know the pedigrees or family trees for their animals (Figure 16-11), it is possible to accumulate information from family Bibles, hospital records, and personal interviews to construct a human family tree. If some unusual trait appears in the family such as extra fingers or toes, or albinism, it can sometimes be traced from generation to generation. Members of the family may remember who had the trait, or they may have kept some sort of record of it, perhaps in the family Bible.

In a pedigree chart a number of symbols are customarily used, such as squares for males, circles for females, and a horizontal connecting line to denote a mating. Children from the mating are represented by

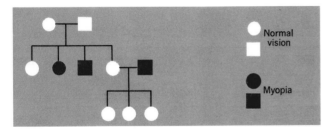

Figure 16-12 *Pedigree chart for myopia. Squares represent males, circles are females, a horizontal connecting line is a mating, and children from the mating are shown by suspended squares and circles. Based on this chart, is myopia a dominant or a recessive condition?*

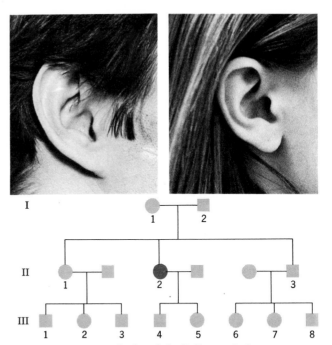

Figure 16-13 *Attached earlobe (left) a trait due to two recessive genes compared to free earlobe (right). Bottom, a pedigree chart for a recessive trait. As indicated on the chart, the trait appears only when an individual inherits both recessive genes.*

suspended squares and circles. Individuals showing the trait are then indicated by solid symbols and ''carriers'' of it by a partially filled-in symbol. A variety of additional symbols are available for showing twins, stillbirths, and so on (Figure 16-12).

After the pedigree is assembled, a geneticist then studies it and attempts to determine the type of inheritance pattern that is involved. Is it a *dominant* trait that appears in one or more members of each generation or a *recessive* one that rarely shows up? Does the trait appear in one sex more than the others? Does it skip a generation? From these observations the geneticist may obtain enough clues to tell a family how the trait is inherited. This information may be of general interest only, or it may be of major concern to the family in the case of a serious hereditary disease.

A classic example is the appearance of hemophilia, a serious disorder in which blood fails to clot, in the descendants of Queen Victoria of England. Of her nine children, one son was a hemophiliac, two daughters were carriers, and a third daughter was a possible carrier. Because records are usually available on the descendants of famous persons, the inheritance of hemophilia can be traced through six generations of this royal family. By studying the pedigree some geneticists believe that the mutation for this disease occurred in an X chromosome of Victoria's father, and that Victoria inherited the mutant gene and was thus a carrier. She did not actually have the disease, but she passed the gene for it on to one son and to some of her daughters.

Although pedigree charts are useful, they have drawbacks. Records may be erroneous or incomplete. Some hereditary diseases show different inheritance

patterns in different families. There are several types of hemophila, for example, and their inheritance patterns are not alike. Finally, recessive genes are difficult to trace in a pedigree chart because they are masked by their dominant partner. Only when an individual inherits two recessive genes does the trait actually appear (Figure 16-13).

For the interested reader, a listing of human hereditary conditions and their modes of inheritance is provided by a sourcebook entitled, *Mendelian Inheritance in Man: Catalogs of Autosomal Dominant, Recessive, and X-linked Phenotypes,* 5th ed., by Victor A. McKusick, Johns Hopkins Press, 1978.

Locating Genes on Human Chromosomes

The present emphasis in human genetics is on locating genes on the various chromosomes. This is a formidable task, because the geneticists must work with 46 extremely small bodies (chromosomes) that are visible only during certain stages of cell division (see Figure 15-12).

The first strategy was to identify genes that might be located on the X chromosome. Because males have only one X chromosome, which does not correspond with the Y chromosome, all of the genes on the male's X chromosome should be expressed, including the recessive ones. In females, who have two X chromosomes, recessive genes would usually be masked by their dominant counterparts on the other X. Hence certain X-linked recessive traits should appear more often in males than in females. Based on this reasoning, the gene for color blindness was assigned to the X chromosome. In fact, it was the first gene to be assigned to a specific human chromosome. This hypothesis is confirmed also by pedigree analysis.

Many additional genes have been assigned subsequently to the X chromosome, including those for certain types of hemophilia and muscular dystrophy. Altogether, more than 100 genes have been located on the X chromosome; their precise sites can often be determined by crossing-over studies.

The Y chromosome in man has only two known genes according to some human geneticists. One of these genes controls the formation of the testes, and the other has to do with the synthesis of certain antigens.

Establishing the location of genes on the nonsex chromosomes is a far more difficult task, because there are many more autosomes than sex chromosomes and each gene on one chromosome has its counterpart on the other. This research has required statistical techniques, data from crossing-over studies in family pedigrees, and data from cell hybridization studies. Despite the difficulties, we now know the location of at least one gene on each chromosome in the human karyotype. A few chromosomes have been extensively mapped, and more than 1200 gene locations are known at the present time.

Cell Hybridization

In 1960 a group of cell biologists in Paris found that when a mixture of mouse cells was grown in a culture medium for a period of time, a new type of cell appeared. It evidently resulted from a fusion of two cells, because its chromosome number was almost double that of the other cells. Moreover, some of the *hybrid* *cells* were capable of reproducing. The phenomenon led to many additional hybridizing experiments, such as fusing mouse and rat cells, and mouse and hamster cells. It was also found that if a strain of a certain inactivated virus (Sendai virus) was included in the culture medium, the hybridizing took place at a much higher rate and was more successful. (A "successful" hybrid is one that can reproduce itself.)

Eventually someone found that it was possible to obtain successful hybrids from a mixture of mouse and human cells. When the hybrid cell first forms, it contains a set of mouse chromosomes and a set of human ones. Also, and more importantly, both sets of chromosomes are functional for a time and produce their normal gene products such as enzymes. The mouse and human chromosomes can be distinguished from each other under the microscope by the use of special dyes.

Although someone with a vivid imagination could envision a mouse-human cell growing into a grotesque half mouse-half human creature, fortunately, nothing like this happens (except in horror films). Instead, the cell undergoes a series of cell divisions. Each time it ejects some of the human chromosome, but not in any particular order. Eventually only a few remain. Because these remaining human chromosomes produce different proteins than do the mouse chromosomes, the chromosomes that carry these genes can be identified. This technique is especially valuable in identifying *linked* genes, that is, genes on the same chromosome. Two genes that are consistently present together can be considered to be linked. By means of this technique, more than 50 genes have been assigned to 18 different human chromosomes in recent years.

An important offshoot of this procedure is the ability to distinguish similar chromosomes from each other. A special fluorescent dye shows up banding patterns, like light and dark rings around each chromosome, and can be used to identify it. Thus, the human chromosomes remaining in a hybrid cell after a series of mitotic divisions can be identified accurately.

Basic Concept
Data concerning human inheritance has been derived from pedigree analyses, and by locating genes on chromosomes from the results of crossing-over studies, statistical analyses, and from cell hybridization experiments.

Figure 16-14 *The short legs of these Ancon sheep resulted from a mutation in sheep with normal length legs.*

THE ORIGIN OF NEW GENES: MUTATION

In 1791 Seth Wright, a sheep raiser in Massachusetts, observed that one of his male lambs had much shorter legs than the other sheep (Figure 16-14). It occurred to him that short-legged sheep would not escape so easily over the low stone wall around his pasture land. He therefore set about to develop a flock of short-legged sheep by using the short-legged male as breeding stock.

Farmer Wright was evidently working with a *mutation*—the appearance of a new characteristic that was thereafter inherited. Well over a century was to elapse before geneticists learned much more about this important event. But long before mutations were understood, livestock growers undoubtedly took advantage of other new traits that suddenly appeared in their flocks or herds.

Today, with our knowledge of DNA, we can define *mutations* as changes in gene structure. Because a gene is a chemical code, there are numerous possible alterations in the chemical makeup of DNA that would change the code and therefore produce a difference in the protein produced by the gene—a mutation, in other words. Such a mutation might be observed as a change in the structure of some part of the body (extra fingers, for example), or detected as a change in body chemistry, as in cystic fibrosis.

Mutations can occur in either body cells or in sex cells. Mutations in body cells (somatic mutations) are not passed to an individual's offspring and thus are of no significance genetically. Somatic mutations usually affect a relatively few cells in an organism's body. Thus they rarely do any harm—unless you consider, as some investigators do, that cancer cells are the result of somatic mutations.

To be inherited, a mutation must occur in germinal tissue—eggs or sperm. Such mutations are termed *germinal mutations.* When the word *mutation* is used in a discussion without specifying whether it is somatic or germinal, it is assumed to mean the germinal type.

Studies indicate a few general characteristics of all mutations: They are rare events, they are unpredictable in occurrence, the mutant gene is usually recessive to the normal gene, and the mutation is often detrimental to the host. The reason mutations are usually harmful is that organisms are already well adapted to their environment (otherwise they would not survive). Almost any random change in their genes is likely to make them less well adapted. On the other hand, because mutations are usually recessive, they do not affect the organism in which they first occur. In fact, they generally do not appear until an organism inherits two recessive mutant genes some generations later.

Basic Concepts

Mutations are changes in gene structure that are passed on to future cell generations. They may be somatic or germinal.

Mutations are rare, unpredictable, usually recessive, and usually harmful to the organism.

Types of Mutations

Mutations may be classified according to the type of

alteration they produce in the structure of the genetic material:

Point Mutations. These are relatively small changes that occur during DNA replication. They involve such errors as loss or repetition of single nucleotide units, misbonding between units, and replacement of one nucleotide by another. Point mutations are the most prevalent type of mutation, accounting for such human traits as albinism and sickle-cell anemia and a host of mutations in other organisms.

Frame-shift Mutations. These are mutations caused by additions or deletions of base pairs in the DNA. When mRNA is transcribed from the mutated segment of DNA, its normal triplet sequence will be thrown off by one base. By disrupting the grouping of bases, the transcription process will "read" an entirely different sequence of triplets for the rest of the length of the gene. When this new sequence is translated into a protein, the protein may be nonfunctional and therefore disruptive to the cell. A type of abnormal human hemoglobin is known to result from this kind of mutation.

Chromosomal Mutations. These mutations result from changes in the structure of chromosomes. During cell division, chromosomes undergo many movements, some of which may lead to abnormalities in their structure. A portion of a chromosome may be broken off and lost (*deletion*); a segment may be duplicated so that its genes are represented twice (*duplication*); or a chromosomal segment may get turned around (*inverted*) (Figure 16-15). The most severe mutation is the loss of all or a major part of a chromosome, a mutation that is usually lethal.

Basic Concept

Mutations may be classed as point mutations, frame-shift mutations, and chromosomal mutations. The latter type produces the most severe abnormalities.

Causes of Mutations

The cause of mutations in wild populations of organisms is unknown, despite numerous studies by many biologists. In the view of some investigators, mutations

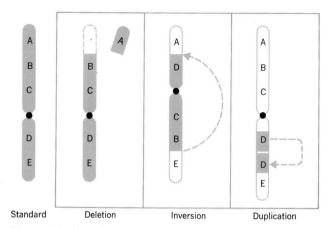

Figure 16-15   *Three kinds of chromosomal mutations. In deletion, a portion of a chromosome is lost. If the order of genes is changed, it is termed an inversion. Duplication may repeat certain genes as shown by the letters.*

may be either *spontaneous* or *induced* by mutation-causing substances (*mutagens*) in the environment. Changes in DNA structure, such as point mutations, could simply be molecular accidents that happen during DNA replication once in a great while; hence they would be spontaneous. However, point mutations and frame-shift mutations can also be caused under laboratory conditions by the use of certain chemicals.

The study of induced mutations began with the pioneer work of H. J. Muller, a geneticist, in 1927. He showed that X rays could induce mutations in *Drosophila* flies. This discovery opened a new era in the study of mutations and mutagenic agents. Since then, many additional mutagens have been identified and studied, falling into two broad categories. One is high-energy radiation, such as X rays, ultraviolet light, and other forms of radiation. The other is mutagenic chemicals, such as mustard gas, nitrous acid, certain dyes such as acridyne orange, and even caffeine.

In recent years, geneticists have sounded a warning that a number of substances that we commonly use or dump into the environment are mutagenic. These include pesticides such as DDT, dieldrin, and aldrin; certain food additives such as the sweetners cyclamate and saccharine; some of the common dyes, including commercial hair dyes; tobacco tars; and soot. Adding to this biological alarm is the recognition that most mutagens are also carcinogens, and vice versa. The two effects are so often associated as to be almost synonomous. This is the basis, in part, for the statement in Chapter 12 that some investigators believe that all human cancers are induced by environmental mutagens. In fact, the *only* cause-and-effect relationships known for any human cancers are those associated with tobacco tars, soot, a few other chemicals, and high-energy radiation, all of which are powerful mutagens.

One of the strongest mutagenic agents in laboratory use is high-energy radiation, referred to hereafter as *ionizing radiation.* Ionizing radiation consists of tiny, rapidly moving particles or energy in the form of waves. As these particles or waveform energy pass through living tissue, they strike atoms, causing electrons to be liberated. These electrons in turn strike various molecules and split them into additional ionized fragments. In this manner a pathway of ionization is created that can be very damaging to cells. Proteins, RNA, DNA, and even entire chromosomes can be affected. If exposed to sufficient amounts of radiation the organism may suffer *radiation sickness.* The most rapidly dividing cells, such as those in the skin, digestive tract, and blood-making tissues are the ones most affected. Symptoms such as anemia, ulcerations of the intestines, internal bleeding, and low immunity to diseases are common. But even if there are no immediate symptoms, certain long-term changes occur. The incidence of cancers such as leukemia is greatly increased. Genetic changes (mutations) also occur although these do not usually appear for several generations.

The best documented evidence for the effects of massive doses of radiation on human beings comes from the studies of the Atomic Bomb Casualty Commission in Japan. Atom bombs were exploded over Hiroshima and Nagasaki in 1945, and the radiation effects on the people there have been carefully documented ever since. The most notable disorders found among the survivors have been thyroid tumors, leukemia, increased incidences of solid tumors, and chromosomal abnormalities. A careful study of the first-generation offspring from the survivors has shown no evidence so far of any hereditary abnormalities that could be attributed to radiation-produced mutations. On the other hand, if these descendants have recessive mutations, it may be many, many generations before they show up.

Figure 16-16   *A group of sun worshippers. This popular pastime carries the danger of increasing the probability of skin cancer.*

Today our knowledge of radiation-induced mutations and cancers clearly indicates that individuals should avoid all unnecessary radiation. Our most common sources, and greatest menaces, are X rays and excess exposure to sunlight. An individual should be very cautious about having X rays of any sort unless absolutely necessary for medical reasons. Sunbathing (Figure 16-16) is another potentially dangerous activity. Studies long ago showed that persons in occupations that required constant exposure to sunlight had significantly greater risk of skin cancer in later life. The term "redneck" originally referred to farmers who literally had red necks from constant exposure to the sun as they worked in the fields. One of their occupational hazards is an increased risk of skin cancer of the neck region. If a tanned skin is a desired cosmetic attribute, as many people seem to believe, then perhaps a better alternative to sunbathing is to use one of the lotions that stains the skin brown. This lessens the risk of skin cancer because the outer cells are dead anyway, and one can always remove the "tan" with a good scrubbing if a lighter skin tone is desired.

Basic Concept

Mutations may be either spontaneous or induced by mutagens. Mutagens include ionizing radiation and a variety of chemicals. Some of the mutagens commonly found in our environment are also carcinogens.

MAN'S GENETIC LOAD

Many, many words have been written about man's genetic load—the accumulation of mutations in our germ cells. Modern medical procedures and drugs have made it possible for great numbers of human beings to survive and reproduce even though afflicted with genetic disorders of various kinds. Eventually, the argument goes, humankind will accumulate such a crushing load of mutant genes that it will become genetically degenerate and go out of existence, or at least be greatly crippled.

There are several arguments, however. First of all, humans show no signs of degenerating; in fact, we live longer and are healthier now than at any time in recorded history. Secondly, human evolution for the past ten thousand years has been more in the realm of social development than in biological change. Hereditary defects do not usually affect social progress to any great extent. In fact, the advancement of medical technology—part of our social evolution—has greatly aided our understanding and treatment of genetic disorders. The time is not too distant when it may be possible to change mutant genes and thereby cure many hereditary diseases.

A final counterargument is that variation (mutation) is the essence of genetic survival in the long run. A population with a wide range of mutant genes in its gene pool has at least a theoretical chance of better surviving future changes than does a population with fewer variations. An example is sickle-cell anemia (due to a mutant gene) that is harmful in some circumstances but is beneficial to populations exposed to malaria.

KEY TERMS

allele	gene	multiple alleles	punnett square
Barr body	independent assortment	mutagen	quantitative inheritance
carrier		mutation	
cell hybridization	interacting genes	pedigree analysis	radiation sickness
chromosomal mutation	ionizing radiation	point mutation	sex-linked gene
codominant genes	law of dominance	polygenic trait	sickle-cell anemia
cystic fibrosis	law of segregation		transmission genetics
frame-shift mutation	monohybrid inheritance		

SELF-TEST QUESTIONS

1. How many genes have actually been found in human cells?
2. How does independent assortment increase hereditary variation?
3. What role did the fruit fly, *Drosophila,* play in the development of genetics?
4. What are some examples of monohybrid inheritance in humans?
5. What is the Mendel's Law of Dominance?
6. Give an example of Mendel's Law of Segregation.
7. Explain the nature of cystic fibrosis and how it is inherited.
8. What is meant by interacting genes?
9. Explain how sickle-cell anemia is inherited.
10. When is it advantageous to be a sickle-cell carrier?

11. If two parents are carriers of the sickle-cell trait, what is the chance that their children will have sickle-cell anemia?

12. In the case of ABO blood types, why cannot an individual carry all three genes?

13. Explain how sex is determined in human beings.

14. What is a Barr body and what is its significance?

15. Explain how sex is inherited.

16. What are sex-linked genes?

17. Why do sex-linked traits show up more often in males than in females?

18. Define quantitative inheritance and give an example.

19. What is a pedigree analysis?

20. Why is the X chromosome valuable as a means of locating genes on specific chromosomes?

21. How many genes are known on the Y chromosome?

22. Describe the technique for producing hybrid cells and the value of this technique is genetics.

23. What is a mutation?

24. What general characteristics do all mutations have?

25. List three types of mutations with an example of each.

26. Name some known mutagenic agents.

27. Are humans developing a lethal load of mutant genes?

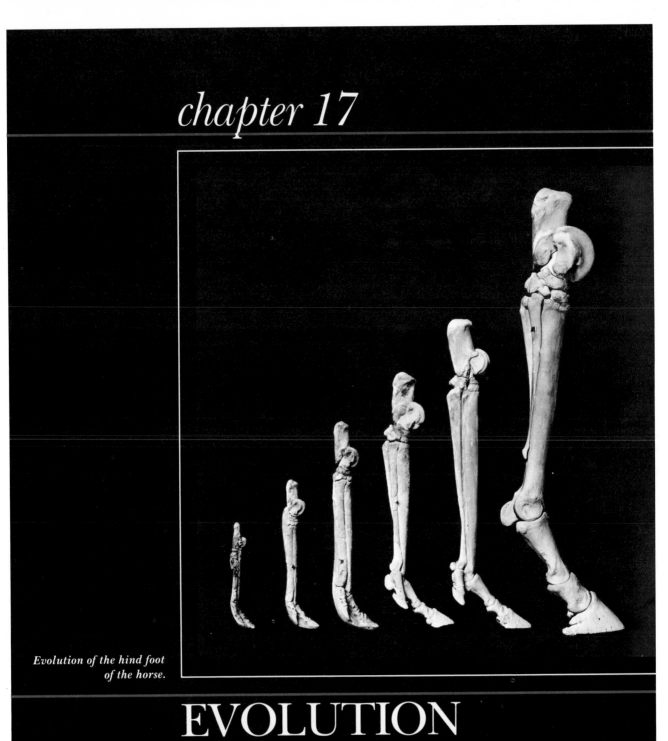

*Evolution of the hind foot
of the horse.*

EVOLUTION
PART I: EVIDENCES

chapter 17

questions to think about

How did life begin?
What evidences support the concept of evolution?
Is it possible to see evolution taking place?

Nothing in biology makes sense except in the light of evolution.

Th. Dobzhansky

The quote at the top of the page is another way of saying that evolution is the major unifying theme in biology. The fossil record, the variations and complexities in the structures and developmental patterns of animals, and the similarities and differences in the basic physiology of organisms make little sense except when viewed as part of the slowly unfolding panorama of evolutionary change through time. Years ago, the concept of biological evolution was highly controversial, and the notoriety associated with that past era unfortunately lingers on somewhat today, especially in the minds of people who do not understand the concept and theory of evolution. Today, however, many disciplines accept without question the basic idea of *evolution,* things changing through time. Astronomers theorize about the evolution of the universe and our own solar system; geologists refer to data concerning the evolution of the earth; anthropologists and archaeologists refer to the evolution of human cultures; and biologists have a large body of evidence to show that most forms of life have changed through time.

Before looking at the kinds of evidence that have led to the widespread acceptance of evolution, we need to set the stage, so to speak, by considering theories about how the earth came into existence and how life may have originated on it.

BEGINNINGS

Origin of the Earth

Astronomers and geologists generally agree that the earth is about 4.6 billion years old. That number is so huge that it is rather meaningless in human terms; perhaps the most important things to know are that the earth *had* a beginning, and that it is extremely ancient.

Evidence from astronomy indicates that the earth formed by the condensing of stellar gases and dust particles. As this took place the mass of material began to heat up and eventually its interior became molten. This melting probably caused the heavier elements (metals) to sink to the core, whereas the lighter materials such as aluminum and silicon floated to the top to form the earth's upper mantle and crust (Figure 17-1).

The molten interior eventually spewed out material to the surface through volcanoes. Included in this material were gases that formed the first atmosphere.

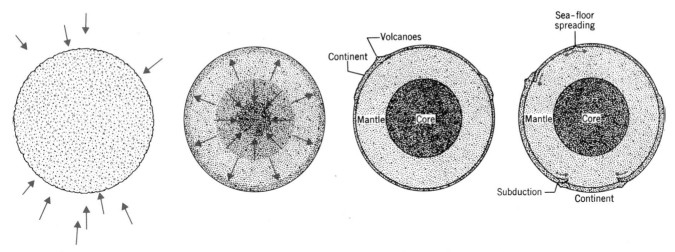

Figure 17-1 *Formation of the earth. According to this hypothesis, the earth began as a condensing of gases and dust particles (left). The interior of the mass became molten with the heavier elements sinking into the core, and lighter elements forming a crust on the surface (middle). Conti-* *nents and oceans formed and an atmosphere enveloped the earth (right). The earth as we see it today is the result of billions of years of weathering, erosion, and other major geological events.*

Until recently it was assumed that this primitive atmosphere contained considerable hydrogen as well as carbon dioxide, water vapor, ammonia, methane, and sulfur-containing gases. Some geochemists now believe that the first atmosphere was more like the earth's present-day atmosphere but without oxygen; that is, with carbon dioxide, water, small amounts of hydrogen and carbon monoxide, and no methane or ammonia.

As the exterior of the earth gradually cooled, the water vapor eventually condensed, forming the oceans. This development set into action the water cycle and the forces that create weather. One major effect of this development was to begin the earthwide wearing down process of erosion and breakdown of rocks that continues to this day.

These matters may seem far removed from present-day life, but the reason for bringing them up is that they provide the background setting for an incredibly momentous event: the origin of life. It is momentous not only in itself, but also in the drastic effects it eventually had on all aspects of the earth's surface and its atmosphere. The advent of life made possible a future time when photosynthetic micoorganisms swarmed in the oceans. Later still, a green mantle of vegetation was to cover great expanses of the land

masses. The photosynthetic action of the first plants eventually transformed the earth's atmosphere, giving it the relatively large oxygen content and low carbon dioxide content that we know today. At the same time elements such as carbon, oxygen, and nitrogen began to cycle through the environment, spending part of their cycle within living cells. All of these events eventually allowed animal life to flourish and cover the earth.

Basic Concept

The earth was formed 4.6 billion years ago by the accretion of dust particles and gases.

Orgin of Life

Dr. Elso Barghoorn's 3.2 billion-year-old fossils, described later in the chapter (Figure 17-8), are our oldest evidences of life. But even though these cells are very primitive prokaryotes, they are still far advanced in structure over what must have preceded them. The surviving fossils unfortunately tell us little about their ancestry. Thus, we are left with a speculative hypothesis about where and how life began. Some scientists prefer to call it a story of "plausible chemical mechanisms," because it is based on well-established chemical

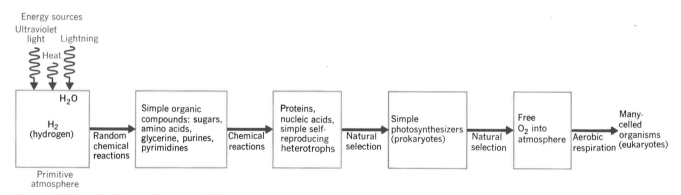

Figure 17-2 *The chemical origin of life concept in the form of a flow chart. Gases of the primitive atmosphere are shown at the left along with the energy sources that caused them to react with one another. Moving from left to right, the chemical substances become increasingly more complex until simple life forms finally evolved. Natural selection then led to simple photosynthetic plants, which in turn provided an energy source for the evolution of eukaryotes and began to add oxygen to the atmosphere.*

processes as well as assumptions about the composition of the early atmosphere.

The story of the *chemical origin of life* begins with the earth's primitive atmosphere. Its exact composition is debatable, as explained earlier, but there is agreement that it contained, in some form or other, the basic elements for making living matter: carbon, hydrogen, oxygen, and nitrogen.

With no protective ozone layer, solar radiation (especially UV light) would have been intense. Ultraviolet radiation, together with the energy provided by lightning and heat, is thought to have triggered random chemical reactions that led to the formation of a variety of simple organic compounds such as amino acids and sugars. Gradually these smaller building blocks got linked into more complex molecules. The occurrence of nucleic acids that could make a useful protein or two, as well as replicate themselves, marked the first self-producing entities. These primitive entities took as their energy supply the simpler organic materials that were floating abundantly in the primitive oceans—thus they were the world's first scavengers. Once a few such groups of self-replicating molecule systems started to duplicate themselves, natural selection would presumably operate to make this population of molecular life better able to reproduce and survive (Figure 17-2). Somewhere along the way, membrane-enclosed bodies (simple cells) such as prokaryotes came into being.

At first the free-floating organic matter would have

been abundant, but once it became an energy source for primitive life, this supply of accidentally formed organic matter in the environment would rapidly have been depleted. Thus the next necessary advance would have been the ability to synthesize an energy supply. Indeed, a simple photosynthetic apparatus evolved more than three billion years ago, which not only provided an energy source (carbohydrate), but also released O_2 into the environment. Some of this oxygen was converted to ozone (O_3) in the upper atmosphere and thereafter acted as a shield against UV radiation. The presence of free oxygen also allowed the evolution of aerobic respiration, an efficient means of energy production. A combination of these events encouraged a rapid expansion of plant and animal life during the Paleozoic era.

Is this concept purely speculative? Not entirely. In 1938, a Russian biochemist, A. E. Oparin, published a book, *The Orgin of Life,* which proposed a chemical origin of life along roughly these lines. This theory aroused considerable interest in the scientific world, and experiments were soon designed to test Oparin's idea that organic substances could be formed in an environment containing no life. In 1953, Stanley Miller at the University of Chicago confined a mixture of gases like those thought to exist on the primitive earth in a glass container. As an energy supply he discharged electrical sparks through the mixture. After a time, amino acids appeared in the container (Figure 17-3).

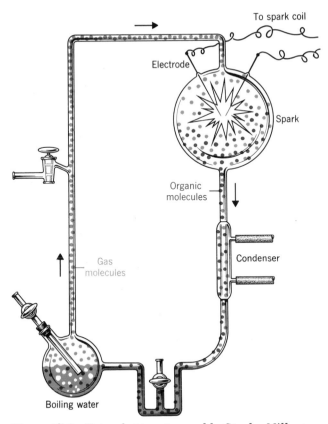

Figure 17-3 *Type of apparatus used by Stanley Miller to show that organic matter arises if energy is applied to a gaseous mixture of ammonia, methane, carbon dioxide, hydrogen, and water vapor.*

Subsequently, similar experiments were performed by Melvin Calvin, Sidney Fox, and others. These experiments indicated that a large variety of organic substances will form in this manner, including the nucleotide units necessary for making DNA or RNA. These experiments do not, of course, prove that such events happened billions of years ago, or that they led to the evolution of living matter. However, they do show that the creation of organic compounds, the building blocks of life, is possible under the conditions thought to have prevailed on the primitive earth.

Supporting evidence comes from some recent findings in astronomy. Meteorites of an unusual type, the stony meteorites, were found to contain carbonaceous particles. Analysis of these particles showed a variety of organic materials, including amino acids and portions of nucleic acids. Certain features of their chemical structure indicate that these compounds probably formed spontaneously rather than as the products of living matter. Also, radiotelescope probes have shown the presence of several kinds of organic molecules in the diffuse gas between the stars, including formalin and cyanide. These molecules are thought to have been formed by the chance collision of atoms. All in all, it appears that the formation of organic matter by nonliving means is not a unique event in our universe. Indeed, many astronomers feel that extraterrestrial life — life elsewhere in the universe — is highly probable. This viewpoint assumes that there are great numbers of planets in the universe, and that life is a likely enough event to have happened on at least a few. So far, there is no observational evidence to support the existence of extraterrestrial life.

What are the alternatives to this concept of the spontaneous origin of life? One is that primitive life came to the earth from another planet, perhaps via meteorites. But even if true, this unlikely event would only change the location of life's origin and remove the possibility of studying it.

Another alternative is the commonly used argument of *creationism:* that the earth and all species as we now see them were created by a supernatural power as stated in the Biblical account of creation in the book of Genesis. A problem with this explanation is that supernatural actions cannot be verified by scientific methodology. In addition, creationism leaves unanswered a vast number of questions about the history of life, such as how to explain the existence of fossils of extinct organisms. Most scientists and other educated people are unwilling to accept the creationist's view of the biological world when an evolutionary viewpoint, supported by evidence from many sources, is available. The quote at the beginning of the chapter makes this point in a simplified way.

Basic Concepts

Life may have arisen by the chance combination of chemicals available in the primitive atmosphere.

The activities of life have changed the surface of the earth and its atmosphere.

EVIDENCES OF EVOLUTION

Fossils

It seems almost contradictory that the most compelling evidence for evolution comes from organisms no longer living and indeed extinct in many cases. Fossils are our only *direct* evidence of life of the past (Figure 17-4). The observation that ancient fossil forms differ from present-day organisms is our strongest evidence that organisms change through time.

Viewing a fossil collection in a museum, or walking about in a fossil-rich deposit such as a phosphate mine gives the erroneous impression that fossils are quite common everywhere. In reality, the probability that an organism will become fossilized is very remote. Usually it must be buried immediately after death, before it has time to decay or be destroyed by scavengers. The oxygenless mud in the bottom of a pond is a good site. Very gradually, the hard parts such as bones and teeth are replaced by minerals and withstand the geological changes as the mud is turned into rock. The soft parts of the organisms are usually lost; perhaps an impression may be left where the organisms lay in the mud (Figure 17-5). However, if conditions are just right, mineral replacement can faithfully show every structure within a cell. In other instances, entire organisms have been preserved in ice, tar pits, and amber (fossilized plant resin) for thousands of years (Figure 17-6).

Finding a fossil is of little value unless we can determine how long ago it was fossilized. Once the ages are known for a group of fossils, they can be arranged in the order of their appearance on the earth. By observing the details of their structures and the times during which they lived, it is possible to deduce their evolutionary history. The wealth of knowledge we now have about the evolution of horses is a classic example of how much information can be obtained by studying the fossil record of a group. By accumulating information like this for many forms of life, paleontologists and geologists have been able to place all major groups of plants and animals in a geological time table similar to that shown in Table 17-1.

Fossils are most often dated by knowing the geological ages of the rock strata in which they were found. The technique fails, however, if the rock's age is not known or if the strata have been disturbed by a change in the earth's crust. Therefore an alternate method based on "radioactive clocks" has become a widely

Figure 17-4 *Reconstruction of the skeleton of a dinosaur,* **Stegosaurus stenops.** *Fossil bones such as these are our only direct evidence of life that lived in the past.*

Figure 17-5 *An imprint fossil of an ancient palm leaf.*

was that the time scale was too short. However, radioactive dating has established that the earth is evidently much older than was thought a century ago. This factor is essential to the modern concept of evolution. The age of the oldest known rocks, 4.5 billion years, indicates that evolution had an immense period of time in which to operate.

Although we usually think of fossils as relatively large objects, techniques have been developed for viewing single-celled fossils. Dr. Elso Barghoorn, a botanist at Harvard University, made thin slices of chert rock from ancient Precambrian formations and found the remains of primitive bacteria and algae (Figure 17-8). These cells are the oldest known fossils, putting the date of the earliest evidences of life near 3.2 billions of years ago.

If fossilization had never taken place in the earth, our knowledge about evolution would be greatly restricted. We depend heavily on the fossil record for knowledge of most of the major evolutionary advances that took place in the plant and animal kingdoms. This point is emphasized by the basic hindrance to a better understanding of our own evolution: the scarcity of fossil remains of early human beings. Our knowledge in this area will advance almost in proportion to the number of new discoveries of fossil remains.

Basic Concepts

Fossils are direct evidences of past life.

Radioactive dating methods aid in accurately placing fossils in a geological time scale.

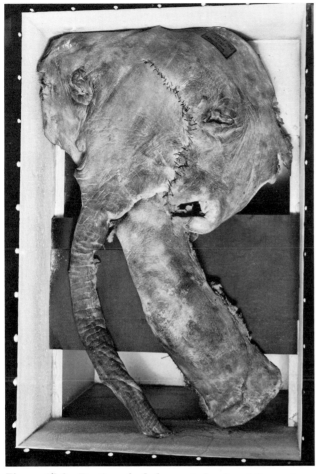

Figure 17-6   *Portion of a baby mastoden found frozen in the ice in Alaska.*

used and valuable technique. These substances, *radioisotopes,* undergo radioactive decay at a constant rate to form other elements. By measuring how much radioactive decay has taken place in a fossil, we can calculate how old it must be (Figure 17-7). In a sense, these radioactive substances offer us radioactive clocks that tell time in tens of thousands or millions of years. For fossils of relatively recent origin, carbon-14 is often used. For extremely old fossils, radioactive potassium or rubidium may be used because of their extremely slow decay rates.

Another valuable aspect of *radioactive dating* has been to establish more accurately the age of the earth. One of the early objections to the theory of evolution

Comparative Structure and Development

If no fossil record were available, the next strongest evidence of the kinship among organisms would be similarities in anatomical structure and embryonic development of organisms today. In fact, this is the basis of the most commonly used method for studying the relationships between organisms or solving problems about relationships. It may be the only technique for gaining information about groups whose fossil record is poor or lacking altogether.

The assumption underlying this method is that organisms or groups that have developed more recently from a common ancestor should resemble each other more closely in their development and adult structures

TABLE 17-1. Geological Time Table

Era	Period		Biological Events			Age in Millions of Years
Cenozoic	Quaternary	Recent				0.0015
		Pleistocene	First man	Age of Mammals		2
	Tertiary	Pliocene				10
		Miocene				25
		Oligocene				38
		Eocene				55
		Paleocene				65
Mesozoic	Cretaceous		Age of Reptiles	Great dinosaurs	First mammals and flowering plants	135
	Jurassic				First birds	185
	Triassic					225
Paleozoic	Permian		Extinction of enormous numbers of Paleozoic species			280
	Pennsylvanian		Great coal-forming swamps	First reptiles		320
	Mississippian					340
	Devonian		Age of fishes (First amphibians)			405
	Silurian		Earliest insects and land plants			430
	Ordovician		First vertebrates (fish)			500
	Cambrian		Beginning of good fossil record of plants and animals			600
Eras of precambrian time						
			Oldest primitive plant fossils			3,300
						3,600
						5,000

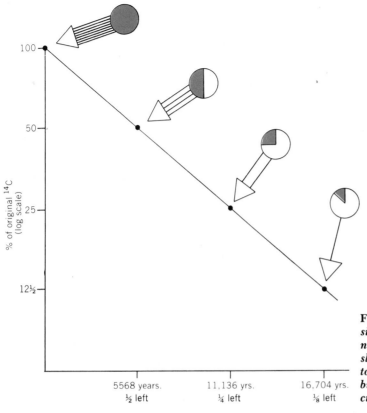

Figure 17-7 *The proportion of ¹⁴C remaining in fossils of different ages. At the time of the death of an organism, its ¹⁴C begins to decay and the ratio of ¹⁴C to ¹²C slowly changes. By measuring the ratio of these two isotopes in a fossil and relating it to the half-life of ¹⁴C, a biologist can calculate the age of the fossil fairly accurately.*

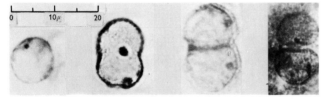

Figure 17-8 *Fossilization at the microscopic level. Stages of cell division in a green alga from one billion year old chert rock of Australia.*

than groups that have been distinct longer. The vertebrates—fishes, amphibians, reptiles, birds, and mammals—provide many good examples. You would probably not be surprised to hear that the anatomy of an ape and of a man are quite similar, as shown in Figure 17-9. The body proportions are somewhat different, but the bones, muscles, blood vessels, and other body parts are extremely similar. And only an expert could tell ape and human embryos apart. These are the

kinds of similarities one expects to find between two closely related forms.

When organisms are not so closely related, like a bird and a human being, for example, the search for similarities and their significance becomes more difficult. However, it is known that there is a general body plan around which all vertebrates are organized. And seemingly divergent or unrelated body forms can be viewed as special adaptations of the more general plan. Vertebrate forelimbs exemplify this unusually well. The basic plan of the vertebrate forelimb is based on a five-digit appendage with hand bones, wrist bones, radius, ulna, and a humerus. The same general musculature, blood vessels, and nerves form part of this general plan. If you compare the parts of the forelimbs of a variety of vertebrates, the basic vertebrate limb plan can be seen even though the limb is modified for a different function in each organism (Figure 17-10).

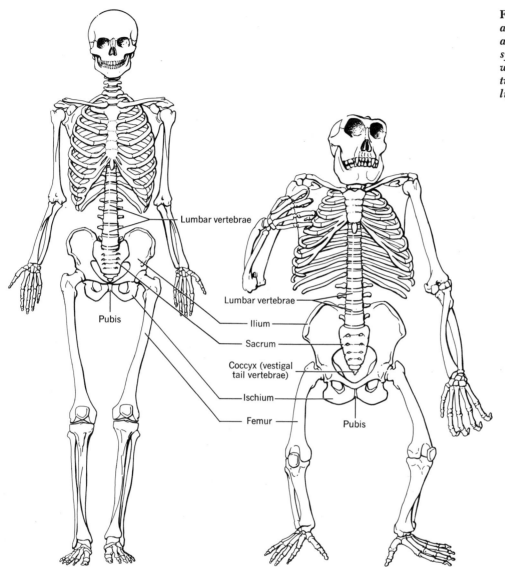

Lumbar vertebrae

Pubis

Lumbar vertebrae

Ilium

Sacrum

Coccyx (vestigal
tail vertebrae)

Ischium

Femur

Pubis

Figure 17-9   *The skeletons of a man and an ape are similar as shown here. Other body systems are also much alike as would be expected between two closely related forms of life.*

Some of these modifications (adaptations) may be extreme. For example, in the bird's limb some bones are lost and others are fused.

In the case of animals with extremely specialized limbs, such as the bird or the horse, it may be necessary to turn to another source of evidence to understand the adult structure. Study of the developing embryo is often valuable, because the general architecture of the body appears first, and only later does specialization of the parts take place. The early bird embryo and horse embryo both show a five-digit appendage. As embry-

onic development continues, however, these appendages are gradually modified to become the highly specialized limbs we see in the adults (Figure 17-10). Comparing the development of embryos of different vertebrates provides many other evidences of their basic kinship. All early vertebrate embryos appear almost identical for a while, having gill pouches, a large bulbous head where the brain is developing, and a tail (Figure 17-11). Development of the major internal organs such as the brain, heart, and gut is also similar for a time. Only later, as in the example of the special-

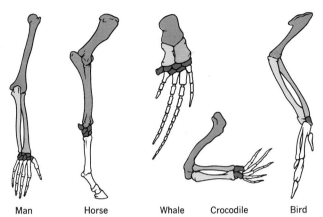

Figure 17-10 *Some vertebrate forelimbs. The basic anatomic plan is the same for the different animals even though each limb is modified for a special function. Homologous bones are shown in the same shade.*

Man   Horse   Whale   Crocodile   Bird

ized limbs, does each embryo reveal where it is heading. The fish embryo forms gills, the turtle embryo starts a tiny shell, the horse's hooves begin to form, and so on.

A nineteenth-century German embryologist, Ernst Haeckel, was so impressed by this unfolding of events in embryos that he proposed an intriguing theory called the biogenetic law. It contains the catchy statement that "ontogeny recapitulates phylogeny." The theory in essence states that the developmental stages of an embryo (its ontogeny) repeat the evolutionary history of its group (its phylogeny). In application it means that a mammalian embryo goes through a fish-like stage, an amphibian stage, and a reptilian stage before it develops into a mammal. Thus, if we carefully study the stages of embryonic development in all major groups, we can learn most of the details of their evolutionary history. There is some validity to the biogenetic law, but only in a very general sense. All vertebrates have similar embryos early in development. The embryos then progress from this general form to highly specific forms. Also, the more similar the adult forms, the later their differences appear in embryonic development. Perhaps the greatest value of this concept is to reemphasize the kinship of vertebrate groups.

Structures that develop similarly and are based on a common plan, as the vertebrate limb, are said to be *homologous* to one another. Homologous structures have similar origins but often have quite different functions, as we have seen already in the limbs of a

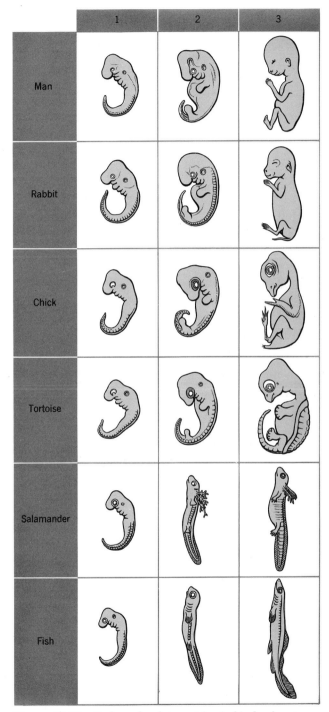

Figure 17-11 *Embryo development in six kinds of vertebrates. At first the embryos are almost identical, having gill pouches, tails, and large, bulbous heads. Only later does each embryo indicate its destiny.*

horse as compared with those of a bird. Homology is an indication of relationship.

To confuse this issue, relatively unrelated organisms may evolve similar-looking structures as adaptations to performing similar functions. For example, a porpoise is a mammal and only distantly related to fishes. However, its body form is much like that of a fish; both have similar body forms as an adaptation to swimming through a dense medium (Figure 17-12). The pectoral fin of a porpoise (a mammal) and the pectoral fin of a shark (a fish) are both used for propulsion, but the basic limb plan of the porpoise is entirely different from that of the shark (Figure 17-13). Features like this that serve the same functions but are in distantly related organisms are said to be *analogous* structures. If two structures appear to be analogous rather than homologous, biologists interpret this to indicate distant relationship. This interpretation is reinforced when other anatomical systems and development patterns are compared in the two organisms or groups.

Basic Concepts

Organisms are considered related if they share structures based on the same basic anatomical plan. These are called homologous structures.

Comparative embryology often shows homologies that cannot be seen in adult structures.

Similar-looking structures may develop in unrelated animals as adaptations for performing similar functions. These are termed analogous structures.

Comparative Physiology

A study of the various chemical constitutents and reactions in many types of organisms provides another strong source of evidence of basic kinships.

As stated in an earlier chapter, carbon, hydrogen, oxygen, and nitrogen comprise 99 percent of the elements in all protoplasm. Carbohydrates, lipids, proteins, and nucleic acids are the organic constituents, and water is the most abundant compound.

ATP is the energy source for performing the various types of work in *all* cells in *all* organisms. To produce ATP, nearly all organisms utilize the same cellular respiration pathways and use the same kinds of raw materials. In fact, this is such an evolutionarily old and universal process that one of its major chemicals,

Figure 17-12   *Similar body forms of a shark (fish) and a porpoise (mammal).*

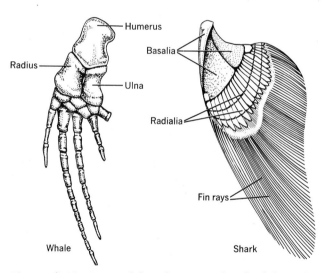

Figure 17-13   *Pectoral fins of a mammal and a fish. Both are used for swimming but their internal anatomy is quite different; hence they are analogous instead of homologous organs.*

cytochrome c, can be used in studying relationships among various organisms.

In theory, one useful way to study kinship or "genetic distance" between two organisms would be through a complete analysis of the nucleotide sequences in their DNAs. Presumably closely related organisms would have more similar sequences (genes) than distantly related ones. This type of analysis is not possible at present except with short pieces of DNA. However, the concept can be used in other ways. One technique is to obtain DNA from an animal and convert it into short, single-stranded pieces. When these are mixed with similar strands from another animal, a certain amount of rejoining of strands (matching of complementary nucleotides) takes place. If the genetic distance is not far between the organisms, the matching is fairly complete; if it is farther, less matching takes place. For example, when human DNA is compared with chimpanzee and monkey DNA, there is more binding between human and chimp DNA than between human and monkey DNA. This confirms what we suspected about these relationships based on other types of evidence.

These DNA analyses are quite difficult to perform. Fortunately, there is an indirect method of getting at the same informtion, using the products produced under the direction of genes—proteins. Related forms of life should have more similar proteins (more nearly identical amino-acid sequences) than less related ones. The kinds of protein that have been extensively examined for this purpose are cytochrome c and various hemoglobins. Both are small proteins, easily obtainable, and found in many organisms. Cytochrome c, with its 104 amino acids, is an important chemical in cellular respiration. It is almost universal in occurrence, and hence very old from an evolutionary standpoint.

The amino-acid sequences in cytochromes from more than 30 organisms are known. These include a variety of vertebrates, insects, and plants. Any variation in the amino-acid sequence is assumed to represent a mutation in the DNA that provided the code for the protein. Thus, a comparison of cytochromes indicates genetic distances, proportional to the numbers of differences in the amino-acid chains.

To give a few examples, cytochromes from a human and a Rhesus monkey show only one difference in amino acids. A monkey and a dog differ in 10 amino acids; a tuna and a chimp differ in 21. These amounts of difference are approximately what we would expect of these animals from our knowledge of their relationships. Table 17-2 shows some additional comparisons.

Recently a detailed comparison of human and chimpanzee genes was made by analyzing the amino-acid sequences in a series of proteins from the two

TABLE 17-2.   A Table of Cytochrome c Comparisons in Different Forms of Life. To obtain the number of amino acid differences in the cytochromes of a dog and a chicken for example, find the dog line at the left and the chicken column at the top. The intersection of the line and the column provide the answer, 10 in this case.

	Rhesus Monkey	Dog	Rabbit	Chicken	Pigeon	Snapping Turtle	Rattlesnake	Bullfrog	Tuna Fish	Fruit Fly	Screwworm Fly	Wheat	Neurospora	Baker's Yeast
Human being	1	11	9	13	12	15	14	18	21	29	27	43	48	45
Rhesus monkey		10	8	12	11	14	15	17	21	28	26	43	46	45
Dog			5	10	9	9	21	12	18	23	21	44	46	45
Rabbit				8	7	9	18	11	17	23	21	44	46	45
Chicken					4	8	19	11	17	25	23	46	47	46
Pigeon						8	18	12	18	25	23	46	46	46
Snapping turtle							22	10	18	24	24	46	49	49
Rattlesnake								24	26	31	29	47	48	48
Bullfrog									15	22	22	48	49	47
Tuna fish										25	24	49	48	47
Fruit fly											2	47	41	45
Screwworm fly												45	41	45
Wheat													54	47
Neurospora														41

organisms, and by matching DNA strands, as described previously. The conclusion was that all of the proteins examined were more than 99 percent identical, and that their DNAs were highly similar. We are left with the intriguing thought "Why are chimps and humans as different as they seem to be if their genes are practically identical?" The answer suggested by the researchers of this problem is that somewhat different regulatory mechanisms function in the two species due to mutations in their gene control systems. The genes themselves are very similar, but they are turned on and off in different sequences in the two organisms.

Basic Concepts

The basic chemical constituents and reactions in organisms provide many clues to their evolutionary relationships.

The genetic distance between organisms can be accurately determined by analyses of their macromolecules including DNA, and proteins such as cytochrome c and hemoglobin.

Evolution Observed

Most evolutionary events take place slowly over rather great periods of time; hence our knowledge of them must always be circumstantial to a large degree. It would be helpful to have some examples of evolution, observable changes in a species in response to a changing environment, that have occurred in historical times. Fortunately, a few such events have occurred.

Probably the best documented example is that of *industrial melanism:* the increase in dark-colored forms of moth populations and some other organisms in response to increased industralization in Great Britain and Western Europe during the past century or so. Originally the vegetation in these areas was relatively light colored, especially the lichen-covered tree trunks. A predominance of the moths were also light colored. However, soot from coal-burning factories eventually coated the vegetation until the landscape became almost uniformly dark. Now, it was hypothesized, the light-colored moths would be seen more easily, and thus captured more frequently by predators than their dark-colored relatives (Figure 17-14). This is precisely what has happened. Light forms, which once were very common, are now relatively rare in the industrial Midlands of England. Because of well-kept

records, it is known that 48 species of moths have become entirely melanic (dark) and about 250 others are much darker, since the start of the Industrial Revolution. In one of the few experiments ever performed in studying evolution, an English biologist, H.B.D. Kettlewell, raised large numbers of light and dark moths, then marked and released them in soot-covered woodlands. By means of motion picture film and recapture data, Kettlewell showed conclusively that predation by birds was much greater on the light-colored moths than on the dark ones. The reverse situation prevailed when moths were released in a nonpolluted woodland where light coloration was more of an advantage than dark in escaping the notice of predators. Thus the hypothesis concerning the change in the genetic nature of these moths by the action of natural selection (predation in this case) was reinforced by Kettlewell's experiments.

Another example of "evolution observed" is provided by the domestication of animals. All domestic animals had their origins in a wild ancestry and they have been changed by humans for our own benefit. Traits we consider desirable have been emphasized by breeding and others have been weeded out. This is human selection instead of natural selection. Significantly, the species that have been domesticated the longest have the most numerous domestic varieties (see Table 17-3). For example, dogs have been domesticated for over 8000 years and have about 200 varieties (Figure 17-15). Cats, domesticated for about half that time, have only 25 varieties. Varieties are interpreted as evolutionary changes. A similar pattern is also found in relation to domesticated plants.

Within very recent times we have seen the occurrence and spread of a number of drug-resistant and pesticide-resistant bacteria and insects. If a population of bacteria is exposed to an antibiotic drug, the bacteria usually die. Sometimes, however, one or two members of the population may happen to have genes that make them immune to the drug. These antibiotic-resistant forms survive, reproduce, and thus give rise to a population that is not affected by the antibiotic. The same sequence of events also happens with insects exposed to certain types of pesticides. This is another example of a very rapid type of natural selection and evolution.

Recent evolutionary changes in human beings are

Figure 17-14 *The dark form of the moth shown here demonstrates industrial melanism. Natural selection favors the dark moth against its dark background.*

TABLE 17-3. The Domestication of Principal Domestic Animals (After Dobzhansky, 1955)

Species	Domestication	Number of Varieties
Pigeon	Prehistoric	140
Donkey	Prehistoric	15
Guinea pig	Prehistoric	25
Dog	10,000–8000 B.C.	200
Cattle	6000–2000 B.C.	60
Pig	5000–2000 B.C.	35
Chicken	3000 B.C.	125
Horse	3000–2000 B.C.	60
Cat	2000 B.C.	25
Duck	1000 B.C.	30
Canary	1500 A.D.	20

Source: Table modified from paper by K. F. Dyer, Evolution Observed—Some Examples of Evolution Occurring in Historical Times. *Journal of Biological Education*, Volume 2, pp. 371–338, 1968.

more difficult to document because our knowledge about human evolution is not that well established. But one example that is clear-cut is that of sickle-cell anemia, discussed in Chapter 16. In parts of Africa where malaria is common, about 40 percent of the members of some tribes are carriers of the sickle-cell gene. This frequency indicates the adaptive advantage of being a carrier of the sickle-cell trait in a malarial region. By comparison, the incidence of sickle-cell carriers among blacks in the United States about 12 generations later is far lower. In the United States this condition is not advantageous in any way.

Basic Concept

Examples of evolution during recorded history include industrial melanism, domestication of animals and plants, the occurrence of drug- and pesticide-resistant organisms, and sickle-cell anemia in man.

Figure 17-15 *The results of human selection for different breeds of dogs.*

KEY TERMS

analogous structure	cytochrome c	homologous structure	ontogeny recapitulates phylogeny
chemical origin of life	evolution	industrial melanism	
creationism	fossil		radioactive dating

SELF-TEST QUESTIONS

1. Why is evolution said to be the major unifying theme in biology?
2. What is the general meaning of the term *evolution?*
3. What is the age of the earth?
4. How old are the oldest known fossils?
5. List the major steps in the chemical origin of life hypothesis.
6. What type of experimental evidence lends support to the above hypothesis?

7. How do meteorites and information from outer space support the chemical origin of life idea?

8. What evidence do we have that life exists on other planets?

9. What is our only direct evidence of life in the past?

10. Discuss ways that fossilization might take place?

11. What are two basic methods for dating fossils?

12. How can the structures of organisms be used to indicate relationships?

13. What is meant by homologous structures?

14. How can embryology be used to unravel complicated homologies?

15. Discuss Ernst Haeckel's "Ontogeny recapitulates phylogeny."

16. What are analogous structures?

17. What aspects of the chemical composition of organisms indicate their basic kinship?

18. What are some techniques for determining actual genetic closeness or distance between two organisms?

19. How does industrial melanism demonstrate evolution taking place within a relatively short period of time?

20. Describe how domestication of animals represents artificial or human selection.

21. What is an example of a recent evolutionary change in human beings?

chapter 18

Scene in the Galapagos Islands. Charles Darwin observed the unusual animals that have adapted to surviving in this stark environment.

EVOLUTION
PART II:
MECHANISMS AND THEORIES

chapter 18

questions to think about

What is the role of adaptation in evolution?

Why is it imperative for species to change over long periods of time?

What role does adaptive radiation play in evolution?

Why it is so difficult to study human evolution?

What is meant by natural selection?

In what ways did Darwin's work and theories influence men's ideas about nature?

---

*The unfit die — the fit both live and thrive
Alas, who says so? They who survive.*

Sarah N. Cleghorn

Biologists have long been intrigued by the various types of adaptive coloration in animals. Animals commonly have color patterns that blend them into their surroundings. Some, like the ptarmigan, a bird found

ADAPTATION

The term *adaptation* refers to the modifications of the structure, functions, or behavior of an organism that better enables it to survive in its environment. The study of adaptations provides valuable insights into evolution, because nature is never capricious with adaptations. Knowing this, we can quickly get caught up in the fascinating game of asking, "What function does *that* adaptation serve?"

Structural Adaptations

Structural adaptations refer to all aspects of the form and anatomical features of organisms that presumably make them better able to survive. They typically show up in body shape, coat color or patterns, limb structures, and teeth (Figure 18-1). They often tell us a lot about what an animal eats, how it moves, and where it lives.

(a)

(b)

(c)

(d)

(e)

(f)

Figure 18-1 *Structural adaptations in a variety of animals.* (a) *Giraffes. Their seemingly ungainly body enables them to browse on tree foliage that other herbivores cannot reach.* (b) *Young sea lion. This animal moves awkwardly on land but in the water is a graceful creature that is skilled at catching prey.* (c) *Wandering albatrosses on their breeding grounds in the Antarctic. Their long slender wings enable them to range far at sea for months at a time.* (d) *A striped skunk in its defensive posture. Its conspicuous coat pattern helps make the animal's presence noticeable.* (e) *A crocodile skull showing its peglike teeth for holding prey until it can be swallowed.* (f) *A horse skull showing its front teeth adapted for cropping vegetation, and the flattened molar teeth for grinding.*

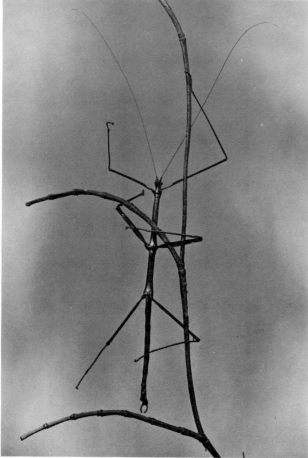

Figure 18-2 *Examples of cryptic coloration. The katydid's wings resemble leaves, and the walking stick insect looks like a twig.*

in the far north, change color with the seasons. The ptarmigan is white during the winter and brown during the summer. Many insects resemble leaves or twigs (Figure 18-2). These types of adaptations are known as *cryptic coloration.*

Warning coloration is common among animals that are especially venomous or dangerous in some way. The easily seen coloration of wasps, the bright colors of a coral snake, and the conspicuous pattern of black and white on a skunk advertise the animal's presence rather than conceal it.

Perhaps the most interesting use of color is *mimicry.* In this type of adaptation, one species mimics the coloration of another (the model). The model is usually a species that is repellent to predators because it tastes bad or is dangerous (Figure 18-3). Several species of harmless snakes mimic the coloration of the poisonous coral snake, and are therefore molested by fewer predators. The monarch butterfly, distasteful to birds, is mimicked by the viceroy butterfly.

Adaptations such as man's manipulative hands or large brain are obviously beneficial. On the other hand we can find many features in our favorite animal (us) whose functions elude us. For example, what adaptive function is served by oil glands, armpit hair, and a relatively hairless body? No one knows for sure.

Human adaptations to high altitudes, cold, and heat have been studied rather thoroughly. Peruvian Indians living in the Andes Mountains have a significantly larger chest and greater lung capacity than do coastal dwellers, a response to less oxygen at the high altitudes.

In relation to temperature, man's body size seems to conform, on a global scale, to a general rule that applies to other warm-blooded animals. *Bergmann's Rule* states that body size is larger in cold climates. This adaptation decreases heat loss, because large-sized bodies have proportionately less surface area in relation to their volume than do smaller bodies. If human body dimensions are compared over large areas such as China, Europe, or the United States, one can observe a gradual descrease in average size from north to south. (Eskimos are an exception.)

The variation in skin pigmentation in various groups of people is another structural feature with several possible adaptive functions. Thus, the dark skin typical of equatorial inhabitants prevents sun-

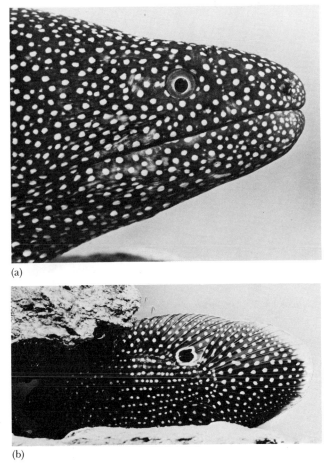

(a)

(b)

Figure 18-3   *An example of mimicry.* **(a)** *The head of the Spotted moray eel, a predatory fish that lives in reefs.* **(b)** *The tail portion of another fish, the Spotted Plesiops, mimicking the moray eel and thereby probably gaining protection against attacks by other fishes. Notice how the spot on the dorsal fin resembles the eye of the eel.*

burn, is more resistant to skin cancer, and has physiological functions described below.

Physiological Adaptations

Physiological adaptations are those related to the functioning of various parts of the body, including all of its internal chemical activities. Examples that we have mentioned in other chapters include sweating, the role of CO_2 in regulating breathing, the functions of glucagon and insulin in controlling blood sugar, and, indeed, virtually all homeostatic mechanisms.

In addition, many structural adaptations are accompanied by physiological adaptations. For example, people who live near the equator, an area of intense sunlight, have more skin pigmentation than those in temperate regions. This is an anatomical adaptation that probably serves one or more physiological functions. One of these functions may be related to vitamin D synthesis in the skin. When sunlight strikes the skin, it triggers a reaction that converts provitamin D into active vitamin D. The vitamin is necessary for proper bone formation and maintenance. A deficiency of the vitamin produces rickets, a disabling disease characterized by bent legs and spinal curvatures. On the other hand, it is also a vitamin that can accumulate in the body and become harmful in high concentrations. Dr. W. F. Loomis, a biochemist at Brandeis University, believes that this function played a major role in the evolution of pigmentation in the skin. In equatorial latitudes, natural selection would favor black skin (more pigmentation) that would shield the deep layers of the skin where vitamin D is made. In northern latitudes, selection would favor white skin so that maximum production of vitamin D could be obtained from available sunlight.

A more recent hypothesis is that dark skin protects against the breakdown of important substances in the skin by ultraviolet radiation in sunlight. One of these substances is folic acid, a member of the vitamin B complex. Folic acid is necessary, for example, for normal pregnancy, development of the fetus, and delivery of a healthy infant. Experiments have shown that the vitamin is rapidly destroyed in blood plasma samples exposed to ultraviolet light. Also, patients receiving ultraviolet light treatments for skin disease develop abnormally low levels of folic acid. Therefore, dark pigmentation helps to shield the deeper layers of the skin from the intense solar radiation experienced in the tropics.

Behavioral Adaptations

The various instinctive activities that are characteristic of different species constitute behavioral adaptations. Like structural adaptations, behavioral adaptations help organisms survive in particular environments. Behavioral adaptations are commonly associated with such activities as establishing territories, courtship and

(a)

(b)

Figure 18-4 *Behavior patterns of the common hognosed snake. In (a) the snake has flattened its neck area like a hood and is hissing audibly. If this bluffing behavior does not succeed in driving away a predator, the snake may roll over and pretend to be dead as in (b). The mouth gapes open and the tongue hangs out in a realistic "death" scene.*

Figure 18-5 *African honey guide bird.*

mating, raising young, finding food, and escaping predators. As you might guess, they are interwoven with functional and structural adaptations. Thus, the insect that mimics a part of a plant must perch on the plant in order for its mimicry to be effective. The bee-mimic fly not only looks like a bee but hovers over flowers in the same manner as a honeybee.

The common hognosed snake, sometimes called a spreading adder, uses an unusual combination of behavioral and anatomical features to discourage predators. When approached, the snake flattens its head and hisses audibly. It often makes false strikes. If these behaviors do not intimidate the predator, the snake rolls on its back and appears dead (Figure 18-4). The illusion is spoiled, however, because if you attempt to right it, the "dead" snake flops back over. Adaptations are not always perfect!

Bird behavior has been studied extensively and offers numerous examples of unusual behavioral adaptations. The African honey guide feeds on bees and bees' wax but cannot open the nest by itself (Figure 18-5). To get help, it attracts the attention of a human being or other large animal by its peculiar flight motion and churring call notes. In this manner it may guide the animal to a bee's nest. If the follower opens the hive to obtain the honey, the bird feeds on the bits of honey and wax that are left strewn around. Another aspect of this adaptation is the presence of special protozoans and bacteria in the honey guide's digestive track to break down the wax. Without these helpers, the bird could not digest the wax.

Human behavior is certainly adaptive, although it is highly conditioned by the culture in which we live. A crucial evolutionary development, as a human behavioral adaptation, was that of tool-making. The adaptation is considered to be one of the major features that helped us along our unique evolutionary pathway.

The adaptations we observe in organisms today evolved long ago and usually do not change appreciably over long periods of time. Thus, studies of the anatomical, physiological, and behavioral adaptations of a group of organisms provide valuable clues and insights into the group's evolutionary history.

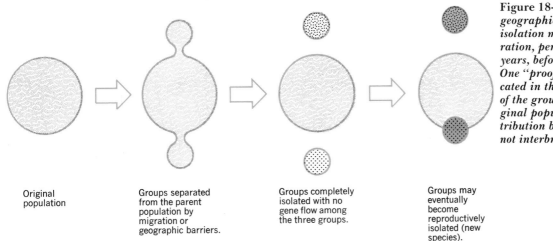

Figure 18-6 *Speciation by geographic isolation. The isolation must be of great duration, perhaps a million years, before new species arise. One "proof" of this is indicated in the last frame. One of the groups overlaps the original population in its distribution but their members do not interbreed.*

Original population

Groups separated from the parent population by migration or geographic barriers.

Groups completely isolated with no gene flow among the three groups.

Groups may eventually become reproductively isolated (new species).

Basic Concept

Adaptation is a major evolutionary feature. Three types are structural, functional, and behavioral.

SPECIES FORMATION

Abundant geological evidence supports the concept of an ever-changing, evolving earth. Over eons of time the surface of the earth, its climate, and the composition of the atmosphere have undergone massive alterations and will, indeed, continue in this manner into the future.

This geological evolution is highly significant in that it has forced most organisms to also change (adapt) or else become extinct. The fossil record of any portion of the earth usually tells us that extinction often did occur, but that some organisms were able to adapt and change through time. The changing of a species into one or more different species is termed *speciation*. Because speciation is a basic event in the process of evolution, biologists have devoted a large amount of time and effort in studying the mechanisms by which it takes place.

In biology, the term *species* refers to a population of organisms that is reproductively isolated (interbreeds only with its own members), is usually recognizable by distinctive anatomical features, and occupies a specific geographical area. The population itself is often com-posed of a group of subpopulations, each adapted to the slightly different environments that usually exist within the major geographical range of the species. These interbreeding subpopulations within a species are known to biologists by a variety of terms such as races, varieties, local populations, or subspecies. An example is the common eastern raccoon. It occupies all of the eastern United States from Maine through the Florida Keys. At least ten subpopulations (subspecies) are recognized, each differing slightly in size or coloration from the other subspecies.

Many biologists believe that subspecies represent an early stage in the formation of new species. Several events would have to take place for *speciation* to happen — that is, for the creation of a distinct species. The first step is the isolation of a population from other subgroups of the species by some sort of a barrier. Usually this is a geographical feature such as a mountain range, a large river, a desert, or some other inhospitable environment. This isolation is necessary so that the population can no longer share its genes with other subgroups. Other barriers to gene flow may eventually arise, such as those involving ecological or behavioral differences. If the isolation time is long and if the populations become sufficiently diverse, then a new species may form (Figure 18-6). The best test of the success of speciation occurs when the barriers are changed so that the populations again meet. A lack of interbreeding (no hybrids can be found) indicates that a distinct species has indeed been formed. The new species is now reproductively isolated from its ancestral stock. This process, or something similar to it, is thought to account for the formation of most of the species we see today.

Many anthropologists and biologists believe that the different groups of human beings found around the world are the equivalent of subspecies of the species *Homo sapiens.* The human races we see today may represent a case in which speciation failed to take place. One interpretation is that our ancestors evolved into a lot of different populations, but these groups did not remain isolated long enough to form separate species. No race or group of humankind today is reproductively isolated from any other group.

It should be noted that biologists are by no means in agreement about the definition of a species or about how new species arise. The diversity of life on the earth is so great that some biologists feel that many different processes of speciation must have occurred. In addition they feel that the term *species* as applied to prokaryotes like bacteria is not equivalent to the species of higher animals or plants because many prokaryotes do not reproduce sexually. This is likely to be a never-ending biological debate.

ADAPTIVE RADIATION

Divergence
The process of speciation, extended through enormous periods of time, often leads to an important evolutionary event termed *adaptive radiation.* When a group of organisms enters a new environment—one not already occupied by other organisms—it tends to *disperse* and become adapted to different parts of the environment. With the occurrence of barriers and sufficient isolation, these organisms are led by natural selection down a variety of adaptive pathways. One of these pathways might control different feeding techniques: One offshoot of the group may become adapted to feeding in tree foliage, another to ground-feeding, and still another to capturing food in lakes and streams. These feeding roles may involve very different morphological, physiological, and behavioral adaptations. Consequently, the basic ancestral group

has undergone an adaptive *divergence* in respect to feeding habits.

Remote islands and isolated continents in the past provided unused environments awaiting the chance arrivals of plants and animals. The few who made such journeys, blown by storms or perhaps surviving on rafts of matted trees, found areas with no competitors and numerous unused living places. Their consequent adaptive radiation pathways have intrigued biologists for a hundred years.

Among the hardy colonizers that reached the volcanic Hawaiian Islands about a million years ago were some small finchlike birds. Having no competitors, these colonizers spread over the islands and underwent an almost explosive speciation (adaptive radiation) to take advantage of the unused ecological situations around them. Some retained a thick finchlike bill for seed eating, some became woodpeckerlike and fed on wood-dwelling insects, others evolved slender warbler-like bills for feeding on small insects and berries, some became parrotlike with bills for eating flowers and fruits, and still others developed long slender beaks for obtaining nectar from flowers. Today this group, or what is left of it, is known as the Hawaiian Honeycreepers (Figure 18-7). Their adaptive radiation took place in a relatively short period of time as evolutionary time is usually considered—less than a million years. A similar set of events occurred in the Galapagos Islands situated 600 miles off the coast of Ecuador. These islands were also colonized by finches and a few other animals that radiated into numerous feeding types. Because Darwin was so impressed and influenced by the strange animal life and speciation that had taken place there, these finches are now called Darwin's finches.

Evolutionary Convergence
Many times during the course of adaptive radiation relatively unrelated organisms have become adapted to similar environments. Obvious examples are birds and bats, or whales and fishes. Superficial similarities may be large, but the basic structure of the organisms betrays their distinct ancestries.

Whales and fishes live in the same general environment and their general body form is much the same.

Figure 18-7 *The results of adaptive radiation in the Hawaiian Honeycreepers. This amazing array of bill shapes arose from a common finch ancestry. As a consequence of this evolution, honeycreepers are able to take advantage of many different sources of food from seeds to flower juices.*

Their respiratory systems show marked differences, however. When the lung-breathing ancestors of whales first entered the water, a system such as gills would have been most useful. Yet such a system did not develop in whales. Instead, mutations occurred that modified the way the whale utilizes its oxygen. The whale's lungs remained the area of gaseous exchange,

and these can accommodate rather large supplies of air. More drastic adaptations occurred in the operation of the circulatory system and in the ability of certain cells to tolerate low oxygen supplies. When a whale dives, its heart rate drops and most of the blood is shunted to the nervous system, with little going to the extremities. The sensitive nervous system receives sufficient oxygen to prevent damage and to allow respiratory movements to be temporarily suspended. Cells in the extremities receive little oxygen and must tolerate a low oxygen supply.

This type of adaptation to life in the water does not appear to be as advantageous to the organism as the development of gills; but lungs were already present and gills were not. Obviously, evolution must function only with structures that are available or with those that happen to develop because of mutation or genetic recombination.

Some Major Divergences

Examples of major radiations that gave rise to entirely new classes of organisms are seen with the conquest of land by amphibians, the replacement of amphibians by reptiles, and then the eventual dominance of mammals over the reptiles (Figure 18-8). Each group was dominant for many millions of years, but each was eventually replaced by a group with more successful adaptations. For example, the evolution of the amniotic egg, with its internal protective membranes and external protective covering, gave reptiles a great advantage over amphibians: complete independence from the water where most amphibians must deposit their soft eggs. With this and other adaptations, reptiles became supreme. For the next 200 or so million years they underwent a spectacular adaptive radiation, giving rise to the many kinds of herbivorous and carnivorous dinosaurs, fishlike forms, spectacular flyers like the recently found fossil of a pterosaur with a 15-meter wingspan (but no teeth!), and many others (Figure 18-9). All of these magnificent creatures were doomed to extinction for reasons still not clearly known. Only a tiny remnant, the modern reptiles and birds, remains today.

As reptiles declined at the close of the Mesozoic era, one of their offshoots was a group of small primitive

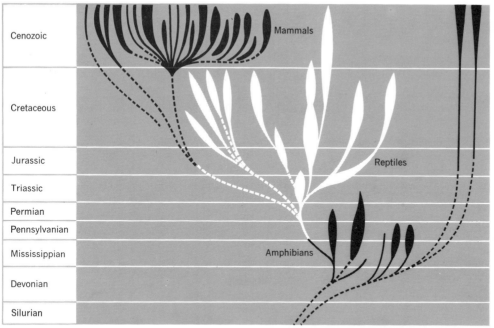

Figure 18-8 *Replacement of amphibians by reptiles and reptiles by mammals in different geological time periods. Each group was successful for an extensive period of time but eventually gave way to another group with better adaptations.*

mammals. These animals slowly took the place of the vanishing reptiles and eventually entered another round of adaptive radiation. Like that of the reptiles, it produced fliers, runners, swimmers, herbivores, carnivores, and scavengers. The present-day mammals are the consequence of this radiation.

The study of adaptive radiation tells us a lot about the major features of evolution. One of these is that *extinction* and *replacement* is the eventual fate of most groups of organisms. Another is that natural selection is frequently pushing organisms down the path of *specialization*. Specialization can be an evolutionary trap because it is always a one-way road—an organism cannot retrace its evolutionary steps back to the generalized form because this would require precisely the same mutations and conditions that produced it in the first place, but in reverse order!

Basic Concepts

Adaptive radiation refers to the adaptive pathways followed by groups in their evolution. It may act to produce divergence or convergence.

Extinction and replacement is the eventual fate of most groups of organisms.

HUMAN EVOLUTION

The evolution of the animal we most want to know about has turned out to be the most difficult to study. To begin with, the human fossil record is fragmentary and therefore incomplete. A few new human-type fossils are found every year, but they add to the complexity of the human evolution story rather than clarify it. Even the chief investigators themselves are not in agreement on how to interpret the human fossil record. Consequently, the information and ideas presented in this section of the text will likely change before this edition of the text is published.

Primates apparently had their origin at the beginning of the Cenozoic era approximately 70 million years ago (see the Geological Time Table in Chapter 17). Primates, as you recall from Chapter 2, are distinguished by a large cerebrum, stereoscopic vision, flattened nails, prehensile appendages, an opposable thumb, and a well-developed collarbone.

Our account of human evolution begins in the Miocene period, which began about 25 million years ago. Fossils from this period indicate the beginning of the human-ape lineage, the hominoids. From this lineage came the great apes (see *Dryopithecus* in Figure 18-10)

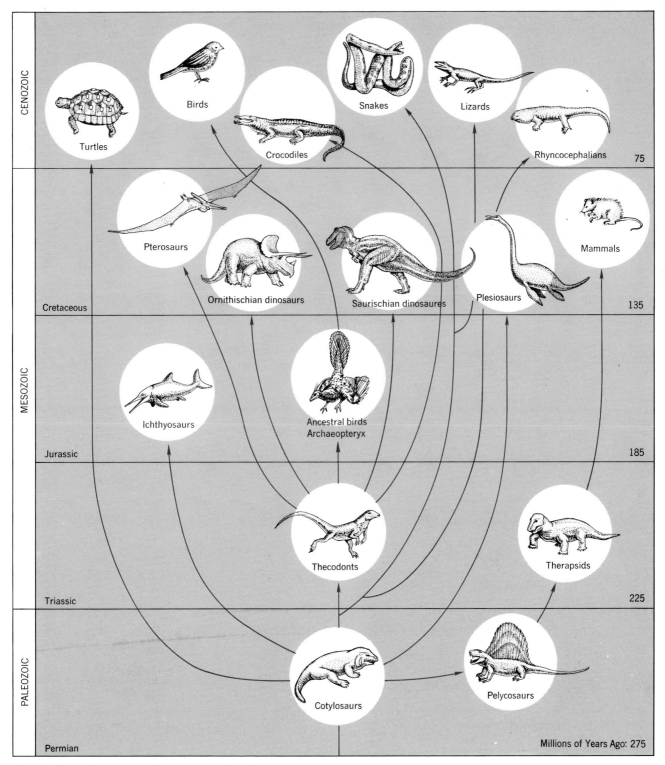

Figure 18-9 *Adaptive radiation in reptiles. As shown, reptiles became adapted to all major habitats and literally dominated the earth for over 200 million years.*

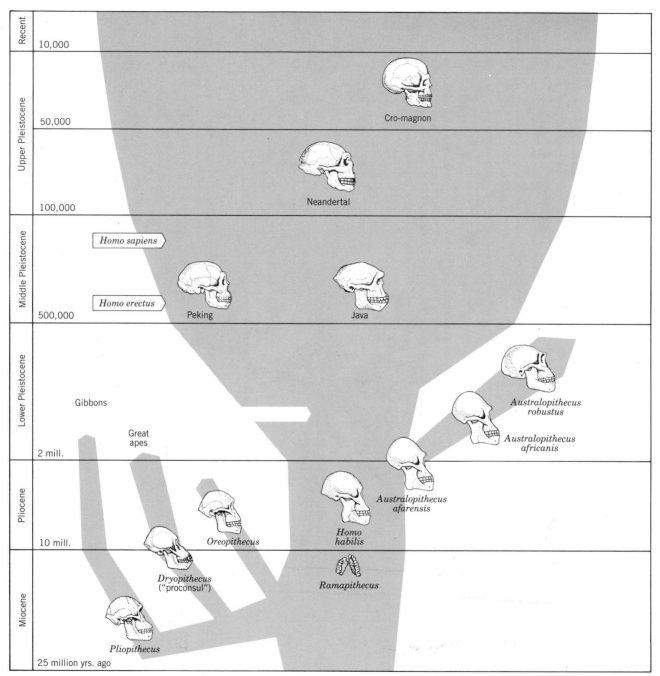

Figure 18-10 *A generalized evolutionary tree for humans. As indicated, human evolution appears to involve a bewildering array of fossil forms instead of a single lineage. It also includes side branches such as* **Australopithecus.**

and another group of fossils, in the genus *Ramapithecus.* The latter group is thought to represent the beginning of human ancestry because they had humanlike teeth.

Ramapithecus lived about 10 million years ago throughout Africa and Asia. They were apelike creatures, probably tree dwellers feeding on fruits and nuts. Opinion is divided as to whether they were bipedal.

From *Ramapithecus,* we jump far up in time to the lower Pleistocene period two to three million years ago. From the well-known years of work of the Leakey family in the Olduvai Gorge of South Africa, plus the fossils found by other anthropologists, we are introduced to the little South African ape man, *Australopithecus.* These creatures were four to five feet in height, less than 100 pounds in weight, bipedal, probably lived in open countryside, and may have used crude pebble tools (although this is debatable). Their brains were only about half the size of the brains of modern humans, but the external features of the brain, as determined from brain casts, were definitely humanlike rather than apelike. From all available evidence, it seems that *Australopithecus* was part of the human lineage in some fashion (Figure 18-11).

In a remarkable story of self-sacrifice and determination, the Leakeys spent 30 years searching in the arid, bleak Olduvai Gorge for the beginnings of mankind. In 1959 Mary Leakey finally discovered a humanlike lower jaw. This was the first australopithicine fossil found in the gorge. Additional fossils eventually turned up, and some of these were interpreted by the Leakeys as being different species of *Australopithecus.* To further complicate this search for human beginnings, a skull was found by Richard Leakey in 1972 that was contemporary with *Australopithecus* but had a larger brain. The new fossil was named *Homo habilis* and considered by the Leakeys as being in the direct lineage with modern humans. *Australopithecus* was relegated by the Leakeys to the position of being a side branch rather than in the main line of human evolution (see Figure 18-10).

In 1974 a young American paleoanthropologist, Donald Johanson, had the good fortune to find the first partial skeleton of an australopithecine in Ethiopia. It was dated by surrounding rock material as over 3.5 million years old and thus much older than the fossils found by the Leakeys. From the skeleton it was determined that the individual was a female and walked upright. Johanson nicknamed this remarkable fossil "Lucy." After a detailed study of Lucy's remains

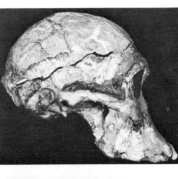

Figure 18-11 *Skull and reconstruction of* **Australopithecus,** *the South African ape man that lived 2 to 3 million years ago.*

and comparisons with other australopithicine fossils, Johanson and a colleague, Tim White, gave Lucy the scientific name *Australopithicus afarensis.* Furthermore, Johanson and White concluded that Lucy's humanlike

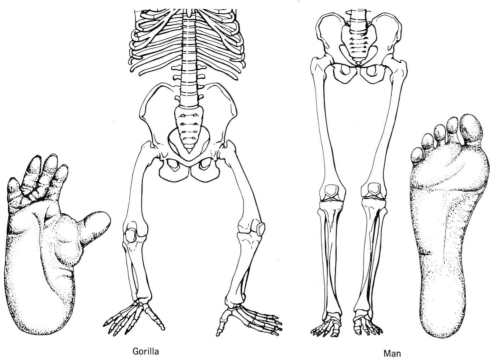

Gorilla   Man

Figure 18-12 *Some of the anatomical changes that came about with bipedalism in humans. Note the elongated limb bones, shortened pelvis, and reshaping of the foot.*

skeletal characteristics and ancient age prove that *Australopithecus* is in the mainstream of human evolution and not a side branch as proposed by the Leakey's.

The Leakeys disagree with this interpretation and the age established for Lucy. As an interesting sidelight to this scientific controversy, in 1981 skull fragments were found not far from Lucy's burial site that are at least 4 million years old; they are thought to be the same species as Lucy. If correct, this pushes our ancestry even further back in time. Hopefully, additional fossil discoveries will clarify this fascinating story of our beginnings.

The evolution of an upright stance (*bipedalism*) was one of the crucial events that allowed humans to emerge as they appear today. This adaptation began more than 3 million years ago; in fact, fossil footprints from Africa indicate that the precursors of early man walked upright at least 3.5 million years ago. This development involved many anatomical changes in the body, including elongation of the hind limbs, shortening and broadening of the pelvis, changes in the hip musculature, strengthening of the hip and knee, and

reshaping of the foot (Figure 18-12). The upright gait was possibly the most important anatomical change that set man apart from other apelike animals. In addition the upright stance, of course, freed our early ancestor's hands for tool-making. With this ability, humans became efficient hunters and gatherers of food.

Another important facet of early human evolution, many anthropologists believe, was the forming of pair-bonds between males and females. This behavior would have improved reproductive success by allowing the female more time to care for her offspring while the male provided food and protection. The male in turn had continual access to a reproductive partner. Such an arrangement may have led to the creation of family groups, small cooperative working groups, and an emphasis on productive social behavior.

By middle Pleistocene, 400,000 years ago, an advanced type of human became widespread. Examples are Java man and Peking man, now called *Homo erectus.* They were cave dwellers, fire users, tool makers, and hunters. They approached modern man in brain size.

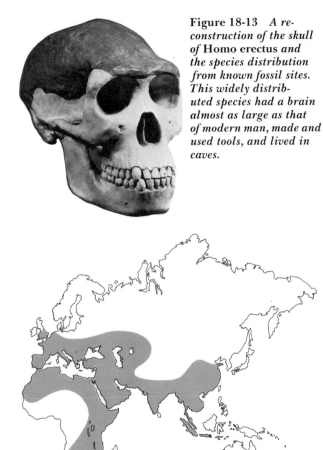

Figure 18-13 *A reconstruction of the skull of* Homo erectus *and the species distribution from known fossil sites. This widely distributed species had a brain almost as large as that of modern man, made and used tools, and lived in caves.*

The wide distribution of their fossil remains attests to their success in spreading over wide areas of Europe, Asia, and Africa (Figure 18-13).

All known fossils after middle Pleistocene are placed in the species *Homo sapiens,* indicating their identity with present-day human beings. One major group, *Neandertal man,* inhabited caves in Europe and the Middle East until about 40,000 years ago. Although often depicted as the brutish stereotype of a cave man, their skeletal characteristics are similar to those of modern man; however, their bones were somewhat heavier and they had more powerful muscles. Indications are that they were extremely physically active peoples. The skull of Neandertal man indi-

cates a brain somewhat larger than that of modern man.

About 35,000 years ago, Neandertal man was replaced by an essentially modern *Homo sapiens* known as Cro-Magnon man. Evidence does not clearly indicate whether the Neandertals were wiped out by the Cro-Magnons or if they simply evolved into these modern peoples. For approximately the past 10,000 years, our evolution has been mostly cultural, because our biology has changed very little. Unlike biological evolution, we can control cultural evolution, if we wish, and direct it into pathways beneficial to all humanity. So far our attempts have been elementary. We repeatedly engage in disruptive actions such as wars and mindless exploitation of the earth's resources. It is both depressing and frightening that our cruelest, most inhumane behavior is directed at members of our own species. Some behavioral biologists such as Konrad Lorenz propose that the human capacity for violence — that is, aggressiveness — stems from our evolutionary past. If our ancestors were hunter-apes, as some claim, then we may retain some of the behavioral features that allowed us to survive: the ability to hunt for prey, to use weapons, to defend territory, to dominate in relationships with our own kind, and so on. Desmond Morris, an English zoologist, develops this viewpoint in his popular book, *The Naked Ape,* as have other popular writers such as Robert Ardrey. In view of our anatomical and physiological heritage, it is not unreasonable to assume that some of our behavior also evolved. However, this assumption has little supporting evidence at this time.

In contrast to these views, paleoanthropologists such as the Leakeys believe that prehistoric man was innately a gentle, cooperative, food-sharing creature and that this behavior was important in early human evolution. Such anthropologists also argue that human aggression is a learned trait that persists because it is so frequently reinforced. For example, many societies encourage and reward competitiveness (a kind of aggressiveness) in their members. Misdirected or unchanneled aggressiveness easily turns into violence and social disorder, especially when people face the frustrations of overcrowding, poverty, and unresponsive social and political organizations.

At this point there is scant evidence whether our ancestors were basically peaceful or aggressive. This is but one of many reasons for continuing to delve into man's evolutionary past. At any rate, we arrive on the modern scene with a creature who is determined to dominate the earth and use it for his or her own benefit. Now we are even beginning to explore space and to seek avenues to other planets. As Shakespeare so aptly stated long ago, "What a piece of work is man!"

Basic Concepts

Hominid evolution began in the Miocene period with *Ramapithecus* about 30 million years ago.

Major factors in human evolution have been the development of an upright stance, tool-making, pair-bonding, and an increase in brain size.

DARWIN AND NATURAL SELECTION

To most people, the name Charles Darwin brings to mind a famous personage who was responsible for proposing a theory about evolution. What most people do not know is that this famous man's personal life and behavior were sometimes so bizarre that one wonders how he accomplished so much.

Biographers tell us that Darwin was quite introverted and shy, and suffered some type of undiagnosed, disabling illness through most of his adult life. He was frequently bedridden for weeks with a variety of ailments which included headaches, upset stomach, weakness, and so on. Despite his chronic illnesses, he maintained a steady correspondence with a number of prominent scientists of his day and poured out a succession of scientific papers and books. This is all the more impressive when you consider that he was not formally associated with any academic institutions or museums, as were most of his scientific contemporaries. In fact, he was never employed by anyone except for the time he spent as a naturalist (unpaid) on the exploratory voyage of *H.M.S. Beagle* (1831–1836). When he returned from this rather arduous voyage, he married his cousin, a member of the Wedgwood family that produced the well-known Wedgwood china. With the aid of generous financial support from his father and his wife's wealth, Darwin settled down to lead the life of a nineteenth-century English country gentlemen

Figure 18-14   *Charles Darwin as a young man.*

(Figure 18-14). As appealing as this sounds to you and me, this life-style evidently did not bring Darwin much peace of mind because he was ill off and on for the rest of his life. Despite the illness, he became a brilliant scientist and wrote many impressive works in biology.

The five-year voyage of the *Beagle* constitutes quite an adventure story in itself (imagine five years with no television or air conditioning!), and Darwin's account of the trip, *the Voyage of the Beagle,* is probably the most interesting of his writings. *H.M.S. Beagle* was a square-rigged sailing vessel converted from naval duty for use as an exploratory and survey ship. She was uncomfortable, only 90 feet long, with 5-foot headroom below decks, overcrowded with crew members, inadequately stocked with food and water, and generally badly equipped for a five-year cruise to remote parts of the world. The *Beagle* was charged with obtaining information about uncharted coastlines in various parts of the world and collecting biological and geological specimens (Figure 18-15). Nonetheless, thanks to the skill and leadership of its captain, *H.M.S. Beagle* performed

Figure 18-15 *(Top) A water color painting of* **H.M.S. Beagle,** *the small sailing vessel on which Darwin spent five years. (Bottom) Map of the voyage of H.M.S.* **Beagle.**

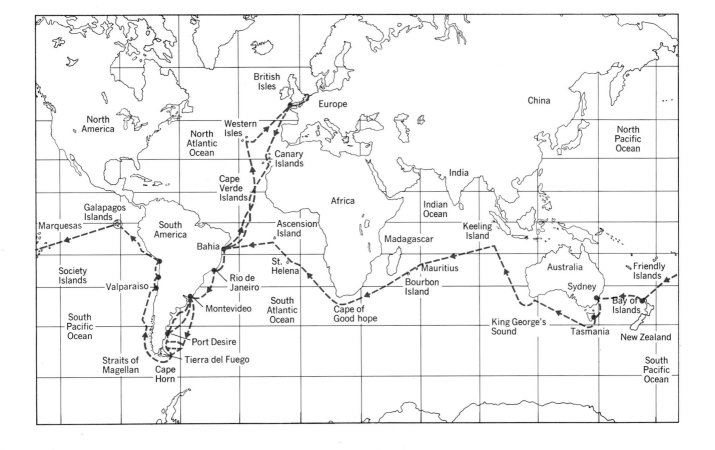

these duties and eventually returned its young naturalist, Darwin, to England.

During the round-the-world voyage, which included stops in South America and the Galapagos Islands, Darwin collected numerous specimens, visited strange lands, and kept a careful diary. All of this information and experience formed the background against which he was to formulate his theory of how evolution took place. The writings of other scientists also influenced his thinking, but the data obtained during his long trip provided a unique fund of information that he often used.

In retrospect, we find that Darwin made two valuable contributions to the concept of organic evolution. First, he provided numerous observations to support the contention that evolution had taken place. The idea of descent through time preceded Darwin, but Darwin provided the first systematic body of confirming evidence. Secondly, he proposed a theory for the mechanisms involved in the process of evolution. This was the contribution for which he became famous and, for a time, very controversial. His theory of natural selection is still considered basically correct; we will examine some modern refinements of it a little later.

Basic Concept

Darwin made two major contributions to the concept of evolution:

1. Evidences that evolution had occurred.

2. A mechanism to explain how evolution took place.

DARWINIAN NATURAL SELECTION

Darwin's theory of natural selection was published in 1859 in a book entitled, *On the Origin of Species by Means of Natural Selection*. The following account is a condensed version of Darwin's main points.

Variation

Members of a population always exhibit a considerable amount of variation with respect to almost any observable trait. Some members are lighter or darker in color, some are larger, some smaller, some are more aggressive in their behavior, and so on (Figure 18-16). In modern terminology we would say that for many traits there is a mean (average) value around which individuals tend to vary.

Figure 18-16 *The variation in physical traits in this group of children of different racial backgrounds is obvious.*

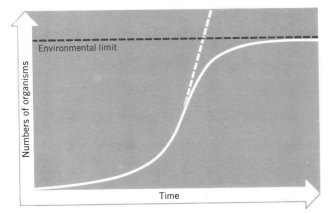

Figure 18-17 *Struggle for existence. A population grows until limited by environmental resources. The population then levels off as shown on the graph. Excess numbers of offspring may be produced in each generation, but most of these will die struggling and competing for available resources.*

Darwin was familiar with such variations because he had made and studied collections of many kinds of plants and animals.

Struggle for Existence

Organisms in a population frequently produce excess numbers of offspring, and yet the population usually remains uniform in size. This implies that many offspring die in some sort of a struggle to exist on whatever resources are available (Figure 18-17).

The overproduction of offspring is a familiar biological phenomenon that has been documented many times for many kinds of organisms. As Darwin pointed out, even in slowly reproducing creatures, such as elephants, if all of the offspring always survived and eventually reproduced, elephants would soon become unpleasantly numerous.

Darwin claimed to have been influenced in his viewpoint about the struggle for existence by the writings of Thomas Malthus, an economist. Malthus pointed out that the human population was increasing at a faster rate than its food supply. Inevitably, he believed, this condition would lead to a struggle for existence, and only forces such as famine, disease, and war would act to check the population growth.

It is of some interest to note that Malthus's ideas about the unchecked population growth impressed not only Darwin but others as well down through the years. Even today the Malthusian doctrine is cited as sort of a doomsday prophecy if humanity does not soon exert better control over human population growth.

Competition and Survival of the Fittest

With many offspring struggling and competing for existence, some will survive and some will die. Those who possess variations that best fit them to the prevailing conditions have the best chance of surviving. The fittest, being likeliest to survive, are therefore the likeliest to pass their traits on to the next generation. Darwin thought it appropriate to call all this process *natural selection*, because it is the environment (nature) which selects those that are fittest for survival.

If, as Darwin envisioned, natural selection functioned generation after generation for long periods of time, species would gradually change through time. Darwin proposed that natural selection applied through eons of time on a slowly but constantly changing planet, accounted for the profusion and the particular adaptations of all the life forms we see on the earth today.

Using this theory, biologists had a basis for reasonable explanations of most characteristics of any organism. The monkey's tail, the zebra's stripes, or the human hand all represent features that enable these organisms to survive better in the struggle for existence. In the course of evolution, these features were selected by nature from a pool of variants in a slow sort of trial-and-error process (Figure 18-18).

Basic Concept

Major points in Darwinian Natural Selection include:

1. **Variation within members of a species**
2. **Struggle for existence**
3. **Competition and survival of the fittest**

THE DARWINIAN REVOLUTION

Interest in evolution and its possible mechanisms was already high when *Origin of Species* appeared in 1859. It was read and discussed by many people in addition to scientists, and it quickly became the center of many

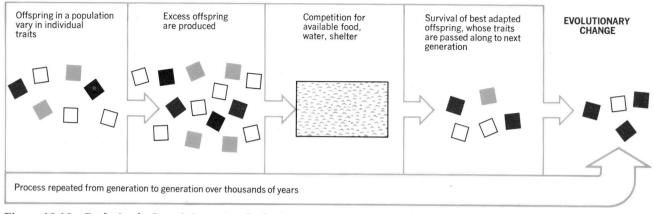

Figure 18-18   *Evolution by Darwinian natural selection.*

debates. As is often the case with new ideas, the arguments pro and con often became based on emotion rather than on intellect. Finally, the dust settled, so to speak, and scientists generally accepted the fact that evolution had occurred and the concept of natural selection as the best available explanation for it. The acceptance of Darwin's work and the implications of his concepts had a profound effect on many existing ideas about man and nature. This effect has been termed the Darwinian Revolution. According to Dr. Ernst Mayr, a renowned evolutionist and taxonomist at Harvard University, some of the major effects were as follows:

1. People recognized that the earth was far older than the age given it by creationists (6000 years). This fact, established by geologists such as Charles Lyell, provided the vast time scale that would be necessary for the occurrence of evolutionary events as Darwin perceived them.

2. A rejection of the idea that the earth underwent periodic catastrophes that wiped out everything (as in the story of the flood and Noah's ark for example). Instead, according to geological evidence, the earth has more or less uniformly and slowly evolved. All of the kinds of geological events that occurred in the past are still occurring today to some degree or other.

3. A rejection of the idea of evolution as a steady progression toward perfection. Evolution by natural selection is a continuing process of adaptation to ever-changing conditions, which therefore can never have any ultimate goal or result. Instead of leading to perfection, it often takes plants and animals down a dead-end pathway to extinction. The so-called "tree of life," in other words, is more like a great branching, unkempt shrub than the tall upward spire of a Douglas fir.

4. A rejection of creationism as the only explanation for the diversity of life and for the origin of life itself. The concept of evolution implies that at some point of time, living matter evolved from nonliving matter.

5. A change from the viewpoint that man is the center of the living world and the endpoint of evolution to the view that he is just one of numerous organisms in the evolutionary stream.

6. Although not listed by Dr. Mayr, there was also a replacing of *Lamarckianism* by natural selection. Lamarck, a famous French biologist who preceded Darwin by a century, had proposed that organisms evolved in response to needs they encountered in the environment and that acquired characteristics could be inherited. According to his widely accepted concept, things that happened to an organism during its lifetime would be passed along to its offspring. For instance, if a person exercises regularly and builds up large muscles, he will pass his well-developed musculature on to his

children. Although you and I know that this is not genetically possible, it is nevertheless an easy intellectual trap to fall into. For example, if you accept the idea that monkeys evolved tails as a consequence of climbing in trees, then you have fallen into a Lamarckian trap. The more accurate way of looking at it is this: Monkeys that evolved longer tails by chance genetic variation found themselves better able to climb in trees.

Basic Concept
The Darwinian Revolution influenced humans' concepts about the following:

1. **Age of the earth**
2. **Catastrophism**
3. **Evolution as a perfecting mechanism**
4. **Creationism**
5. **Man as the center of life**
6. **Lamarckianism**

MODERN NATURAL SELECTION

After 1900, the rapidly developing field of genetics added some new dimensions to Darwin's natural selection concept. For this reason, biologists sometimes refer to the present-day concept, which includes all that we have learned from genetics, as *Neodarwinism.* Going back to our old outline of natural selection on page 329, we can now add the following modifications.

Variation
Darwin and others knew that variation existed, but no one at that time understood the sources of genetic variation. Variation is now attributed to events that occur during meiosis and recombination of gametes, and to mutation.

We have said that mutations are unpredictable and usually disadvantageous. How, then, can they ever help the cause of evolution? The answer lies in the great amounts of time available. Once in a great while, a mutation takes place that happens to be beneficial and leads to better adaptiveness. Unfavorable mutations are quickly weeded out; the occasional favorable one becomes the basis for evolutionary change.

The important thing is how a mutation affects the gene "pool" of a population rather than its effect on a single member. Members of a population vary according to the genes they happen to inherit from the gene pool of the population. A higher proportion of mutant genes in the pool will cause more variation in the population than a low frequency of mutant genes.

Struggle for Existence
Darwin's view that larger numbers of offspring are produced than usually survive is generally true. In addition, it is now known that a species may have built-in mechanisms for limiting its numbers. Some organisms, when crowded, bear fewer offspring, reproduce less often, or may even devour part of their litters.

Competition and Survival of the Fittest
Competition, as now viewed, rarely involves direct physical combat. More often it is a matter of using available resources or acquiring reproductive partners more efficiently. The competitors may never even see one another.

The fittest organisms are now considered to be those most successful at reproducing the most offspring. "Survival of the best reproducers" would be a better description than survival of the fittest. The best reproducers, for whatever reason they are best, leave the largest number of offspring and, therefore, the greatest number of genes in the gene pool. For this reason, their traits are found more often in subsequent generations. It is not merely the ability to survive, but the ability to survive and reproduce that really counts.

Applying this concept of natural selection functioning generation after generation, we see that gene pools, and thus populations, change slowly through time. However, the outcome of the process is totally unpredictable for two reasons. Mutations are random and unpredictable, and the probability that a beneficial mutation will occur precisely when it is needed is slight. However, over vast eons of time, coincidences that are unlikely in the short run eventually do occur. Secondly, changes in the environment of the earth itself do not follow a predictable pattern. Life forms continue to adapt to the ever-changing environment, which more or less leaves "selection by nature" up to the whims of nature (Figure 18-19).

Basic Concept

Modern natural selection involves the following ideas:

Variation is attributed to genetic events.

There is a struggle for existence and competition.

The fittest are defined as the best reproducers.

Offspring vary due to genetic events such as crossing over, recombination, and mutations

↓

Excess numbers offspring often produced

↓

Competition occurs, often subtle and usually involving improved reproductive means

↓

Survival of the best reproducers

↓

Repeat process over eons of time

Figure 18-19 *Evolution by modern natural selection.*

KEY TERMS

adaptation	Cro-Magnon man	*Homo sapiens*	specialization
adaptive radiation	cryptic coloration	Lamarckianism	speciation
Australopithecus	Darwin	natural selection	structural adaptation
behavioral adaptation	divergence	Neandertal man	struggle for existence
Bergmann's Rule	*Dryopithecus*	neodarwinism	survival of the fittest
bipedalism	extinction	physiological adaptation	variation
competition	*Homo erectus*	*Ramapithecus*	warning coloration
convergence	*Homo habilis*	replacement	

SELF-TEST QUESTIONS

1. (a) Define the term *adaptation.*
   (b) List three types of adaptations and give examples of each.
2. Explain why skin color is an anatomical adaptation.
3. In what way could skin color also be a physiological adaptation?
4. What is the relation between speciation and geological evolution?
5. Define the term *species.*
6. What are thought to be the major steps in speciation?
7. Define adaptive radiation.
8. What is meant by the term *divergence* in relation to adaptive radiation?
9. Explain how the Hawaiian Honeycreeper birds exemplify adaptive radiation.

10. What is the difference between divergence and convergence in terms of evolutionary trends? Give examples to support your answer.

11. Why is our knowledge of human evolution so meager?

12. What is the significance of the fossils in the genus *Ramapithecus?*

13. Discuss the importance of *Australopithecus* in human evolution.

14. Explain the importance of bipedalism in human evolution.

15. On what grounds might you defend the idea that humans evolved from a hunter-ape background?

16. On what grounds might you defend the idea that humans evolved from peaceful, cooperative ancestors?

17. What was bizarre about Darwin's behavior when he was an adult?

18. What did the voyage of *H.M.S. Beagle* contribute to Darwin's knowledge?

19. Name the two major contributions Darwin made to the concept of evolution.

20. Discuss three major points in Darwin's natural selection theory.

21. What influence did Thomas Malthus have on Darwin's ideas?

22. Name five major effects of the Darwinian Revolution.

23. State how modern natural selection differs from the Darwinian concept.

24. Why is the outcome of natural selection totally unpredictable?

chapter 19

THE BIOSPHERE

chapter 19

questions to think about

Why is the earth a highly suitable place for life?

What roles do mineral cycles play in the biosphere?

What is an ecosystem?

How is life supported in an ecosystem?

Can nonhuman animals show altruistic behavior?

---

*The waters of ancient seas set the pattern of ions
in our blood. The ancient atmospheres molded
our metabolism.*

George Wald

Has it ever occurred to you that the earth is a highly
suitable place for living things? The temperature is
about right, there is light for photosynthesis, plenty of
water, and the necessary chemicals to sustain life. Is
this just a fortunate coincidence, or is there some
deeper significance to the earth — life relationship?

The answer is that chance and coincidence are
partly responsible, but that the presence of life itself
has profoundly affected many features of the earth and
its atmosphere as we know them today. To understand
this we need to look at some of the physical characteristics of the earth, how these affect life, and the consequences of this earth-life interaction.

THE BIOSPHERE

The term *biosphere* refers to the part of earth on which
life (*bios*) exists in the soil, water, and air at the surface

of our planet. Three basic conditions make the biosphere possible. One is an abundant supply of solar
energy, the ultimate energy source of all life on the
earth. Another is ample liquid water, a necessary ingredient for life. Finally, the third is the existence of a
variety of chemicals in the form of elements, ions, and
minerals that will sustain life.

Solar Energy

It is certainly understandable why many ancient cultures worshipped the sun: It is quite literally the giver
of life. Radiation from the sun reaches the earth as a
mixture of short and long wavelengths, partly filtered
by the atmosphere. The wavelengths important to life
are infrared (heat), visible light, and the ultraviolet.

By coincidence, the earth's distance from the sun is
approximately right to support life as we know it. If the
planet were too close, the sun's infrared and ultraviolet
radiation would be too intense. If too far, the infrared
and visible wavelengths would be too weak.

The *visible light* of the sun is, first of all, the energy
source in photosynthesis. The chlorophyll in leaves is a
complex organic molecule capable of converting certain wavelengths of light into chemical energy. This
energy in turn takes part in several sets of reactions in

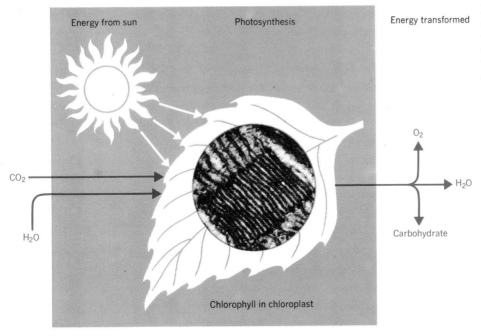

Energy from sun   Photosynthesis   Energy transformed

CO_2

H_2O

O_2

H_2O

Carbohydrate

Chlorophyll in chloroplast

Figure 19-1   *Energy conversion by photosynthesis. Chlorophyll in leaves converts carbon dioxide and water into carbohydrate, using sunlight as an energy source.*

the leaf that are crucial to all living things. One of these reactions is the production of oxygen from the splitting of water molecules. In photosynthesis, oxygen is a byproduct, but this oxygen is the only source for replacing the oxygen consumed in animal and plant cell respiration. A second crucial set of reactions leads to the linking of carbons and hydrogens to create simple carbohydrate molecules (Figure 19-1). This carbohydrate, or some form of it such as starch, is the energy supply for the entire living world. For this reason, the photosynthetic organisms of the world—of all types, from algae to higher plants—are given the ecological term of *primary producers.*

Light, of course, also makes vision possible, and virtually all forms of animal life have evolved visual receptors of some sort. The perception of colors and patterns is involved in functions such as protection (as in mimicry), and courtship and mating behavior.

The 24-hour cycle of daylight and darkness also regulates animal and plant activity to a great degree. Some animals are active only at night, some in daylight hours; a few, such as bats, stir about mostly at twilight. Plants undergo photosynthesis only during daylight, of course, but some of their other activities may be

adapted for nighttime. Quite a few plants open their flowers only at night, for instance, making use of the pollinating potential of nocturnal insects such as moths.

Other important responses to light are the seasonal changes that animals and many plants show in response to the changing number of daylight hours in a 24-hour period. This *photoperiodism* most often affects reproduction. Many plants produce flowers and fruit only when growing in latitudes where the days are long and the nights short, or vice versa. Flowering is controlled by plant hormones that in turn are formed only when the photoperiod is right. The most dramatic response of plants to photoperiods is seen in the broad-leafed trees that form the familiar forests in the eastern United States. Each fall these trees lose their leaves, and the lush green forest quickly changes into almost a biological desert of tree trunks and bare branches (Figure 19-2). The animal inhabitants must either depart for the winter of have some special adaptations for surviving in this barren winter forest. There are few other types of natural communities that experience such a complete annual change.

In animals the breeding cycle is usually keyed to

Figure 19-2 *An oak-hickory deciduous forest in the winter season presents a barren appearance. Do you see any food that an animal might eat?*

geographic distributions and can range over larger expanses of the earth. Several kinds of evidence indicate that at least some of the dinosaurs may have been warm-blooded, which in turn would explain the presence of dinosaur bones in areas that evidently were quite cold in Mesozoic times. But being warm blooded has a metabolic price tag on it: It takes more energy to maintain a constant internal temperature, and a warm-blooded animal requires more food than does a cold-blooded creature of the same size.

Ultraviolet (UV) light is ecologically important in at least two respects. First, it is lethal to many forms of life, including viruses, bacteria, and fungi. However, UV light would be far more harmful than it is if the earth were not surrounded by a layer of ozone (O_3) in the atmosphere that absorbs much of it. Second, the wavelengths that do reach the earth's surface are necessary for the synthesis of vitamin D in the skin, hair, or feathers of many birds and mammals. When sunlight strikes the skin, it triggers a reaction that converts provitamin D into vitamin D. The vitamin is necessary for proper bone formation and maintenance. A deficiency of the vitamin produces rickets, a disabling disease characterized by bent legs and spinal curvature.

seasonal day length in such a way that the young are born at an opportune time. The function of this adaptation is obvious, but how does the animal's body distinguish one season from another? This question has been studied extensively in birds, whose reproductive organs actually grow larger as day length increases and then decrease in size during the short days of autumn. The retina of the eye appears to be the receptor that triggers the release of the reproductive hormones.

The *infrared* part of sunlight warms the earth, making a considerable portion of it inhabitable. Of all the life forms that have evolved, only two groups, birds and mammals, have internal temperature-regulating mechanisms. That is, only birds and mammals are "warm-blooded." All the others fare quite well with body temperatures approximately the same as their surroundings. A major advantage of warm-bloodedness is that such animals are less restricted in their

Available Water

Another basic condition for a successful biosphere is an abundant supply of water in a liquid state. This necessity can be summarized by a quote from G. Evelyn Hutchinson, a zoologist at Yale University: "All actively metabolizing organisms consist largely of elaborate systems of organic macromolecules dispersed in an aqueous medium." In other words, life consists mostly of complex systems of large molecules in a watery environment. In addition we might add that all cells and tissues are bathed in water, and water is the transporting medium within plant and animal bodies.

On a broader scale, water forms major habitats for living things, in the form of lakes, rivers, and oceans. Because oceans cover more than two-thirds of the earth's surface, it is not surprising to find that 97 percent of the earth's water is salt water. And, of the remaining 3 percent, three-fourths of it is bound up as ice and snow at the poles; nearly all of the remaining water is found in fresh-water bodies. Only a tiny

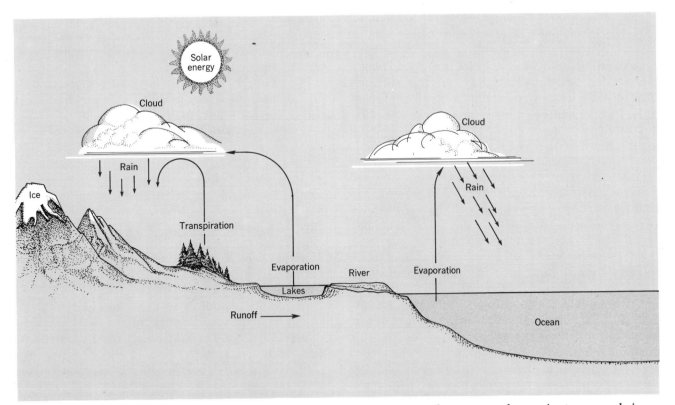

Figure 19-3 *Water cycle. Water vapor enters the atmosphere by evaporation from bodies of water and transpiration from plants. This water then eventually returns in the form of precipitation. Excess precipitation may end up as ice or snow or else return to the sea via streams and rivers. There is a general worldwide balance between evaporation and precipitation.*

amount of water actually circulates through the atmosphere in the form of atmospheric moisture, rain or snow, but this small amount functions in a water cycle that creates what we call weather. The vital water cycle influences many aspects of plant and animal life (Figure 19-3).

Water is a dense medium that does not hold very much oxygen. Whereas air holds about 210 ml of oxygen per liter, sea water contains only about 5 ml of O_2 per liter, and fresh water holds only 7 ml per liter. Consequently, aquatic creatures have specialized breathing devices such as gills to enable them to extract oxygen efficiently from as much water as possible. It is not surprising to find that many kinds of fishes have lungs as well as gills, and come to the surface to gulp air when the O_2 level drops in their surroundings.

Terrestrial forms of life face the problem of preventing undue loss of water from their bodies, because air tends to be very drying. This is accomplished with physical adaptations. Special body coverings such as scales or thick skin help protect the moist interior, and kidneys carefully regulate the elimination of water from the body.

Chemicals and Exchange Mechanisms

Organisms and their cells live by maintaining a constant exchange of elements, ions, minerals and gases. These substances and their importance in living tissues are listed in Chapter 2.

The exchange mechanisms that make these materials available to living organisms are often complex as well as important. First of all, materials may be in a solid, liquid, or a gaseous state. Carbon, for example, exists in solid states (as in a fossil fuel or in limestone), in a gas, CO_2, and dissolved in water as carbonic acid,

H_2CO_3. To be useful to an organism the material must often move from one state of matter to another. Thus, carbon must enter the gaseous state as CO_2, perhaps from the combustion of a fossil fuel, before land plants can use it in photosynthesis. The movement of elements from one state of matter to another is referred to as crossing interface boundaries. Life would be impossible were it not for the frequent exchanges of materials across interface boundaries between solid, liquid, and gaseous states of matter.

A second type of exchange mechanism involves movement of materials into the cells of organisms, that is, through the cell membrane itself. In nearly all instances the material must first dissolve in the fluid around the cell before it can be transported through the cell's membrane. Thus, for ions, minerals, or gases to move from soil into a plant's roots, they must be dissolved in water in order to get through the root cell membranes. The same condition is met when an animal eats food. The food must be digested in order to get its constituents into a liquid state, so that they can cross the cell boundaries (another solid state) into the animal's bloodstream (liquid state), and so on. Gaseous exchange is yet another example, because a gas such as oxygen must enter into a liquid state before crossing respiratory membranes. It is no accident that water, with its great solvent abilities, is vital to many of these exchanges.

Basic Concept

The earth is conducive to life because of three basic conditions found in the biosphere:

1. **Adequate intensities of solar energy**
2. **Abundant liquid water**
3. **Mechanisms of interface exchanges between different states of matter**

SOME CONSEQUENCES OF THE LIFE-EARTH INTERACTION

Cycles

The biosphere we see about us today, and indeed are part of, resulted from millions of years of interactions between living forms and the inorganic or nonliving part of the earth. The biosphere is a huge, complex system whose dynamics are known only very generally.

Among the consequences of the long evolution of the biosphere was the establishment of many types of cyclic phenomena concerned with the utilization of energy and materials.

One category of these cyclic events important in the biosphere is those that involve elements which move from the atmosphere, lithosphere (soil) or hydrosphere (water) into living things and then back into these environments. These are called the *biogeochemical cycles.* All of the elements that function in animals and plants probably follow some sort of cyclic pathway; the details for a number of these elements have been worked out including carbon, oxygen, nitrogen, phosphorus, and calcium. Two of these are described here.

Carbon Cycle. If you were asked, "Where do you find carbon?" your answer would have to be "Everywhere." A small amount is in the atmosphere as CO_2, a considerable amount exists in the tissues of all living things and in dead organic matter, and immense amounts are found in sedimentary rocks and the fossil fuels in the earth.

The most active part of the carbon cycle (Figure 19-4) involves the photosynthetic activity of plants on the land and in the hydrosphere. These plants take in CO_2, and through photosynthesis, they use this carbon to build the carbon chains in carbohydrates. These compounds then become the basic energy supply for both plant and animal cells. As living things use the energy, via cellular respiration, some carbon is released back into the environment in the form of CO_2. Some carbon is temporarily locked in the tissues of the plant or animal, but most of this, too, will return when the organisms die and are decomposed by microorganisms. On land, forests are the major CO_2 consumers, and they also tend to tie up a lot of carbon in the form of wood (cellulose). This carbon will not return to the cycle until the wood burns or is decomposed.

In the sea, marine plants are the major users (fixers) of carbon. The microscopic plants provide food for other small organisms (*zooplankton*), which are gathered up by larger organisms in various intricate food webs in the hydrosphere. Again, CO_2 is returned to the ocean as a consequence of cellular respiration and the action of decay organisms. Marine plants do not stockpile carbon in wood, but vast amounts of it end up in inorganic forms such as calcium carbonate ($CaCO_3$) in

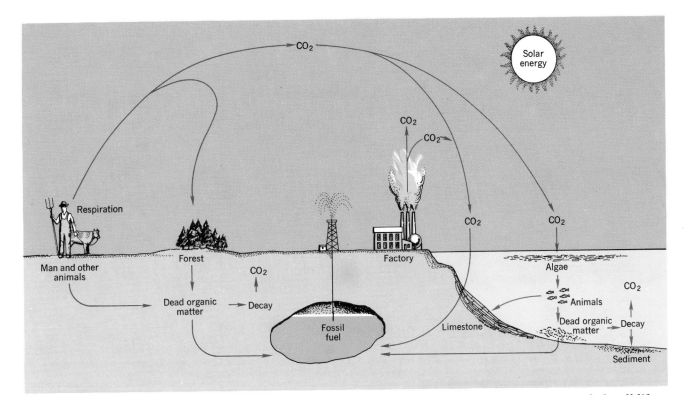

Figure 19-4 *Carbon cycle. Carbon is added to the environment (in the form of CO$_2$) as a consequence of cellular respiration and combustion. The process of photosynthesis constantly converts part of the CO$_2$ into organic material that becomes the basic energy supply for all life. When organisms die, decomposer organisms help return the carbon to the earth and eventually the atmosphere. Some carbon tends to stockpile in fossil fuels and limestone.*

the shells of shellfish and in sediments on the sea floor. The oceans also help maintain a balance in atmospheric CO$_2$ because CO$_2$ readily goes into solution in sea water, and waves and wind return CO$_2$ to the atmosphere.

Finally, to complete our picture of the carbon cycle, huge amounts of carbon are locked into sediments such as limestone and into fossil fuels. This carbon is destined to remain there until brought to the surface of the earth by geological upheavals or by the intervention of man.

Nitrogen Cycle. The nitrogen cycle (Figure 19-5), like the carbon cycle, also involves an element that circulates from the atmosphere into living organisms, and then back into the atmosphere; however, many details of the cycle are different.

Nitrogen is an essential component of the amino acids that make up proteins. Because 79 percent of the atmosphere is nitrogen gas, there would seem to be a great abundance of nitrogen for making these vital organic substances. However, the problem is that nitrogen gas, N$_2$, is inert; that is, it does not readily form compounds with other elements. Hence the bulk of life forms on earth cannot use N$_2$ directly. Most of the "fixing" of nitrogen into usable compounds is accomplished by specialized groups of microorganisms; smaller amounts are fixed by lightning, volcanic action, and by humans in the manufacture of chemical fertilizers.

One group of nitrogen-fixing bacteria in soil converts nitrogen into nitrate, NO$_3$, which is readily utilized by plants in making amino acids and proteins.

The most important of the nitrogen-fixing bacteria live associated with the roots of a few dozen species of plants, primarily legumes, the pod-bearing plants such

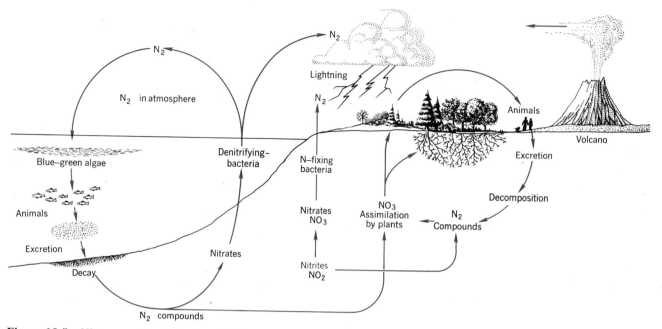

Figure 19-5 *Nitrogen cycle. A portion of the large supply of inert nitrogen in the atmosphere is converted into usable compounds mostly by specialized nitrogen-fixing microorganisms, and to a lesser extent by lightning and volcanic action. The major source of nitrogen-containing materials comes from the waste products of organisms and from the action of decomposers. Denitrifying bacteria return nitrogen to the atmosphere.*

as soybeans, peas, beans, and alfalfa. The bacteria interact with root tissues of the host plant, forming nodules where their activities are carried out (Figure 19-6). This is a mutually helpful arrangement, because neither the legume nor the bacterium can fix nitrogen alone. The presence of nitrogen-fixing bacteria makes these plants extremely important both to the nitrogen cycle and to agriculture. Legumes enrich the soil by releasing nitrogen compounds into it, and at the same time they often provide a highly nutritious food crop for people or pasture crop for livestock.

Perhaps the major source of nitrogenous materials is provided by waste products of animals (feces and urine) and from the decomposition of dead animals and plants. Ammonifying bacteria convert the nitrogen-containing wastes into ammonia (NH_3) and ammonium (NH_4). Other specialized bacteria convert the ammonia into nitrites (NO_2) and then nitrates (NO_3). Nitrates are then assimilated through the roots of many plants and used in their metabolism.

All of these activities would tend to stockpile nitrogen in the soil if it were not for another set of microorganisms, the denitrifying bacteria. These organisms break down nitrogen-containing substances and release N_2 gas back into the atmosphere. And so the cycle continues.

In reviewing these cycles, note that microorganisms and plants play key roles in both. Animals are relatively unimportant, although human activities, are increasingly infringing on these cycles. Industry adds large amounts of CO_2 to the atmosphere, and large amounts of nitrogen are used in chemical fertilizers. The long-term impact of human activities on these and other cycles is not yet known.

Basic Concepts

The elements necessary to sustain the biosphere circulate in characteristic and complex pathways called biogeochemical cycles.

Microorganisms and plants play key roles in these cycles.

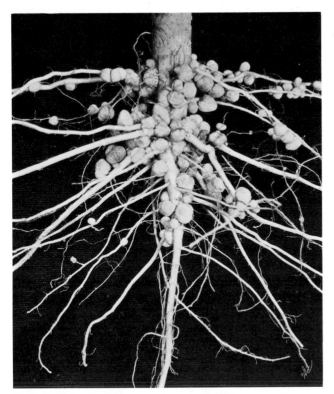

Figure 19-6 *Nodules on soybean roots. These nodules contain nitrogen-fixing bacteria and are beneficial to soybeans and other legumes.*

THE ECOSYSTEM CONCEPT

In recent years, ecologists have found it useful to view portions of the biosphere as ecological systems, that is, as units of plants and animals interacting with each other and their surroundings. The shortened term *ecosystem* has subsequently come into use.

An ecosystem includes all living things and their physical surroundings in a large geographical unit such as a forest, grassland, or lake, that is self-sustaining when provided with sunlight. Plants use the sunlight for photosynthesis, which in turn provides the basic food supply for all of the life in the ecosystem. The plants, animals, and microorganisms are so well adapted to each other and to their physical surroundings that they tend to persist in a distinct assemblage (a grasslands, for example) for extended periods of time.

All ecosystems are made up of certain basic components (Figure 19-7).

Abiotic (Nonliving) Components

These are the elements, minerals, water, gases, and soil that make up the surroundings of organisms. Most of these abiotic components take part in vital cycles in the ecosystem, as we have already seen.

The Producer Component

The term *producer* refers to all of the green plants in an ecosystem, because through photosynthesis they produce an energy supply for themselves and all other life in the system. These are also often termed *autotrophs*, "self" feeders, because they make their own food supply. Producers provide the food base that directly or indirectly supports all of the nonphotosynthesizers. This food base supplies not only an energy source but also minerals and other nutrients.

In land communities, the producers are the trees, shrubs, grasses, and herbs. In water environments, the most important producers are algae. Producers depend heavily on many abiotic components for their success. Quite obviously, a soil rich in minerals and receiving adequate rainfall will support a lush population of producers. And a scarcity of any single crucial abiotic substance, such as nitrogen or water, will greatly inhibit the growth of producers and thus affect the nature of the whole ecosystem.

The Consumer Component

Consumers constitute the animal life of an ecosystem, obtaining their nutrients and energy by subsisting on other life forms. For this reason they may be termed *heterotrophs*, "other" feeders. Some feed directly on plants and hence are *primary consumers* (herbivores). Rabbits, cattle, and many insects are typical examples in terrestrial ecosystems. Herbivorous fishes such as carp and mullet are examples in aquatic communities. In either case, the survival and numbers of herbivores are linked closely to the producer component and tend to fluctuate with it.

Secondary consumers feed on herbivores (or on each

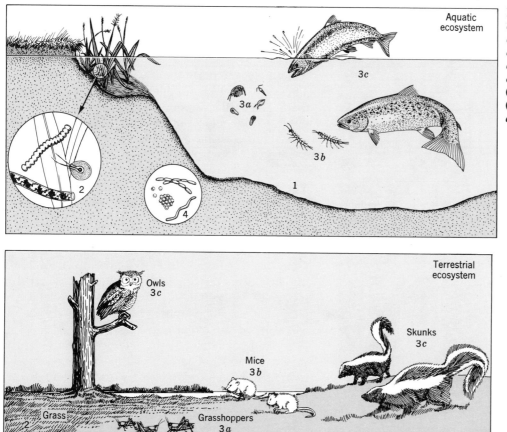

Figure 19-7 *Examples of aquatic and terrestrial ecosystems to show (1) abiotic materials, (2) producers, (3a) primary consumers (herbivores), (3b) secondary consumers, (3c) tertiary consumers, and (4) decomposers.*

other at times) and thus are called carnivores or predators. Dragonflies, weasels, hawks, bass, and sharks are examples. In some ecosystems it is possible to have *tertiary consumers,* such as a shark that feeds on king mackeral, which in turn feed on small plankton-eating fish.

Other subtypes of consumers include omnivorous animals such as opossums, catfish, roaches, and human beings, which feed on both plant and animal matter, and scavengers, such as vultures and burying beetles, which eat the remains of other organisms.

The Decomposer Component

Decomposer organisms are consumers, such as bacte-

ria and fungi, that break down the complex compounds of "dead" plant and animal matter in order to obtain nutrients. Because they use external digestive processes, their action releases simpler substances, such as minerals, into the environment; therefore, these organisms play major roles in important phenomena such as the nitrogen cycle. In addition, of course, they are essential in disposing of the remains of plants and animals. The decomposers are only a small part of the total mass of living matter in an ecosystem, but their functional importance far overshadows their mass.

This brief description very generally summarizes the major components of any ecosystem. This concept is a useful one because it fits our observational experience of nature, and because it also lends itself to various kinds of ecological studies. For example, ecologists can measure the various abiotic components and attempt to determine their influence on the ecosystem. They can analyze the producer component and try to calculate how much organic matter it produces per year. They can determine "Who eats whom or what," by analyses of stomach contents and by field observations; in this way some of the various feeding relationships in the ecosystem can be worked out.

Basic Concept

An ecosystem consists of four major components: abiotic materials, herbivores, carnivores, and decomposers. It functions as a self-contained unit as long as it is supplied with sunlight.

ENERGY FLOW IN ECOSYSTEMS

Food Chains and Food Webs

The terms *producer, consumer,* and *decomposer* obviously refer to the production and use of food materials; indeed, these processes dominate the functions of all ecosystems.

A simplified way to view this energy flow through an ecosystem is to think of it in terms of numerous food chains. For example, one such chain might consist of grass, eaten by grasshoppers, which in turn are eaten by mice, which form part of the food supply for skunks. We could diagram this as:

$$grass \rightarrow grasshoppers \rightarrow mice \rightarrow skunks$$

As a student you are probably very familiar with this food chain:

$$grass \rightarrow cow \; (hamburger) \rightarrow hungry \; student.$$

The disadvantage of using food chains to represent energy flow is that they oversimplify the actual events as they occur in nature. When biologists analyze the stomach contents of consumers in a food chain, they usually find the remains of a variety of organisms. Mice, for example, eat many kinds of insects in addition to grasshoppers. A food chain does not represent this accurately.

A more realistic picture of energy flow relationships would be a food web that indicates the interrelationships of all the food chains in a community. As shown in the generalized food web for a forest community (Figure 19-8), foliage provides food for more than one herbivore, mice are eaten by more than one consumer.

There are two general types of food chains or webs: grazing food webs and detritus (decomposer) food webs. Grazing food webs involve use of living plant matter such as leaves or fruits as already described. Detritus food webs involve organic matter that falls to the forest floor and is consumed by bacteria, fungi, and invertebrates. Studies have shown that there is usually a considerably larger energy flow through the detritus web than through the grazing web, contrary to what one might expect.

Relatively complete food webs are known for only a few ecosystems, but these provide us with enough information to make some broad generalizations. For example, a relatively mature, stable community such as a forest will have a complex food web with long chains of consumers. If part of the web is damaged by the removal of a few members, the remainder of the web can compensate for the loss and continue to function. In contrast, a simple web with short chains, like a crop planted by humans, is quite fragile. A slight change in the climate or an overabundance of an herbivore—an insect pest, for example—can become catastrophic. This knowledge should make us cautious about trying to simplify natural ecosystems.

Ecological Pyramids

Food webs are too complex to describe or illustrate easily, and they do not express the quantitative aspects of the feeding arrangements. A graphic way of showing this quantitatively is by constructing *ecological pyramids.* A horizontal bar is used for each major component, with the length of the bar being proportional to the quantity of that component. For example, if we count the numbers of grass plants, grasshoppers, mice, and skunks in a portion of a community, it would be possible to construct a *pyramid of numbers,* as shown in Figure 19-9a. These data form a pyramid because each successive tier contains fewer organisms than the one below it. Such a pyramid has limited value because it does not contain any information about the size or

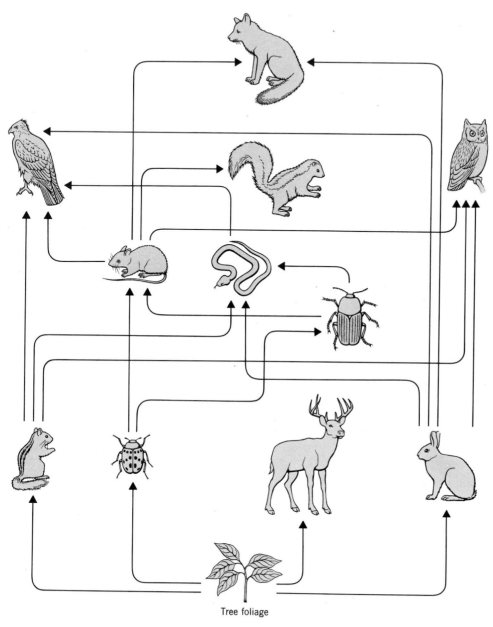

Tree foliage

Figure 19-8 *A food web based on a hardwood forest in New Hampshire. Arrows indicate direction of energy flow.*

energy content of each level. An alternative is a *pyramid of mass,* obtained by weighing the producers and consumers from a portion of the ecosystem (see Figure 19-9b). This type of pyramid shows us that an enormous amount of grass is required to support a relatively small mass of owls, and that each stage in the pyramid has far less mass than the preceding one. Pyramids of mass are still somewhat misleading because some organisms may have a lot of mass that contributes very little as food material. Thus, the woody material of a tree adds considerable mass to a producer level but contributes little usable energy to

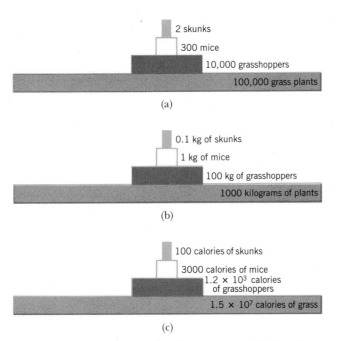

Figure 19-9 *Three types of ecological pyramids (based on imaginary numbers).* **(a)** *Pyramid of numbers;* **(b)** *Pyramid of Mass;* **(c)** *Pyramid of Energy.*

the consumer level of the pyramid. Similarly, in a marine community, some clams with heavy shells provide little food.

A third kind of pyramid, the *pyramid of energy,* overcomes many of the objections noted above and better illustrates some of the actual energy relationships within a community of organisms. To make this pyramid, ecologists must determine the energy value of various members in the food web. In doing so, they can eliminate the trunks of trees or the shells of bivalves if they do not seem relevant to the study. Not only is this pyramid a more accurate representation of the energy picture, but it also allows us to compare different ecosystems. Figure 19-9*c* shows a pyramid based on data derived from the energy content of its members.

In all of these pyramids, each level is considerably smaller than the one below it. Why does this condition prevail? The answer lies in the principles of energy dynamics. In simplified form, one of these energy laws states that energy can be transformed from one type to another, as solar energy to chemical energy, but these conversions are never 100 percent efficient. Some of

the energy, in other words, ends up in the form of heat and is lost to the system.

Basic Concept

The existence of an ecosystem depends on the flow of energy through it from producer to herbivore to carnivore. This energy flow may be expressed as a food chain, a food web, or as an ecological pyramid.

The Energy Flow Model

In order to sum up all the major effects of the energy flow through an ecosystem, ecologists often turn to a model called an *energy flow diagram.* Such a model can show how much sunlight falls on the system per unit of time, how efficiently the solar energy is converted into organic matter by the producers, how efficiently the consumers and decomposers use the organic matter, and how much energy is lost to the system as heat.

Figure 19-10 shows this in a diagrammatic way. Note that the producers occupy the largest energy rectangle in the drawing. The amount of organic matter formed by the producers as a result of photosynthesis is termed the system's *gross production* (G.P.). Some of the production maintains the plants themselves in cellular respiration. What is left over becomes the *net production* (N.P.). Some of the net production is used for growing new producers, but 10 to 20 percent is available for use by herbivores. Note in Figure 19-10 that the figure representing herbivores is far smaller than the producer rectangle.

The transfer of energy from herbivores to carnivores follows the same rule: from 10 to 20 percent of the net production of the herbivores is utilized by second-level consumers. As you can see, the energy loss in going from one component level to another is quite large. This loss represents the heat produced in cellular respiration and what is lost in each energy transfer from one component to another. Finally, as the diagram shows, decay organisms utilize energy from both producers and consumers.

Most of the nutrients that have passed from level to level can be recycled or used again. However, all of the energy will eventually be dissipated as heat is lost from the system.

Net production levels are important, and they have been calculated for many ecosystems. Some of these

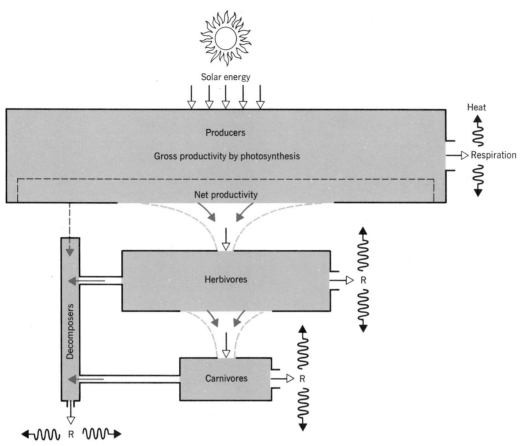

Figure 19-10 *Energy flow diagram. A large amount of energy is lost going from one level to another in the form of heat. Also, energy is not recycled, and hence a new supply must be available constantly.*

figures are noted in Figure 19-11. Note that tropical forests often have much higher N.P.s than do temperate forests, as you might expect with the lush growing conditions of the tropics. It is somewhat surprising to note the very low N.P. of ocean and coastal ecosystems, in view of popular writings that depict the oceans as swarming with seafood waiting to be harvested.

Basic Concepts

The total organic matter formed by a producer is its gross production. The amount remaining after supplying the producer's energy needs is the net production.

Consumers utilize 10 to 20 percent of the N.P. of the level on which they feed.

Nutrients recycle in ecosystems, but energy flow is a one-way path.

INTERACTIONS BETWEEN SPECIES IN ECOSYSTEMS

Predation

Predation refers to the use of one organism by another for food, and it is probably the most common type of interorganismic energy transfer in nature. Herbivores are not commonly referred to as predators, although in a sense they do "prey" on plants. The food chains, webs, and pyramids described in the previous section include a large number of predator-prey interactions.

Hunger would seem to be the major motivating force that compels predators to seek prey. This is probably true in many instances, but examples are common, especially among large carnivores, where the predator kills more prey than it can eat. A number of organisms such as crows, some owls, foxes, hyenas, and

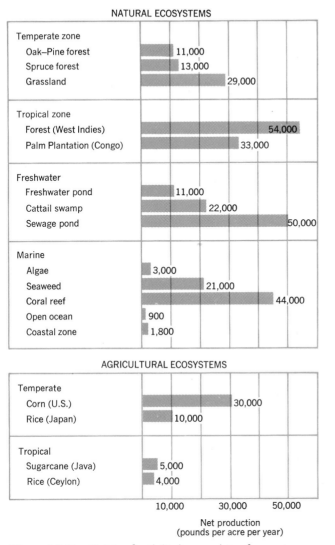

NATURAL ECOSYSTEMS

	Net production
Temperate zone	
Oak–Pine forest	11,000
Spruce forest	13,000
Grassland	29,000
Tropical zone	
Forest (West Indies)	54,000
Palm Plantation (Congo)	33,000
Freshwater	
Freshwater pond	11,000
Cattail swamp	22,000
Sewage pond	50,000
Marine	
Algae	3,000
Seaweed	21,000
Coral reef	44,000
Open ocean	900
Coastal zone	1,800

AGRICULTURAL ECOSYSTEMS

	Net production
Temperate	
Corn (U.S.)	30,000
Rice (Japan)	10,000
Tropical	
Sugarcane (Java)	5,000
Rice (Ceylon)	4,000

Net production
(pounds per acre per year)

Figure 19-11 *Net productivity in a variety of ecosystems.*

mountain lions store dead prey in a cache for later use. Moles are known to paralyze earthworms and store them for a winter food supply. Many birds and other animals catch prey in order to feed their young rather than to satisfy hunger.

Not all predators catch their own prey. Many are known to scavenge food killed by others. Hyenas, for example, get about one-fourth of their meat by this method (Figure 19-12), and all large African carnivores steal meat from other predators.

The methods that predators use for finding and catching prey are exceedingly varied, ranging from trial-and-error seeking, stalking, ambushing, communal hunting, and scavenging as described above. Trial-and-error hunting is exemplified by owls that shake tree branches to flush out small birds, cattle egrets that follow cattle or other large animals to grab insects that might be stirred up, or an octopus that pounces on any object that remotely resembles a crab.

There are numerous predators that obtain prey from an ambush, either by hiding or by camouflage (looking like a part of the environment). Web spinning spiders often wait in hiding until an insect becomes entangled in their webs. The ant lion, an insect commonly called a doodle bug, lies beneath the sand at the bottom of a small pit. If an ant crawls into the pit, the ant lion bursts out of the sand and captures it.

One of the most fascinating predator-prey interactions relates to the subduing of dangerous prey. A classic example is the attack of a mongoose on a cobra (Figure 19-13). The mongoose is very aggressive and makes the cobra strike at it repeatedly. Eventually the mongoose is able to bite into the neck or back of the snake and kill it. In Africa, small groups of Ground Hornbill birds attack the highly dangerous Black Mamba snake and repeatedly peck at it until it is subdued. Wolves work in a team to bring down a moose because a single wolf would not be a match against this large animal.

Human beings have often attempted to manipulate predator-prey populations for their own benefit, sometimes successfully but sometimes to their detriment. A successful application, now a classic in biology, occurred in the late 1800s in California. A scale insect, accidentally introduced from Australia, became a serious threat to citrus trees. As a control measure, its natural predator in Australia, the ladybird beetle, was brought into California and released. The ladybird beetles fortunately destroyed nearly all of the scale insects and have since kept them in check.

Some of our other experiments with predator-prey manipulations have not turned out so well. The mongoose was introduced long ago into many Caribbean islands such as Jamaica, because the early English settlers imagined that the area was infested with poisonous snakes that had to be destroyed. The mongoose did indeed wipe out nearly all of the native snakes

Figure 19-12 *Two hyenas and a jackal scavenging meat from the remains of an animal killed by other predators.*

Figure 19-13 *A mongoose attacking a cobra. Mongooses also feed on birds and many small mammals.*

(none poisonous), and then began killing birds and small mammals. Today the mongoose is a serious pest on many of the islands and has wiped out forms of wildlife unique to the area.

Basic Concept

Predation is the most common type of energy transfer in nature. Predators capture prey by trial-and-error seeking, stalking, ambushing, communal hunting, and scavenging.

Prey Defenses

Just as predators utilize many methods in capturing prey, prey species have numerous ways of avoiding capture. Many organisms avoid detection by predators because they resemble some part of the environment. Because green is a common color in nature, many insects are green and thereby blend into the color of foliage. Others are brown, resembling tree trunks or dead leaves. Some lizards, many flat fish (like flounders), and cephalopods can quickly change color or color patterns to better match their backgrounds. Seasonal color changes are often found in Arctic animals, such as Arctic hares which have a brown summer coat and white fur during the winter.

Another widely used adaptation is that of disruptive (irregular) markings that change the appearance of the shape of the body. The young of birds that nest on the ground, such as bobwhite quail and many shore birds, have irregular color patterns that effectively blend them into the background. In many adult birds the females are well camouflaged, probably because they sit on the nest and are thus vulnerable, whereas the males are conspicuously colored.

Finally, some forms drape themselves with portions of their environment and therefore become inconspicuous. The sponge crab, *Dromia*, places pieces of sponge on its back, and some sea urchins cover themselves with small shells.

Another type of deceptive mimicry is to resemble an organism (the model) that is generally recognized by the predators as being dangerous or distasteful (see Figure 18-3). This is called Batesian mimicry. Several species of flies mimic bees or wasps in appearance and even make buzzing sounds like their models. Ants are the models for many small mimicking insects. Some beetles actually live in the ant nest and share the ant's food. The brightly colored and highly poisonous coral snakes are mimicked by a variety of other species of harmless snakes.

Not uncommonly, animals may mimic inedible objects. Examples are the walking sticks (insects), twiglike caterpillars, and leaf-mimicking grasshoppers such as the katydid (see Figure 18-2). Many butterflies resemble dead leaves. An especially bizarre mimicry is shown by several species of caterpillars that resemble bird droppings as they lie curled on leaves.

Unpleasant tasting or dangerous animals often advertise these qualities by having bright coloration (e.g., the wasp), by their behavior (buzzing of a bee), or by emitting a chemical that repels predators (the bombardier beetle). A number of caterpillars are covered with stinging hairs and these species are usually brightly colored. The striped skunk uses a variety of signals such as coloration, pawing of the ground, and emission of a powerful chemical to signal its unpleasant nature.

Even if detected by a predator, the behavior of the prey species may confuse the predator. The most common response to discovery is flight, often in a zigzag manner. Many fleeing animals, such as frogs, grasshoppers, and moths, flash a bright color in flight and then suddenly become motionless and seem to disappear.

Prey species may resort to startle displays or behavior when threatened. Many moths flash a bright orange color when they spread their wings, and numerous butterflies have large eyelike spots on their wings (Figure 19-14). These seem effective in bluffing small predators such as birds. An unusual startle behavior is that shown by some toads and the puffer fish. These animals swell up, often to impressive sizes, when molested (Figure 19-14).

Sounds are also widely used as startle mechanisms. Click beetles snap their abdomens vigorously making a loud click and often throwing their bodies some distance through the air. The rattlesnake's rattle sound is well known, as is the sound made by a porcupine shaking its quills just before it charges into a would-be predator (backwards!).

One of the most impressive startle displays is put on by the common hog-nosed snake. This harmless and usually docile snake rears its head and flattens its neck in a cobralike fashion. This action is accompanied by loud hissing, tail shaking, and false striking motions if the predator comes close. If all else fails, the snake rolls over and feigns death, even letting its tongue hang out realistically (see Figure 18-4).

Finally, prey may retaliate against their predator foes by using teeth, claws, bills, and other body parts. Horns, antlers, and hooves are effective defensive weapons, as are the spines of sea urchins, many fish, and porcupines. A number of insects emit unpleasant chemicals when molested.

Although predation may seem to be a cruel and heartless way of life to us, it is a necessary event in an ecosystem's energy flow. It may actually benefit the prey population by acting as a control on population growth. In many instances the victims of predation are the very young, sick, or old members of the prey population. The loss of sick or defective individuals may benefit the gene pool of the species. Often the predator seeks a physically disabled member, as when hyenas spook a herd of wildebeast and watch the fleeing herd for the slowest or weakest members. This is called "testing." Wolves often "test" a would-be

Figure 19-14   *Two examples of startle displays. Some moths (left) flash eyelike spots when disturbed. Other animals such as the puffer fish, swell up and become difficult for a predator to swallow.*

moose victim. If the moose stands its ground instead of fleeing, the wolves quickly give up.

Even though we now better understand the important roles of predators in ecosystems, we persist in trying to eliminate many of the larger ones. In the western United States, especially, it is still considered desirable by many ranchers to kill mountain lions, coyotes, and eagles. This is deplorable in view of the data that clearly show us the important roles of these creatures in their ecosystems.

Basic Concept

Prey attempt to avoid capture by using concealment, camouflage to resemble the environment, disruptive markings, mimicry of a dangerous species, mimicry of an inedible object, effective defense mechanisms, and startle mechanisms.

Competition

Competition is the outcome of organisms needing the same materials from the environment. Plant requirements are sunlight, water, minerals, and growing space. Animal requirements involve food, nesting sites, and mates. In most instances the competition is a subtle interaction rather than a direct confrontation, which the word *competition* tends to convey. Competitors fre-

quently have little or no physical contact with each other. For example, grasshoppers, mice, rabbits, and antelope may all act as primary consumers (therefore, as competitors for grass) in a grassland community. Several carnivores frequently utilize the same prey.

Studies of competition many years ago led to the proposal of a biological concept termed *Gause's competition-exclusion principle*. This principle states that two species having the same environmental requirements cannot live together; competition will force one species to leave or to become extinct.

Frequently two related species *are* found living in the same habitat or community, which seems at first to be contrary to Gause's principle. However, a close examination always reveals that they utilize slightly different food sources or, in the case of plants, may require different nutrients from the soil. Two animals may utilize the same food but at different times of the day, as is the case with butterflies and moths. Related birds such as thrushes may coexist in the same woodland by feeding at different levels in the trees. Thus the Grey-cheeked Thrush feeds in foliage, whereas its close relative, the Veery, feeds primarily on the forest floor.

Basic Concept

Competition is the outcome of organisms requiring the

Figure 19-15 *A mollusk with a commensal bryozoan attached to the shell near the opening of the siphon through which water enters. The bryozoan obtains food in this manner but evidently does not harm or help the mollusk. (Drawing by Richard A. Boolootian, Science Software Systems, Inc.)*

same materials from the environment. Two species with the same requirements cannot live together because competition will force one of them to leave or become extinct.

Symbiosis

Another common interaction in nature is symbiosis. This term literally means *living together*. It applies to any intimate association between two dissimilar organisms, regardless of the harmful or beneficial outcomes of the relationship. The symbiotic association is frequently associated with food acquisition by one or both symbionts but may also provide shelter or support for them. We will consider three types: commensalism, parasitism, and mutualism.

Commensalism. In this kind of interaction, one member (the commensal) benefits and its host is not affected. For example, many bivalved mollusks have filamentlike bryozoans attached to the shell around the area where the mollusk takes in and pumps out water (Figure 19-15). The bryozoans feed on the microorganisms caught in the stream of water; the mollusk is neither helped nor harmed.

In many of the sandy ridge areas of Florida, a small deer mouse, *Peromyscus polionotus,* lives in burrows that it digs in the sand. A common lizard of the area utilizes the opening of the mouse burrow from which it digs a short side tunnel slanting almost to the surface of the soil. When disturbed, the lizard bursts up through the soil as though to escape a predator. The lizard gains shelter by this adaptation, and the mouse is unaffected.

Many plants are adapted to an epiphytic way of life; that is, they grow on the trunks or limbs of other plants, deriving support but not nourishment from them. Residents of the southeastern coastal areas are

Figure 19-16 *The flea, a common external parasite, is flattened for crawling through fur and has legs modified for clinging to hair.*

familiar with the gray, streamerlike, Spanish moss that drapes so many trees. This epiphyte, a flowering plant rather than a moss, does not damage its host unless it becomes thick enough to shade the tree's leaves.

Parasitism. When one symbiont uses another for a food source, the relationship is one of parasitism. The parasite, usually smaller than the host, spends part or all of its life cycle on or within its host, using the food or the tissues of the host for nourishment. Normally the relationship is obligatory in that the parasite must have specific hosts in order to complete its life cycle.

A variety of parasites are adapted to living on the exterior of their hosts. Common examples are fleas and lice on animals, and dodder and mistletoe on plants. This type of parasitism requires many special adaptations. Fleas have laterally flattened bodies for crawling through fur, and legs adapted for clinging to hair (Figure 19-16). Dodder and mistletoe have special roots that penetrate the host plant's vascular tissues to obtain food. Dodder is so closely associated with its host that the production of flower-inducing hormones in the host also causes the dodder plant to flower.

Internal parasites are even more modified and highly adapted for their way of life. Parasitic worms, for example, frequently show a drastic reduction in all body systems except the reproductive apparatus (Figure 19-17). This system is complex, highly developed, and extremely fecund. Frequently the life cycles of internal parasites include a stage of asexual reproduction. Internal parasites are usually very host-specific,

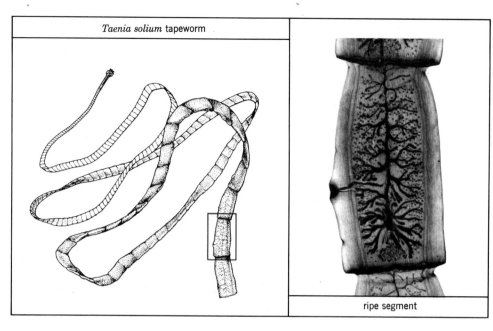

Taenia solium tapeworm

ripe segment

Figure 19-17 *The tapeworm, an example of an internal parasite. Details of one segment are shown in the photomicrograph. The dark multibranched structure is the reproductive system filled with eggs. (R. A. Boolootian)*

most likely as the result of a long evolutionary history of adaptation to a particular host.

Parasitism is evidently a successful way of life, judging by the large number of organisms that utilize it. For example, thousands of species of wasps deposit their eggs in the bodies of other insects. When these eggs hatch, the larvae use the internal tissues of the host for food, but not so rapidly as to kill the host immediately, because this would also destroy the larvae. One large group of wasps uses plants as the hosts on which the larvae feed. Typically this causes an enlarged growth on the plant called a *gall*, the counterpart of a tumor in animal tissues.

There are various degrees of parasitism and some barely fit the definition we used earlier. In "social parasitism," the female Brownheaded cowbird does not build a nest; instead, she deposits her eggs in the nests of other birds. Often the baby cowbird grows more rapidly than the young of the host bird, and the host's own progeny may die for lack of attention. The European cuckoo uses a similar form of social parasitism.

Parasitism has been studied more intently than other symbiotic relationships because of the number of diseases associated with it. Some of the more serious human diseases caused by parasitic organisms include malaria, amoebic dysentery, sleeping sickness, and a variety of bacterial and viral infections.

Mutualism. An interaction in which both symbionts benefit is termed *mutualism.* As in all of these relations, there is a wide range of interactions from voluntary mutualism, such as shrimp that clean fish, to inseparable types, such as the association of fungi and algae to form lichens.

Lichens consist of a mass of fungal filaments in which are embedded cells of green or blue-green algae. The spongy fungal tissue holds moisture that both members may use; the algae are photosynthetic to provide an energy supply and vitamins that the fungus cannot synthesize. Moreover, lichens possess distinctive body forms and functions not shown by either symbiont alone. For example, some lichens live on bare rocks where few other organisms can live. Others produce antibiotics.

In marine habitats, a number of small creatures are involved in a "cleaning symbiosis." At least six species of small shrimp, frequently brightly colored, crawl over fish, picking off parasites and cleaning injured areas (Figure 19-18). This is not an accidental occurrence, because fish are observed to congregate around these shrimp and stay motionless while being in-

Figure 19-18 *A shrimp cleaning a moray eel.*

spected. Several species of small fish (wrasses) are also cleaners, nearly all of them having appropriate adaptations such as long snouts, tweezerlike teeth, and bright coloration. Conspicuous coloration probably communicates that these animals are not prey.

Basic Concept

Symbiosis is an intimate association between two dissimilar species; commensalism, parasitism, and mutualism are three types of symbiosis.

INTERACTIONS WITHIN SPECIES IN ECOSYSTEMS

Up to this point we have considered interaction between *different* species of organisms. Members of the *same* species also interact closely with each other because they require the same items from the environment. This is accomplished by territoriality, social dominance, and altruistic behavior.

Territoriality

Territoriality is an activity exhibited by many animals, especially vertebrates. In general, a male occupies a circumscribed area, drives away other males that enter it, courts the females there, and perhaps raises young. This behavior is particularly noticeable in birds and can be seen in any backyard in the springtime. The male bird frequently uses a singing perch, his song identifying the area to other males and females of his species. In an aquarium one can often observe fish defending certain areas of the tank against the other fish. Even domesticated animals like dogs show territorial behavior, as exemplified by a small dog chasing a much larger one out of its yard — but no farther!

Territorial behavior apparently serves a variety of functions. It obviously spaces out animals over an area, preventing crowding and perhaps food shortages. It reduces conflict between members of the population because the territory's owner communicates his domination by auditory and visual signals. Other males usually avoid an occupied territory.

Social Dominance

Social dominance, or *pecking order* as it is sometimes called, is another behavioral action that helps decrease intragroup conflict. It is most commonly observed in social aggregates such as flocks and herds where members of the group become arranged in social hierarchies. Usually the strongest, largest, or most mature individuals dominate other members, which in turn dominate individuals below them in the hierarchy (pecking order), and so on. In a flock of chickens, for example, the dominant ones eat first at the feed trough and take the choice roosting sites. After the hierarchy is established, most members of the flock accept it, allowing daily activities to proceed with a minimum of strife.

In many cases the dominant members have responsibilities as well as privileges. Thus the dominant members of a baboon pack are the large mature males, and an important part of their function is defending the pack against predators such as leopards. In this case, both dominance and cooperativeness are essential to baboon survival.

Social ranking may have considerable hereditary and evolutionary significance, because individuals at the top of the pecking order do most of the mating and produce the most offspring. By contributing more genes to the gene pool of their population, the dominant members influence its future.

Altruism

If a human being risks life and limb to save another human from danger, we call this heroic behavior an altruistic act. Altruism was thought for a long time to be primarily a human trait. A study of animal behavior shows, however, that altruistic acts are a normal part of the action patterns of a variety of nonhuman animals. A soldier bee, for example, dies after stinging a potential invader of the beehive. Some birds give warning calls when a predator is sighted, alerting their flockmates to seek shelter but exposing themselves to danger in the process. If a baboon troop is attacked by a predator, the largest males in the troop come forth to fight the predator and may die as a result. Numerous additional examples are known.

That this type of behavior could evolve in animals seems contrary to natural selection. If you recall from Chapter 18, selection favors the best reproducers, that is, the ones who contribute the most genes to the gene pool. But if an animal dies in an altruistic act, it cannot pass its genes on. If the strongest and largest male baboon is killed defending the troop, how does this help the baboon troop's gene pool? The answer to this puzzle seems to be a genetic one. In the process of defending the troop and perhaps forfeiting the chance to reproduce, the dominant male may make it possible for his relatives (brothers, sisters, etc.) to reproduce successfully. Because they share copies of many of his genes, he is able to continue his genetic continuity through them. This is called *kin selection*. Thus it seems that altruism in the nonhuman animal world has the selfish motive of preserving one's own genes. Do you suppose this has any bearing on human altruism?

Basic Concept

Interactions within species include territoriality, social dominance, and altruism.

KEY TERMS

altruism	consumer	heterotroph	predation
autotroph	decomposer	mutualism	primary producer
biogeochemical cycle	ecological pyramid	net production	producer
biosphere	ecosystem concept	nitrogen cycle	social dominance
carbon cycle	energy flow diagram	parasitism	symbiosis
commensalism	gross production	photoperiodism	territoriality
competition			

SELF-TEST QUESTIONS

1. Define the term *biosphere.*
2. What three basic characteristics make the earth a highly suitable place for life?
3. In what ways does solar energy make the biosphere possible?
4. Define photoperiodism and give an example.
5. Discuss the role of the infrared portion of sunlight in the biosphere.
6. What protects us from getting lethal doses of ultraviolet light from the sun? How is UV light beneficial to some living things?
7. Discuss some ways in which water is essential to the biosphere.
8. What is the importance of exchange mechanisms to the biosphere?

9. To what does the term *biogeochemical* refer in discussing cycles?

10. How does carbon fit into the definition of a biogeochemical cycle?

11. Describe the passage of nitrogen from the atmosphere into living things and back into the atmosphere.

12. Why are animals not key organisms in most of the cycles?

13. List the four major components of an ecosystem and discuss each of their roles in the functioning of an ecosystem.

14. Describe with examples what is meant by a food chain and a food web.

15. What are three types of ecological pyramids?

16. Describe an energy flow model for an ecosystem.

17. What does gross production refer to? How does it differ from net production?

18. What does it mean to say the nutrients recycle in an ecosystem but energy flow is a one-way path?

19. Define predation and give some examples.

20. What are some ways that prey organisms avoid capture by predators?

21. Explain Gause's competition-exclusion principle.

22. Define the term *symbiosis* and list three types.

23. Define commensalism and give an example.

24. What is meant by parasitism? What evidence indicates that it is a successful way of life?

25. Define mutualism and give some examples.

26. List three types of interactions that occur between members of the same species.

27. What functions are served by territoriality?

28. What is the advantage to a species of exhibiting social dominance?

29. How can altruism be a selfish endeavor?

chapter 20

A wilderness area in the Grand Teton National Park, Wyoming.

THE BIOSPHERE
AND
HUMAN LIFE

chapter 20

It's not nice to fool Mother Nature.

TV Commercial

BIOMES AND THEIR USE

The animal and plant life of the biosphere as we see it today (or study it from past records) appears to exist in large assemblages that biologists call biomes. A *biome* is defined, generally, as a grouping of distinctive plants and animals occupying a sizable geographic area. Biomes are named according to their dominant plant forms or some conspicuous physical feature. An example is the grasslands biome that is distinctive because it consists predominantly of grass plants and forms of animal life which are adapted to this type of vegetation. The geographical limits of land biomes are determined by factors that include amount of sunlight, soil type, rainfall, and year-around temperature conditions.

Prehistoric humans lived in these biomes as, indeed, people do today. The effects of *Homo sapiens* on the various biomes have been awesome, as pointed out in the following pages.

Tundra

The tundra is the vast, treeless, cold biome that borders the Arctic Ocean (Figure 20-1a). Its vegetation is made up of grasses, mosses, and dwarfed woody plants (Figure 20-2). Life is harsh, and permanent residents are few. Notable inhabitants are the small, mouselike lemmings, which erupt in population explosions from time to time; caribou; grizzly bears; and snowy owls. During the short summers, large numbers of ducks and geese nest in the tundra.

Net production in this ecosystem is low and food chains are short; in addition, replacement of damaged vegetation is very slow because of the climate (and the subsoil, which is permanently frozen). One reason environmentalists object to construction projects such as roadways and oil pipelines in this biome is that such manmade damage is virtually permanent.

Early humans had to experience and endure the cold, forbidding tundra as they spread from Asia into North America over a land bridge where the Bering Straits now exist. Perhaps they adapted and survived as the Eskimos do today by constructing snow domes (igloos) from blocks of ice in the winter and sod-roofed dugouts in the summer months (Figure 20-3).

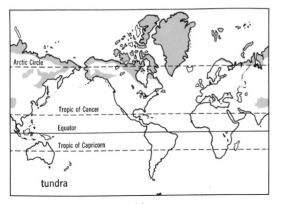

tundra

(a)

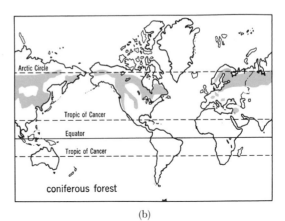

coniferous forest

(b)

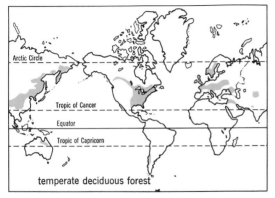

temperate deciduous forest

(c)

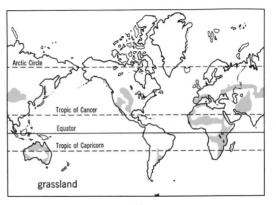

grassland

(d)

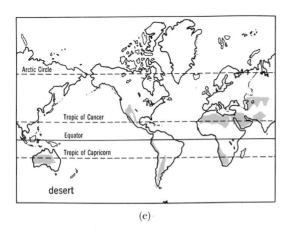

desert

(e)

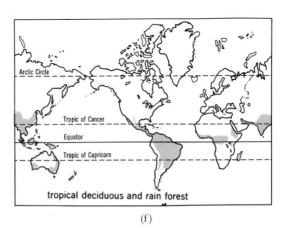

tropical deciduous and rain forest

(f)

Figure 20-1 *Geographic distribution of six biomes.*

Figure 20-2 *Tundra. Dwarfed vegetation and plentiful water are major features of this rather harsh environment. Notable inhabitants include the lemming, caribou, snowy owl, polar bear, canada goose, and arctic fox.*

Figure 20-3   *A dwelling made of pieces of sod cut from the tundra.*

Northern Coniferous Forest (Taiga)

The taiga is a large belt of evergreen forest lying just south of the tundra (see Figure 20-1*b*). The trees of this biome are mostly spruce, fir, pine, and cedar. Their leaf shapes, needle- or scale-like, prevent excessive evaporation of water and freezing. Their flexible branches and conical shapes help them avoid snow damage during the winter (Figure 20-4). Thickets of broad-leafed trees such as birch and alder are also found. The biome is important economically because of its extensive timber and fur resources.

The winter climate of the great northern woods is cold and severe, but the dense growth of trees provides an abundant food supply of leaves, wood, and seeds. This in turn supports a fair-sized consumer population of moose, squirrels, snowshoe hares, and some unusual birds like the crossbill, with its distinctive adaptation for removing seeds from cones.

A portion of the biome extends southward from Alaska into central California as a luxuriant, humid coniferous forest of hemlock, cedar, fir, arborvitae, and the spectacular redwood trees of northern California. Coastal fogs drift through these forests almost continually; consequently they are sometimes called temperate rain forests. Because these beautiful forests are in great demand for recreational use, homesites, and for timber, conflict has arisen over their use. Conservation groups such as the Sierra Club have fought strongly to protect the redwoods from the timber com-

panies, but unfortunately they appear to be losing their battle.

Temperate Deciduous Forests

The term *deciduous* refers to plants that lose all their leaves for a part of the year, a trait common to many of the trees and shrubs in this biome. These trees, including oaks, maples, beech, elm, and hickory, are sometimes spoken of as broadleafs or hardwoods in contrast to coniferous, narrow-leafed softwoods.

Deciduous forest communities once formed a continuous band across eastern North America, Central Europe, large portions of China, and southeastern Siberia (see Figure 20-1*c*). Today much of this area is occupied and utilized by human beings, so that the biome rarely is found in its original state.

Most deciduous forest areas experience a distinct seasonal pattern of cold winters and hot summers, but on the average the climate is moderate. An important factor in the development of the broad-leaf forest climax is the even distribution of rainfall (30 to 60 inches per year) over the seasons.

This biome consists of a variety of community types such as oak-hickory and beech-maple forests, but all share the striking seasonal contrasts of being leafless during the winter and heavily foliaged during the summer. This trait alone has a considerable influence on the rich fauna of these communities, because many of the animals are adapted to the food and protection afforded by the trees.

In its natural state it is rich in species, net production is high, and the structure of the ecosystem is stable. Because of the rich variety of primary producers, much of the animal life is adapted to the food and shelter offered by the trees. Squirrels, woodpeckers, warblers (insect feeders), woodchucks, chipmunks, deer, and bear are typical natives (Figure 20-5).

Grasslands

Where seasonal rainfall distribution is too uneven to support tree growth, the producer population of an ecosystem consists chiefly of grasses (see Figure 20-1*d*). In our country we call the tall grasslands *prairies* and our shorter drier grasslands *plains. Savanna, pampas,* and *steppe* are terms applied to similar grasslands in other parts of the world.

Figure 20-4 *A view of the Taiga or Northern Coniferous Forest composed mostly of firs, spruces, pines, and cedars. Some of the typical animals shown here are the porcupine, showshoe rabbit, gray wolf, evening grosbeak birds, moose, and mule deer.*

Figure 20-5 *Temperate Deciduous Forest. This forest of broadleafed trees and shrubs is rich in species and has a high net production. Its appearance changes drastically in the winter as the leaves drop from the trees (see Figure 19-2). Some typical animals include the treefrog, gray squirrel, yellow warblers, red fox, racoon, salamander, and bobcat.*

Figure 20-6   Grasslands. The soil is deep and rich but uneven rainfall prevents trees from growing here. Common animals include the jackrabbit, pronghorn antelope, coyote, prairie dog, and bison (historically).

Net production in grasslands in high unless the land is overgrazed or abused by farming. Organic matter builds up rapidly in the deep soil, creating a rich humus. For this reason this ecosystem has been extensively used for farming, especially of cereal grains (which themselves are grasses). The drier grasslands over the world are frequently overused as grazing lands.

A major climatic feature leading to the formation of grasslands rather than forests is the uneven seasonal distribution of rainfall. Grassland areas may have as much as 40 to 60 inches of rain a year, but it often falls in torrents interspersed with long dry periods. Trees cannot survive these droughts and the frequent fires that accompany them.

The native animals of grasslands are adapted for running, leaping, or burrowing. Typical animals of the American grasslands are rabbits, pocket gophers, "prairie dogs," antelope, coyote, and in the past, bison (Figure 20-6). Interesting ecological equivalents of these creatures are found in other grasslands in the world, such as kangaroos in place of antelopes, cheetahs in place of coyotes, and Australian marsupial moles in place of pocket gophers.

Ancient dwellers in the grasslands were faced with winds, cold nights, and severe winters. With few plant materials to serve as building materials, they used animal hides stretched over a frame of poles (Figure 20-7). A hide tent gives protection from snow and wind but is portable to support a nomadic way of life.

Deserts

Grasslands generally give way to deserts where rainfall drops to less than 10 inches a year. Intense solar radiation, hot days and cold nights, high evaporation rates, and constant winds are characteristic of the desert biome.

Contrary to popular opinion, very few deserts are devoid of life. Most of them support a characteristic set of arid-adapted plants (xerophytes) and animals (Figure 20-8). Many plants, such as cacti, store water in thick, leafless stems; the shrubs have small, thick leaves that may be shed in times of extreme drought. Much of the desert animal life is nocturnal and spends the hot daylight hours in burrows or shaded nooks. Rodents, reptiles, some unusual birds such as the roadrunner

Figure 20-7 *A tent made of animal hides by plains-dwelling Indians. Such a tent provided protection against the elements and at the same time was portable for the Indian's nomadic way of life.*

(which chases down snakes and lizards — there really is such a bird!), and even a few amphibians inhabit the desert.

The deserts have suffered relatively little from use by humans because they have fairly inhospitable climates. In fact, the deserts of the world are slowly increasing in area. Bordering areas such as grasslands often turn into desert when abused by overgrazing.

About one-third of the land area of the earth is desert (see Figure 20-1*e*), and increasing human population pressure almost demands that more of this biome be made usable for agriculture. Some types of desert soils are quite fertile when irrigated, and this practice is used in many parts of the world. The major drawback is finding a large source of fresh water that can be diverted to the desert area. In the southwestern United States, most of the waters of the Colorado and Rio Grande rivers have been put to this use, but at considerable cost. Many arid lands do not have either the necessary capital or such suitable water sources.

Because many arid lands like those of north Africa border on bodies of salt water, *desalination* would seem to be a logical, and perhaps the only, alternative. The cost of removing salt from seawater is fairly high with

Figure 20-8 *Desert scene in Big Ben National Park, Texas. Less than 10 inches of rainfall a year produce this type of biome. Yucca plants dominate this scene but a considerable variety of plant species is found in various deserts. Typical animals of deserts include the kangaroo rat, lizards such as the fringe-footed sand lizard and gila monster, elf owl, horned lizard, sidewinder rattlesnake, roadrunner bird, and scorpions.*

present technology, at least in the quantities needed for farming. Even so, some Arab countries such as Kuwait are starting to use desalinated water for this purpose.

Desert dwellers of the past often turned to mud masonry for their dwellings. Many desert soils harden into a bricklike consistency after they have been wet, and mud can be applied like plaster to a wooden framework or made into bricks. Thick masonry walls insulate well against the high daytime temperature and cold nighttime temperatures of the desert, and they shield the inhabitants from the intense solar radiation. The ancient cliff dwellers in the far western United States built exquisite villages from sun-dried bricks. Some of these beautiful little dwellings are over a thousand years old, but the dry climate and protection of overhanging ledges kept them well preserved (Figure 20-9). Most of these desert dwellers were farmers, although in most parts of the world desert inhabitants are nomadic.

Figure 20-9 *Portion of a prehistoric cliff-dweller's village at Mesa Verde, Colorado. The thick walls made with sun dried bricks protected the inhabitants from temperature extremes over 1000 years ago. Why do you suppose that these dwellings have windows but no doors?*

Tropical Biomes

It sometimes surprises first-time visitors to the tropics to find that tropical areas are not uniformly covered by impenetrable jungles, as is so often seen in movies and on television. The tropical areas of the world contain a variety of ecosystems, including deserts, grasslands, and various kinds of forests (see Figure 20-1f). However, of these, the *tropical rain forests* are sizable and have many interesting features.

As the term implies, rain forests receive abundant rainfall (over 80 inches a year); temperatures are consistently high, and days and nights are uniform in length. Virtually all of the plants are tall-growing trees whose branches are often covered with other plants called *epiphytes* (plants that use other plants for support). Thick-stemmed woody vines (*lianas*) are also typical. The tree canopy is usually so dense that the forest floor is dark, humid, and open. Tropical forests are made up of large numbers of species of plants but relatively few individuals of any one species. The exceedingly varied animal life shows this same pattern (Figure 20-10).

The climatic conditions of the tropical forest biome sound ideal for farming purposes. Furthermore, their net production is the highest of any biome (Figure

19-11). The problem is that many of the tropical soils consist of a red, porous earth that is poor in organic matter and in most plant nutrients. Tropical plants "hoard" their nutrients, in the sense that when they die, decay is so rapid that their nutrients are quickly taken up by other plants. When a tropical forest is removed for timber or for agricultural purposes its mineral supply quickly leaches down through the porous soil, leaving a hard, unworkable, infertile surface. In fact, this type of soil, called *laterite*, is often used as a road and building material in the tropics because it turns into rock when exposed to air. Native farmers use a slash-and-burn technique to clear a piece of forest, grow two or three crops such as manioc (a starchy root), and then abandon the land and let it return to forest.

As a consequence of the above practices, the wet tropical forests of the world are rapidly diminishing; ecologists fear that only remnants of this magnificent biome will remain a century from now. This would be a biological catastrophe because of the loss of so many diverse plant and animal species. Many of these are new to science and some are potentially useful as sources of new drugs and other products. As in all biomes, every species represents a unique gene pool that can never be reconstructed if the species is eradi-

Figure 20-10 *Tropical rain forest in the lowlands of Ecuador. Vines, epiphytes, and a dense tree canopy typify this humid forest. In tropical forests such as these, animal life includes the vervet monkey, tree boa, three toed sloth, parrots and toucan birds, and centipedes.*

cated. Therefore, such losses must always be avoided if possible.

Dwellers in *tropical forest* areas need shelter from the frequent rainfalls. They usually make a shelter frame with lightweight poles such as bamboo and thatch the roof with leafy materials like palm fronds. The roof is sloped for water runoff. The floor, if any, is raised some distance above the ground for air circulation and to avoid standing water (Figure 20-11). Tropical peoples are often permanent dwellers, either simple farmers or hunters of plentiful game.

Basic Concepts

Biomes are the major ecosystems that comprise the biosphere.

The distinctive life forms of each biome reflect the climatic conditions intensity of solar radiation, and soil types of the areas in which they live.

Marine and Freshwater Biomes

The most extensive, and some of the least productive, biomes are aquatic. Marine and freshwater biomes are distinguished chiefly by the amount of dissolved salts. Sea water contains over 40 elements, largely as ions, and this mineral abundance provides an excellent chemical environment for life. Many of these elements also occur in freshwater, but in far smaller amounts. Bodies of water contain a variety of ecosystems, as do terrestrial environments. Lakes, rivers, marshes, coral reefs, and open oceanic regions all have characteristic complements of life. All the basic ecosystem components are found in aquatic biomes: producers, consumers, and decomposers (Figure 20-12).

As has already been mentioned, algae are the basic producer organisms. *Plankton* (small forms of life with limited means of locomotion) play a large role in the energy flow in aquatic communities, both as producers (phytoplankton), and as consumers (zooplankton). Large consumers, with well-developed swimming abilities, are called *nekton*. These organisms include fish, turtles, birds, and some mammals, as well as squid and sharks in marine habitats. Decomposers include microorganisms and fungi, as in the terrestrial biomes. In addition catfish, mollusks, and crustaceans may serve as scavengers in disposing of bits of organic matter and carcasses of dead organisms.

Figure 20-11 *In hot, rainy climates the raised floor for air circulation and thatched roof for water runoff are highly functional.*

Figure 20-12 *A Spartina salt marsh at low tide on the North Carolina coast. The black, soft mud and Spartina grass support an interesting assemblage of small animals such as fiddler crabs, worms, and marsh birds. This ecosystem occupies extensive coastal areas of the eastern United States.*

Humans, of course, use all of these aquatic biomes extensively for getting rid of sewage, garbage, and industrial wastes. At one time everyone assumed that wastes could simply be dumped into the nearest stream or river. The consequence was that all major river systems in all highly developed countries became the equivalent of sewer lines into the sea. In addition, the earth's rivers have long functioned for travel and commerce, irrigation, and recreation, and as a source of drinking water for nearby cities. The latter two functions have necessarily been discontinued in many rivers.

Pollution of the *oceans* is less apparent because of their vast size. Nevertheless, the large oil spills of recent years remind us that size alone does not make a system immune to pollution. Coastal areas near seaports and cities are commonly polluted with industrial wastes and sewage. And even in midocean, water samples frequently show evidence of human activities, such as cellulose shreds (probably from toilet tissue) and bits of plastic materials.

Oceans are used by humans for transporting materials, for recreation, and as a source of food. Fish are a rich source of protein, although at present they play a minor role in the world's total consumption of protein (about 5 percent). The world's fish catch is at present about 60 million tons per year. Of this amount, three-fourths is consumed directly by humans and one-fourth is used as livestock and pet food.

It is often suggested that seafood should be harvested more extensively to provide more protein for the world's hungry people. However, the present pattern of seafood use suggests that this is not a realistic solution to world hunger. The world's people simply prefer to meet their protein needs from other sources. Eating habits are cultural patterns that are not easily changed. (Would you be willing to eat fish as often as you now eat hamburger and steak?)

Some marine fisheries specialists believe that the world's annual seafood harvest could be increased considerably. However, they warn that such increases should be carefully supervised. So far, the world's marine resources have not been well managed; the motto has generally been to catch as much as you can as fast as you can, before the next person gets it. Consequently, some resources, such as the sardine fishery on the west coast of North America, have been virtually

wiped out, and several species of whales are on the verge of extinction. Only through stringent regulations have certain popular coastal seafoods such as abalone and lobsters survived excess exploitation.

There is a hopeful side to the problem of aquatic ecosystem abuse. The United States now has an extensive program to clean up its river systems and its large inland bodies of water such as the Great Lakes. Cities that once poured their raw sewage into rivers are being required to install better sewage disposal systems, and industries are being forced to curtail drastically their discharge of wastes into the environment. As a result, considerable improvement has already been noted in Lake Erie and in a few river systems, and the once abundant fish life is being restored.

RACHEL CARSON AND THE ECOLOGICAL REVOLUTION

The earth travels on its orbit through space, carrying with it a living cargo surrounded by a protective envelope of atmosphere. The cargo is a self-perpetuating one, at least to the present time, involving the cycles and energy flow we discussed in the previous chapter.

The concern of many thoughtful individuals today is that the human portion of the cargo is threatening the well-being of the remainder by excess reproduction and by abusing its surroundings. Even though human beings are the newest addition to the "cargo," our influence in the entire biosphere during our brief presence has been awesome.

All of this is by now common knowledge, for in the past decade we have lived in an "ecological revolution." But this state of public awareness about environmental problems is surprisingly recent. It is thought to have been triggered by the 1962 publication of a book—*Silent Spring* by Rachel Carson. The book concerned only one problem, the adverse effects of pesticides on the environment, but it was so accurately documented and well written ("must" reading for anyone who professes a concern with environmental problems) that it almost immediately focused the attention of scientists and the public on a truly serious problem. Spokesmen for the pesticide industries condemned the book as inaccurate, unscientific, and emotional; there were even attempts to discredit Ms. Carson's professional qualifications as a biologist to write such a work. Nevertheless, ecologists found that many pesticides were, indeed, dangerous to ecosystems, including their human component. The result was a drive to regulate their use.

One healthy outcome from this controversy was that it made the public and the ecologists aware of numerous additional environmental abuses: A lot of things in addition to pesticides were polluting our water supplies and atmosphere. We were damaging our biomes in many ways, and our whole quality of life seemed to be deteriorating. The overall consequence of this awakening was the ecological revolution that is still with us.

SOME ENVIRONMENTAL PROBLEMS

Highly developed countries such as the United States and the European nations have experienced three major types of serious environmental problems; air pollution, water pollution, and pesticide effects. Other kinds of problems prevail in less industrialized parts of the world. Drought might be a catastrophic problem in a Central American or African country, if, for instance, its people are already on the verge of starvation, whereas air or water pollution would be relatively unimportant. The ecological problems, and the solutions to them, are far different in the developing world from those in our society.

Air Pollution

Even in "clean" environments the atmosphere carries numerous materials: many kinds of dust particles, microorganisms, water droplets, pollens, minerals — and even poisonous gases in the vicinity of active volcanos. Not long ago scientists found that the "smoke" of the Smoky Mountains in the Appalachians is a type of natural "pollution," consisting of a complex of chemicals exuded by the trees themselves (Figure 20-13). But

Figure 20-13 *A typical view in the Smoky Mountains in North Carolina. The haze in the distance is a natural air pollutant produced by the forest trees that cover these beautiful mountains.*

these natural substances rarely damage the biosphere or human beings; rather, it is the actions of humans that produce harmful air pollutants.

Artificial air pollutants are produced mostly from combustion of fuels in industries, from internal combustion engines in vehicles, and from the burning of wastes such as rubbish. Industries such as chemical plants and paper mills also produce air pollutants through various noncombustion processes. (You may have been reminded of this as you gasped your way past a paper mill on the highway.)

Pollutants take numerous forms. The most common ones are *particulate matter* (particles over 0.1 micron in size, such as smoke and ash), and a variety of chemicals in the gaseous state or in tiny liquid droplets called *aerosols*. The most common gases and aerosols are *sulfur dioxide* (SO_2), from the burning of coal and oil; *nitrogen dioxide* (NO_2), from the same sources and from car engines; and *carbon monoxide* and *hydrocarbons*, mostly from engine emissions. Table 20-1 summarizes the sources and amounts of these pollutants.

In terms of what causes the *most* pollution, motor vehicles lead the parade because of their high emissions of carbon monoxide and hydrocarbons, as you can see from the table. Not far behind are industrial activities. Roughly half of industrial pollution is produced by electrical generating plants that burn coal or oil for fuel. Nuclear power plants have a great advantage in this respect, because they are completely "clean" with regard to air pollution. But their drawbacks are that they produce tremendous quantities of heat energy, itself an environmental pollutant; their radioactive wastes are difficult to dispose of; and they present potential danger if an accident should spew quantities of radioactive materials into the environment.

Automobile emissions are an especially troublesome problem because of the nature of the pollutants they produce. Carbon monoxide (CO) is a poisonous gas, because it attaches to hemoglobin and reduces its capacity to carry oxygen. In many major cities during the rush hours, the CO level may reach 300 parts per million parts of air — 10 times the recommended safe level for human beings. Studies indicate a close correlation between accident rates and a rise in CO. Elderly people and sufferers of lung disorders such as emphysema and chronic bronchitis are especially affected by carbon monoxide.

Lead from leaded gasoline is another potentially dangerous pollutant, although its actual damage to human health has not been documented. Very small quantities of lead have severe neurological effects, and oil companies have been directed to eliminate gradually this additive from gasoline.

TABLE 20-1.   Major Air Pollutants—Estimated Emissions of Major Air Pollutants in the United States in 1971, in millions of tons

Source	*Partic- ulates*	*Sulfur Oxides*	*Nitrogen Oxides*	*Carbon Monoxide*	*Hydro- carbons*	*Percent of Total*
Combustion from industries and power plants	6.5	26.3	10.2	1.0	0.3	21
Vehicle engines	1.0	1.0	11.2	77.5	14.7	51
Combustion of trash and rubbish	0.7	0.1	0.2	3.8	1.0	3
Industrial processes other than combustion	13.5	5.1	0.2	11.4	5.6	17
Miscellaneous	5.2	0.1	0.2	6.5	5.0	8
Totals	26.9	32.6	22.0	90.7	26.6	

As shown in Table 20-1, autos also produce sizable quantities of nitrogen oxides and hydrocarbons. These chemicals are not only irritating to respiratory membranes, but they are also the major components in the reactions that produce *photochemical smog*. This unusual reaction occurs when a layer of cool air at ground level is held there by an overlying layer of warm air, a *temperature inversion* (Figure 20-14). Pollutants accumulate at ground level instead of being dispersed, as would normally occur. The action of sunlight striking this complex mixture of hydrocarbons and nitrogen oxides triggers a series of chemical reactions that produce many harmful byproducts (Figure 20-15). Some of these compounds such as ozone and PAN (peroxyacetyl nitrate), are especially damaging to human beings and can also harm vegetation and farm crops. Inversion layers most often occur in basins surrounded by mountains, as in Los Angeles.

The type of smog more commonly found over cities in the eastern United States and in Europe results from the accumulation of sulfur oxides and particulate matter, such as soot, in stagnant air masses over these cities. Sulfur dioxide (from the sulfur in coal and fuel oil) is usually the most common pollutant in this type of smog (Figure 20-15). This chemical is irritating to respiratory membranes and makes breathing difficult. The incidence of bronchitis, sore throats, eye irritations, common colds, and headaches rises sharply in areas such as New York City when smog is present. Among people with serious lung ailments, death rates increase.

Air pollution is a deadly serious affair, as has already been demonstrated tragically in this country and in Europe. In December 1952, heavy smog in London caused approximately 4000 deaths. In Donora, Pennsylvania, in 1948, a smog of several weeks' duration caused hundreds of deaths, as did the smog that hung over New York City in 1962 and again in 1963. Heeding these warnings, some industrialized countries are taking steps to control air pollution. Stringent legislation in London has greatly decreased their infamous "pea soup" fogs, which were really smogs. In the United States, serious efforts are directed toward decreasing the use of high-sulfur coal and fuel oil by industry, and by establishing pollution emission standards for automobiles. The outcome of these efforts is not yet evident, however. Los Angeles residents are still immersed in their photochemical smog, and many cities along the eastern seaboard are almost continually enveloped in a cloud of pollutant haze. Unfortunately

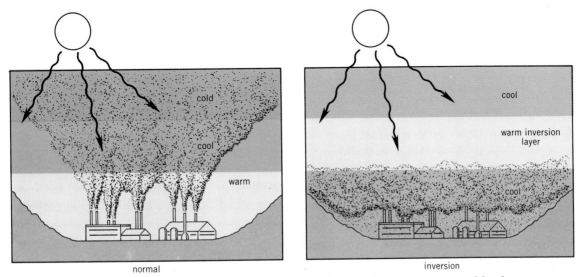

Figure 20-14 *The formation of a temperature inversion holds air pollutants near ground level as shown at the right. (Jeff V. Erickson)*

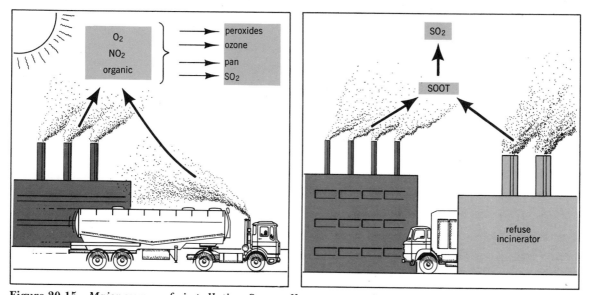

Figure 20-15 *Major sources of air pollution. Some pollutants are modified by sunlight (left) to produce photochemical smog. Sulfur dioxide and particulate matter are produced by industries and the burning of rubbish (right). (Jeff V. Erickson)*

the nation's energy problems may undo some of the small progress that has been made. Due to the increasing demand for electrical power and the high cost of fuel oil, some power plants are reverting from low-sulfur fuels to the more abundant high-sulfur coals.

Related to these problems is the occurrence of a new threat to the environment, that of *acid rain.* For the past several years, the rain and snow falling on the eastern United States, southeastern Canada, and northern Europe have become increasingly acidic. This is evidently caused by sulfur and nitrogen oxides reacting with other substances in the air to form sulfuric acid and nitric acid. This acidic rain in turn contaminates lakes, streams, and public drinking water, dissolves mineral nutrients from the soil, and corrodes buildings. Thousands of lakes in Canada as well as numerous bodies of water in New England and in New York State no longer support aquatic life because the water in them has become so acid. Emission-control legislation by Congress has been proposed, but most industries argue that it is prohibitively expensive to eliminate sulfur dioxide emissions. Many environmentalists feel that corrective measures should be taken immediately in order to control this serious problem.

On the plus side of this rather gloomy picture are various studies on the earth's atmosphere, which show that human activities have not as yet produced significant changes in the composition of air on a worldwide scale. The oxygen content, for example, has not yet been lowered, as some scientists had feared. Carbon dioxide has increased slowly but steadily (13 – 18 percent) since 1850, due mostly to increased combustion from industrialization and destruction of forests all over the planet. At the same time, the global temperature has warmed about 0.4°C. This is thought to be the consequence of the *greenhouse effect:* As carbon dioxide increases in the atmosphere it helps to retain heat from sunlight near the surface of the earth. This is similar to the effect of the glass panes in a greenhouse which allow the sunlight to pass through but reduce the loss of heat from the air in the greenhouse (Figure 20-16). The long-term effects, if any, of this warming trend are not known.

Basic Concepts

Air pollutants take the form of particulate matter and chemicals in the gaseous state or in aerosols.

Motor vehicles cause the most pollution, primarily carbon monoxide and hydrocarbons.

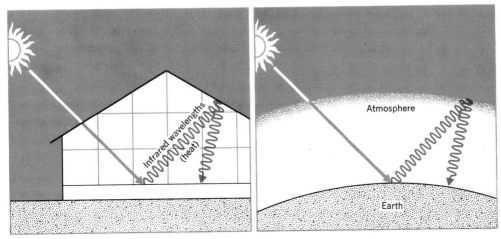

Figure 20-16 *Greenhouse effect. Excess CO_2 in the atmosphere would act like glass in a greenhouse in trapping infrared radiation on the earth.*

Water Pollution

Just as there is not any "pure" air in nature (except perhaps at extremely high altitudes), similarly there is not any pure water. The clearest mountain stream carries a load of minerals, dissolved gases, and organic matter from the plants and animals that live in or near the stream. However, as with air pollution, human activities are the ones that lead to pollution problems in aquatic ecosystems.

Humans put two general types of pollutants into aquatic environments: *enriching substances,* which stimulate excessive growth of microorganisms and algae, and a wide variety of *poisonous materials,* which damage aquatic organisms. In addition, many people would list objects such as beer cans, old shoes, automobile tires, and garbage as *visual pollutants,* even though they neither enrich the water nor poison the life in it.

Enriching substances consist predominantly of human sewage, but they also include fertilizer washed off from farms, animal wastes, and detergents. The nitrates and phosphates in these materials serve as nutrients to the algae in the water and stimulate their rapid growth—an algal "bloom" occurs. Some of these algae may impart offensive tastes or foul odors to the water. Furthermore, as some of this lush crop of algae dies, bacterial and fungal decomposers become abundant, often to the point of depleting the water of free oxygen. Oxygen-dependent organisms in the water, such as fish, suffocate. As they decompose, phosphates and nitrates are released to stimulate an-

other algal bloom, and so the cycle continues. This damaging process is known as *eutrophication.*

Eutrophication in a body of water such as a lake can render the water unfit for any human use. Organic matter dumped into rivers may cause eutrophication to some degree, but the flow of the water may keep it from proceeding to the point of utter oxygen depletion. However, in small rivers and streams that do not carry enough water to handle the load of organic matter, oxygen may be totally depleted.

An additional threat posed by sewage is that of contamination of the water by disease bacteria and viruses, including those that produce hepatitis, polio, botulism, various kinds of dysentery, and typhoid. This problem has reached such serious proportions that cities are being persuaded through governmental legislation and financial aid to install sewage treatment systems or improve existing ones.

A sewage treatment system often involves three stages (Figure 20-17). In the *primary* treatment stage, sewage consisting of wastes from homes and runoff water from streets enters the sewage plant. It is processed through screens and settling tanks to remove solid matter, such as large lumps of unpleasant things and coarse materials such as sand or grit. Then the liquid portion, still loaded with dissolved organic matter, is either directed into the secondary treatment stage or (more often) is poured into a nearby stream or river. Primary treatment does not improve sewage biologically to any appreciable extent, because what

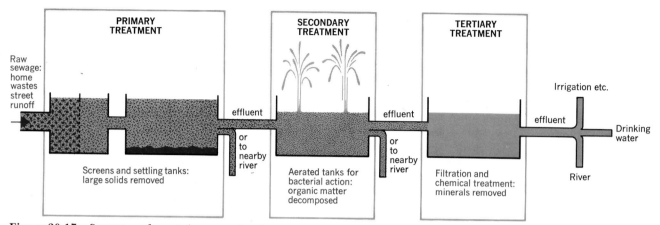

Figure 20-17 *Summary of events in sewage treatment.*

was originally putrescent, foul-smelling, *lumpy* sewage is now putrescent, foul-smelling, *liquid* sewage. It still contains its population of pathogens and minerals and will cause eutrophication in rivers and lakes. Therefore, communities are urged to process the sewage further.

In *secondary* processing, the sewage is held in aerated tanks where microorganisms break down the organic matter into simpler compounds. Once this material is removed, the effluent is filtered and then chlorinated to kill any pathogens. Most sewage plants then return this relatively clean liquid to the environment, where it at least does less damage than does sewage that has received only primary treatment. However, even this secondary effluent carries a considerable load of nitrates, phosphates, and other minerals capable of causing enrichment.

If a community can bear the financial cost, its sewage can enter a *tertiary* treatment stage involving additional filtering and chemical treatment to remove its mineral load. The effluent from this treatment is pure enough to drink and, indeed, is far cleaner than the drinking water used by most cities. Few communities now provide tertiary treatment, because of the cost, but it may become a necessity as a source for water in the future. To an environmentalist, there is a certain appeal to the idea of a city's residents recycling the wastes that they produced.

Poisonous pollutants arise mostly from the discharge of industrial wastes into bodies of water. These are often mixtures of chemicals, petroleum products, acids, metals, herbicides, and pesticides. In the Appalachians, acid water runoff from old coal mines and from new strip mines can be very damaging. Accidental oil spills from ships and storage areas also belong in the poison category. These pollutants can render a river almost sterile. Dumpings of poisonous wastes have caused spectacular fish kills in areas of the Great Lakes, Chesapeake Bay, San Diego harbor, and many other areas. Perhaps a low point was reached when portions of the Cuyahoga and Buffalo rivers were declared fire hazards because of the load of combustible pollutants and debris found in them; a portion of the Cuyahoga River actually caught fire!

A somewhat typical example is the plight of cities on the lower Mississippi River. Industrial development is heavy along the river, and the river becomes progressively more polluted as it proceeds toward the sea. Each town quite literally has to drink the pollutants from the towns and industries upriver from it. By the time the Mississippi reaches New Orleans, still 100 miles from the sea, it contains an enormous assortment of chemical wastes. New Orleans obtains its drinking water from the river. Even after fairly extensive treatment, this water still contains over 30 chemicals, some of which are known to be carcinogenic in laboratory animals. The populace of New Orleans is understandably concerned, but their only solution at present is to drink bottled water from less contaminated sources.

One unexpected and frightening consequence of industrial pollution was first detected in Minamata, a Japanese coastal city, in 1950. Fishermen and their families started to become ill with unexplained neurological ailments, eventually resulting in 17 deaths and 23 permanently disabled victims. Seabirds and household cats suffered the same symptoms. After considerable detective work, the mysterious ailments were found to be caused by mercury poisoning from the seafood that the fishermen, their families, the cats, and the seabirds all used for food. The only mercury source in the area was an industrial plant whose wastes contained small amounts of metallic mercury, but mercury is not poisonous in its elemental form. The mystery was solved when it was found that microorganisms in the sea convert metallic mercury into an organic form, methyl mercury, which is poisonous to living systems. This type of mercury poisoning is now termed *Minamata disease*. This incident has made industries more cautious about discharging mercury into the environment. This is just one dramatic, unexpected consequence of placing a normally rare element in the environment in a concentrated form. Perhaps the same type of event could happen with other elements.

Basic Concept
Water pollutants are classed either as enriching substances that cause eutrophication or as poisonous materials that harm aquatic life.

Pesticides
Human beings inhabit the earth in numbers that are far higher than could be supported by natural ecosystems. This population density is possible because of man's agricultural skills. Agriculture shortens food chains, simplifies ecosystems, and creates enough net production in domesticated plants to support the ever-growing population. One major reason for today's productive agriculture is insecticides. Man's food crops are often *monocultures*—huge plantings of a single species of plant. The problem is that any insects that are adapted to feed on a particular kind of plant can reproduce almost explosively, because they have a virtually unlimited food supply. The plant crop is doomed unless the insect predation can be stopped. Insect poisons have been the best technique for doing this for a long time.

The original insecticides were often made of plant products: pyrethrins from dried chrysanthemum

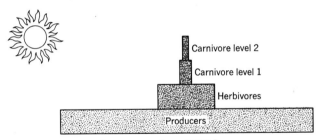

Figure 20-18 *Substances such as DDT become more concentrated in each successive level in an ecological pyramid.*

flowers, rotenone from derris plants (a vine in the East Indies), and nicotine from tobacco plants. These compounds are efficient contact-type insect killers, but they last only a short time in the environment. Other widely used insecticides included compounds containing arsenic, zinc, lead, copper, and mercury. These too had a limited lifespan, and some of them were dangerous to human beings.

A breakthrough in insect control came in 1937 with the preparation of a *chlorinated* organic substance called *DDT*. It effectively killed many insects, seemed to be harmless to humans and other large animals, and it was effective for many months after use. In the tropics, DDT became the major agent in combating the malaria-carrying mosquito; it is credited with saving countless human lives, because malaria is still one of humanity's major killers. Additional chlorinated-hydrocarbon pesticides were developed, including chlordane, dieldrin, and heptachlor. These substances and DDT were used in great quantities in many parts of the world, including the United States. At first they seemed like an unmixed blessing.

Eventually, however, some disturbing facts became evident. For one, these substances work their way up food chains or pyramids from herbivore to carnivore. But more significantly, substances become more concentrated at each successive level (Figure 20-18). This sometimes led to large kills of birds that fed on insects that had eaten DDT on vegetation. DDT was found also to interfere with the reproductive cycles of many birds, especially the top carnivores in their food chains, such as hawks and pelicans. A disturbing note was the discovery that human beings from all parts of the world accumulated at least small amounts of DDT in their fatty tissues. And finally, it was discovered that the chlorinated hydrocarbons were not only being distributed to all parts of the planet, including the Antarctic, but they were also accumulating in the soil, where they were not being broken down. All this evidence suggested that accumulating DDT represented a long-term hazard to human life. DDT use was finally banned from use in the United States and in a number of other countries and the other chlorinated hydrocarbons are also on their way out.

A second highly effective group of pesticides is the *organic phosphates,* such as parathion, malathion, and diazinon. These compounds should more properly be called *biocides,* because in certain concentrations they kill any form of life, including humans. These chemicals are highly effective against insects and are biodegradable in the soil. However, they are highly dangerous: Many human deaths have been attributed to their improper use and handling. The home gardener should be extremely careful when using such products. In view of the dangers involved in the use of chlorinated hydrocarbons and organic phosphates, what alternatives do we have for controlling crop-eating insects? The most promising seem to be *biological controls* of several kinds.

One such method is to introduce a predator that feeds on the insect pest. A classic example is the use of ladybird beetles to control citrus scale. There is a decided risk involved in introducing foreign animals and plants, however, because they may themselves become pests.

Another type of biological control involves insect *pheromones.* This promising technique has been most widely used against the gypsy moth, whose caterpillar is extremely destructive to forest trees in the eastern United States. Chemists synthesized a substance called Disparlure that is similar to the sex attractant used by the female gypsy moth. Spread in the woods and placed in traps, this lure prevents the male moths from finding their lady friends and reproducing. Success with this technique has been minimal so far.

Still another approach is the *sterile male technique.* This method was used with great success against the screw-worm fly that infects wounds on livestock. Flies are raised in containers in large numbers. They are then sterilized with radioactivity before being released over the countryside. The sterile flies mate with wild flies but, of course, produce no young (Figure 20-19).

In the southern United States, this method was so successful that the fly was thought to be extinct in this country. However, severe outbreaks occurred after 1968. The fly still survives in Mexico, and treatments must be repeated, at considerable expense, to keep it out of the United States.

The sterile male technique is a highly promising one because it controls only the target pest. Attempts are being made to apply it to other major insect pests such as the gypsy moth and boll weevils.

Biological controls are greatly preferable to pesticides for several reasons: They are safer to use, do not harm beneficial insects or other nontarget organisms, and are less expensive than chemical pesticides in many instances. Unfortunately biological control methods are known for only a few insect pests, and farmers are still obliged to continue using pesticides that do not discriminate between friendly and harmful insects.

Basic Concepts

Most of the widely used pesticides are chlorinated hydrocarbons or organic phosphates. Excessive use of these chemicals have damaged many ecosystems.

Biological controls, when available, are greatly preferable to chemical pesticides because they affect only the target organism.

HOW DID WE GET INTO THIS MESS?

Ecologists, sociologists, politicians, and many others have debated at length about the causes of our environmental problems. The arguments condense into four factors that are all at least partly responsible and closely interrelated.

1. *Urbanization.* The concentration of huge numbers of people in cities has undoubtedly put stresses on the environment, especially as regards air and water pollution. A city is somewhat analogous to a giant parasite. It continually takes materials from its host, the biosphere, but returns only waste products.

2. *Population Pressure.* The human population is undeniably increasing far more rapidly than the biosphere can support (Figure 20-20). We often seem only a few steps ahead of the doomsday prophesied by Thomas Malthus. The sheer numbers of people have

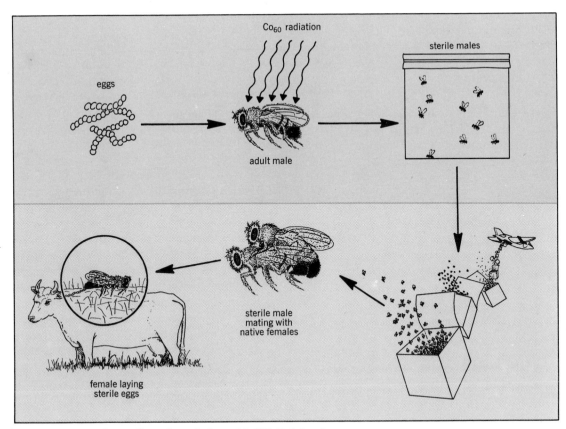

Figure 20-19 *The sterile male technique. (Jeff V. Erickson)*

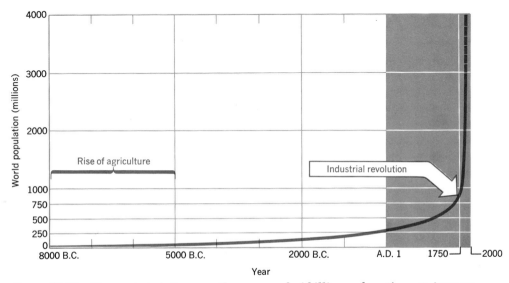

Figure 20-20 *Human population growth now exceeds 4 billion and continues to increase.*

certainly damaged various ecosystems, as we saw earlier in the chapter. Contemporary biologists such as Paul Ehrlich and Garrett Hardin believe that many of our problems stem from excessive population growth. Their popularized writings present many convincing arguments for this viewpoint.

3. *Technology and Wealth.* As science and technology advanced, making possible such achievements as the industrial revolution, nations and individuals began to accumulate great wealth. This process in turn led to still greater exploitation of natural resources, and thus to abuses of the environment. According to Barry Commoner, this factor has been the most influential one in degrading the environment. Part of his argument is based on the observation that wealthy nations use a disproportionate share of the planet's resources compared with their population sizes. Many less wealthy nations have far greater population density (number of people per square mile), but use far less of the earth's resources. We certainly have far more automobiles, television sets, home furnishings, and luxury items than any nation in the world. The manufacture of these items extracts a huge toll from the environment in terms of materials and energy. Barry Commoner's popularized writings such as *The Closing Circle* make these and other points convincingly.

4. *Democracy and Individual Resource Ownership.* Our form of government makes it possible for an individual, or a group of individuals as a corporation, to own areas of land and the resources the land contains. Moreover, these resources often can be used or abused at the discretion of the owners. Forests can be cut, minerals mined, grasslands grazed, and water polluted almost without restraint. Only when it appears that ecosystems are being seriously endangered does government step in, establishing regulations and setting aside areas to be preserved as national parks, forests, and monuments. This philosophy of private ownership is still strong in our society, and conflicts frequently occur when a federal regulatory body such as EPA (the Environmental Protection Agency) attempts to "protect" some portion of the environment from damaging exploitation. Typical examples have been the resistance met in trying to regulate strip mining in various parts of the country and drilling for oil in offshore coastal areas.

Because the problem of environmental degradation appears to be based on a variety of events and situations, there are no simple solutions to the problem. In the United States we have slowed our population growth dramatically. We are now concerned about various kinds of urban problems, we have embarked on a variety of environmental improvement programs, and we are trying to turn our technological expertise towards a wiser use of resources. Progress may seem slow, but these are the only rational procedures that seem available at this time.

Basic Concept

Environmental problems in our country evidently stem from four conditions:

1. **Urbanization**	3. **Technology and wealth**
2. **Population pressure**	4. **Individual ownership of resources**

KEY TERMS

acid rain	epiphyte	nekton	sterile male technique
biocide	eutrophication	northern coniferous forest	taiga
biological control	grassland		temperate deciduous forest
biome	greenhouse effect	organic phosphate	
chlorinated hydrocarbon	laterite soil	pheromone	temperature inversion
	liana	photochemical smog	tropical rain forest
DDT	Minamata disease	plankton	tundra
desert			

SELF-TEST QUESTIONS

1. How does this chapter support the quote, "It's not nice to fool Mother Nature"?

2. Define the term *biome* and list the factors that determine the type of biome found in a particular geographical area.

3. For each of the following biomes tell its major climatic features, principal plant and animal life, and how it is used by mankind: (a) Tundra; (b) Taiga; (c) Temperate deciduous forest; (d) Grasslands; (e) Deserts; (f) Tropical biomes; and (g) Marine and freshwater biomes.

4. What was Rachel Carson's role in triggering the ecological revolution of the 1960s and 1970s?

5. List the three major environmental problems encountered in highly developed countries.

6. What is the difference between natural and artificial air pollutants?

7. What are the most common artificial pollutants? What is the cause of the most pollution?

8. Discuss the type of pollutants produced by automobile emissions.

9. Describe how a photochemical smog sometimes comes about.

10. What produces the type of smog found over cities in the eastern United States and in Europe?

11. Discuss the greenhouse effect and its possible cause.

12. Why is there no "pure" water in nature?

13. What two types of pollutants do humans add to aquatic environments?

14. Describe the series of events that lead to eutrophication.

15. Describe the steps involved in sewage treatment.

16. Why were the original pesticides not as dangerous as the types used today?

17. For what reasons was DDT finally banned from use in the United States?

18. What is the major danger of the organic phosphate pesticides?

19. Explain what is meant by *biological control* of insects.

20. Why is this type of control preferable to pesticide chemical control?

21. List the four factors that seem to be the cause of our present environmental problems.

glossary

Abortion. Expulsion of a fetus from the womb before it can survive outside its mother's body.

Acetylcholine. A transmitter chemical produced at the axonal ends of parasympathetic nerves.

Acetylcholinesterase. An enzyme found in the synapse that breaks down acetylcholine as it accumulates.

Acid rain. Precipitation that has become acidic as a result of air pollutants such as sulfur and nitrogen oxides.

Acromegaly. A condition in adults caused by excess amounts of the growth hormone, somatotropin. Characterized by enlarged hands and feet, protruding jaw, coarse facial features, and excessive body hair.

Acrosome. A membranous sac of enzymes located on the head of a sperm cell.

Action potential. Reversal of electrical charge on a nerve cell membrane. A nerve cell impulse.

Active transport. The movement of substances through a cell membrane by an expenditure of energy.

Adaptation. The adjustments that organisms make to their environment. Adaptations may involve the structure, function, or behavior of an organism.

Adaptive radiation. The many different adaptive pathways followed by a group of organism in its course of evolution.

Addison's disease. Condition caused by insufficient adrenal hormones. Characterized by a bronzed looking skin, decrease in heart size and work capacity, and death if not treated.

Adenine. An organic base unit in a nucleic acid; part of the chemical code.

Adenosine diphosphate (ADP). A complex chemical found in all cells and related to energy output of the cell. ADP results from the removal of a phosphate group from an adenosine triphospate (ATP) molecule.

Adenosine triphosphate (ATP). A complex chemical, found in all cells, that provides the chemical energy cells use in performing any kind of work.

ADP. See **adenosine diphosphate.**

ADP-ATP cycle. The breaking down of small organic compounds in order to convert ADP molecules into ATP molecules.

Adrenals. Endocrine glands on the upper end of each kidney. Produces several hormones with widely varied functions.

Aerosol. Extremely small liquid droplet.

Afterbirth. Placenta and umbilical cord after being expelled from the uterus following a birth.

Albinism. Lack of pigmentation in the skin, hair, and iris of the eyes caused by recessive genes.

Albumin. A protein found in the blood that functions as a major transporter of materials such as ions and some drugs.

All or None Law. A nerve fiber carries a full-strength impulse or none at all.

Allantois. Outgrowth of embryonic hindgut that becomes part of the umbilical cord.

Allele. Any of the contrasting forms of a gene that may occupy the same site on a chromosome.

Altruism. A behavior in which an organism puts itself at risk in order to protect another.

Alveoli. Extremely tiny air sacs through which gases are exchanged in mammalian lungs.

Amino acid. An organic molecule that contains at least one carboxyl group and one amino group. Proteins are composed of amino acids.

Amniocentesis. Removing a small amount of amniotic fluid in order to obtain loose cells from the embryo. Certain diseases can be diagnosed from these cells.

Amnion. The membrane that encloses the embryo in fluid.

Amylase. An enzyme in saliva that breaks down starches.

Anaerobic respiration. An energy-yielding event in which a substance, other than oxygen, is used as a hydrogen acceptor.

Analogous structures. Those that have similar functions and appearance but different embryological origins.

Anaphase. A stage in cell division when the chromosomes move from the center of the spindle to the poles.

Anatomy. The study of the structure of an organism.

Androgens. Collective term for male sex hormones.

Anthropoidea. A suborder of primates that has the general body form of a human being (monkeys, apes, and humans).

Antibiotic drug. A drug that inhibits the protein-making process and therefore the growth cycle of bacteria.

Antibody. A protein produced by a special group of white blood cells in response to invaders such as microbes. Antibodies inactivate such invaders in various ways.

Anticodon. Triplet code of nucleotide bases on a transfer RNA molecule.

Antidiuretic hormone. Hormone from the pituitary gland that regulates excretion of water from the kidneys.

Antigen. Substance that has invaded an organism and provoked the formation of antibodies; usually microorganisms or large molecules such as proteins.

Antihistamine. A chemical released from damaged tissue cells.

Aorta. The large blood vessel that transports blood out into the body from the left ventricle of the heart.

Applied research. The systematic accumulation of observations directed toward the solution of a problem.

Arteriole. An extremely small artery that connects with a capillary bed.

Artery. A blood vessel that conducts blood away from the heart. Usually contains a considerable amount of smooth muscle.

Arthropoda. The largest phylum of animals. Its members have jointed appendages and are encased in an exterior skeleton, among other features. Includes insects, crustacea, and spiders.

Artificial inovulation. Placing an externally fertilized egg back into the uterus to continue its development.

Artificial insemination. Placing sperm into the female reproductive tract by artificial means.

Asexual reproduction. A portion of an organism grows into a replica of the organism. Includes budding, fission, and vegetative reproduction in plants.

Association neuron. A nerve cell that connects a sensory neuron with other neurons.

Astigmatism. A defect in the curvature of the cornea that causes blurred vision.

Atherosclerosis. The buildup of fatty deposits in arteries.

ATP. See **adenosine triphosphate.**

Atrioventricular node. Mass of tissue in the atrial wall that helps coordinate the contractions of the heart chambers.

Atrium. Upper chamber of the heart. Right atrium receives venous blood from the body; left atrium receives oxygenated blood from the lungs.

Australopithecus. Fossil ape man found in the Olduvai Gorge in South Africa.

Autoimmune disease. A malfunction of the immune system that allows the body to damage its own tissues.

Autonomic nerves. Neurons that serve the internal organs of the body, and join the spinal nerves near the spinal cord. Autonomic nerves consist of two antagonistic divisions, the parasympathetic and sympathetic.

Autonomic nervous system. The portion of the nervous system that controls the smooth muscle tissue in organs and glands.

Autosome. A "body" chromosome, in contrast to a sex chromosome.

Autotroph. An organism that is capable of making its own energy supply, usually by photosynthesis.

Axon. Portion of a neuron that carries impulses away from dendrites to some other part of the nervous system or to a muscle or a gland.

B cell. Lymphocyte that is capable of forming antibodies.

Barr body. A tiny speck of chromatin, representing an X chromosome, and found only in the nucleus of cells from females. It is a sex indicator.

Basal cell layer. Layer of cells in the skin that forms new epidermal cells.

Basal metabolism. The energy utilized by the function of the internal organs when the body is at rest.

Base pairing. The specific linking of adenine to guanine and thymine to cytosine.

Basic research. The systematic accumulation of observations not directed at immediate practical applications.

Basilar membrane. Thin membrane in the inner ear on which rests long rows of hair cells.

Batesian mimicry. Imitation of an organism that is recognized by predators as being dangerous or distasteful.

Beads-on-a-string theory. A theory of chromosome structure which proposes that a chromosome consists of DNA associated with globular clusters of histone proteins.

Behavioral adaptation. The actions of organisms that better enable them to survive. The hissing of the hog-nosed snake is an example.

Bergmann's rule. Concept that body size is larger in warm-blooded animals in colder climates than in warm ones.

Bile. Secretion from the liver and gall bladder that separates fats into extremely small particles.

Bio-assay test. A technique that uses living tissues or organisms to test for the presence of substances such as hormones or antibodies.

Biocide. A chemical that kills any form of life.

Bioethics. The application of ethics to biomedical problems.

Biofeedback. A conditioning technique that allows an individual to willfully control certain autonomic nervous system functions.

Biogenetic law. Concept that the developmental stages of an embryo repeat its evolutionary history.

Biogeochemical cycle. Cycling of elements from the earth or atmosphere into living things and back into the earth. Microorganisms often play crucial roles in these cycles.

Biological clock. A hypothetical mechanism in the body (brain) that controls certain biological rhythms.

Biological control. Any technique for controlling harmful pests by use of means that affect only the target organism.

Biological principle or law. A theory or hypothesis that has been verified repeatedly and appears to have wide application in biology.

Biology. A branch of science that studies the phenomena of life.

Biome. Assemblage of distinctive plants and animals that occupies a sizeable geographic area.

Biosphere. The part of earth on which life exists: soil, water, and air at the surface of the planet.

Bipedalism. Locomotion in an upright body position on the hind limbs.

Birth control pill. Mixture of a synthetic estrogen and progestin in pill form that prevents ovulation if taken on a regular daily schedule.

Blastocyst. Small, hollow ball of cells resulting from repeated cell divisions of a zygote. Synonomous with blastula.

Blastula. Hollow ball of cells resulting from cleavage of a zygote.

Blood. A liquid tissue that circulates through the blood vessels.

Blood protein. Any one of a variety of proteins found in blood. Three important ones are albumin, the globulins, and fibrinogen.

Blood sugar. Another name for glucose, the most commonly used energy source in animals.

Bomb calorimeter. A device in which a unit of food is burned in order to measure its energy content.

Bowman's capsule. Cuplike structure that encloses a glomerulus. Part of a nephron.

Brain stem. Portion of the brain made up of the medulla, pons, midbrain, thalamus, and hypothalamus.

Breech birth. Baby born buttocks first, making the delivery more difficult.

Buffer. A chemical that helps maintain a certain acid-base level in a solution.

Caecum. A saclike portion of the large intestine. It may be large and functional as in some herbivores or small and nonfunctional (an appendix) as in many carnivores.

Calorie (with an upper case C). Amount of heat required to change the temperature of a kilogram of water from $14.5°C$ to $15.5°C$. also called a kilocalorie.

Calorie (with a lower case c). Amount of heat required to change the temperature of one gram of water $1°C$.

Calorimeter. A device for determining the rate of energy used by an organism.

Camera eye. The vertebrate-type eye, so called because it is generally similar to a camera with its shutter (iris), lens, and film (retina).

Cancer. An unrestrained growth of cells in an organism.

Capacitation. Interaction of sperm with the chemical environment of the female reproductive tract.

Capillary bed. A network of tiny, interconnected, thin-walled blood vessels found throughout body tissues.

Carbohydrase. An enzyme that digests carbohydrates.

Carbohydrate. A major organic substance in cells, consisting of a carbon chain to which are attached hydrogen atoms and hydroxyl groups.

Carbon cycle. A biogeochemical cycle in which carbon moves from the atmosphere into plants, then into animals and eventually back into the atmosphere.

Carcinogen. Any substance or energy form that causes cancer. Examples are tars in cigarette smoke and X rays.

Carcinoma. Solid tumor type that makes up about 85 percent of human cancers.

Cardiac tube. The structure that functions as the heart in early developmental stages of an embryo. It later grows into a four-chambered heart.

Cardiovascular fitness. Preventing the buildup of fatty deposits inside the body's arteries by means of regular, vigorous, physical activity.

Carnivore. A flesh-eating organism.

Carrier (genetic). Heterozygous state, as Aa.

Carrier-facilitated diffusion. Movement of molecules through membranes with the aid of special protein carriers.

Catalyst. A substance that promotes a chemical reaction without itself being changed or used up in the reaction.

Cataract. An opaque area in the lens of the eye.

Cell. The basic structural unit of all forms of life.

Cell hybrid. Cell resulting from the union of cells from different organisms. See Hybridoma.

Cell life cycle. The stages through which a cell may go in its life cycle: growth phase one, synthesis phase, growth phase two, and mitosis.

Cell membrane. A thin boundary structure made of a complex of lipids and proteins.

Cellular respiration. A series of chemical reactions in cells that derives energy from short carbon-chain nutrients for making ATP molecules.

Cellulose. A complex carbohydrate used by plants as a structural material. Wood, for example, is made of cellulose.

Central nervous system. The spinal cord and brain.

Centrifuge. A machine capable of spinning materials rapidly in order to separate substances of different densities.

Centriole. A structure in the cell composed of tiny tubules arranged in a circle. It plays a role in forming the mitotic spindle.

Centromere. Point of attachment of a spindle fiber to a chromosome.

Cerebellum. A large portion of the brain that coordinates motor activities of the body.

Cerebral cortex. The gray matter that forms the outer layer of the cerebral hemispheres.

Cerebrum. Made of two large hemispheres, many small subdivisions, and covered with a layer of gray matter. It has primary control over all motor activities, is the interpreting center for all the senses, and is the site of learning, memory, and all intellectual activities.

Cervix. Lower portion of the body of the uterus that joins the vagina.

Chalone. Chemical found in many tissues; inhibits cell division.

Chemical bond. The energy that binds atoms to one another.

Chemical origin of life. Hypothesis that life arose by the combination of elements that were present in the earth's primitive atmosphere.

Chlorinated hydrocarbon. A compound containing carbon, hydrogen, and chlorine; the chemical basis of a category of pesticides such as DDT.

Chloroplast. A plant organelle that contains chlorophyll.

Cholesterol. A fatty material found throughout the body in cell membranes, myelin sheaths, and used in making steroid hormones.

Chorion. Term applied to the outer cell layer of the blastocyst at a certain stage of development.

Chorionic villi. Fingerlike outgrowths of the early mammalian embryo that grow into the uterine wall.

Chromatin. Part of the cell nucleus: bodies composed of intertwined protein and DNA-containing fibers.

Chromosomal mutation. Change in the structure of a chromosome including deletion of a part, duplication of a part, or the inversion of a portion of the chromosome.

Chromosome. A threadlike body, in the nucleus of a cell, containing the hereditary material and a complex of nucleoproteins.

Chronobiology. The study of cyclic events in the body.

Chyme. The term for partially digested food in a semiliquid form in the small intestines.

Cilia. Short, hairlike organelles that protrude from cells. They are made of microtubules.

Ciliary escalator. Action of the cilia in the upper respiratory tract in pushing mucus up into the pharynx.

Ciliary muscle. Small, circularly arranged muscle that helps change the shape of the lens in focusing images on the retina.

Circadian cycle. A rhythmical event that is related to a 24 hour day-night time period.

Circumcision. Removal of the foreskin from the penis.

Citric acid cycle. A stage of cellular respiration in which energy is produced by the removal of hydrogens and carbon dioxide molecules.

Class Mammalia. Vertebrates that possess hair, suckle their newborn with milk, and are warm-blooded.

Cleavage. Initial series of cell divisions by the zygote until it becomes a ball of cells called the blastula.

Clitoris. Small protuberance in the vulva covered with a foreskin. It is very sensitive and functions in sexual excitement.

Clone. An identical copy of a cell or of an organism.

Cloning. Producing identical individuals from body cells.

Closed-type circulatory system. A blood system in which the blood is confined within arteries, capillaries, and veins as it moves about the body.

Cochlea. Spiral shaped structure that houses the inner ear passageways.

Codominance. Hereditary condition in which genes interact with one another rather than being dominant or recessive. The gene for sickle-cell anemia is an example.

Codominant gene. Gene that interacts with another gene to produce an intermediate phenotype.

Codon. A triplet of bases on a messenger RNA molecule.

Coenzyme. A substance required by some enzymes in order for them to become activated.

Coitus interruptus. Withdrawing the penis during intercourse just prior to orgasm.

Collateral circulation. Accessory arteries that supply important parts of the body in addition to the major blood supply for these parts.

Colon. The large intestine.

Color blindness. Defect in color vision caused by a mutant, recessive gene on the X chromosome.

Commensalism. An interaction in which one member benefits and its host is neither helped nor harmed.

Competition. The consequence of organisms needing the same materials from the environment.

Competition-exclusion principle. Two species having the same environmental requirements cannot live together.

Complex carbohydrate. A carbohydrate made of large numbers of simple sugar units linked in long chains.

Compound. A substance formed by the combining of atoms of two or more different elements.

Condom. Sheath for the penis to contain the sperm and semen after ejaculation. Usually made of thin latex rubber.

Cones. Specialized light-sensitive cells in the retina. Cones function in relatively bright light and provide color vision.

Connective tissue. Cells that support the body such as bone, cartilage, ligaments, collagen, and the blood cells.

Consequentialist theory. A school of ethical thought which holds that the goodness of an action is judged by the results (consequences) of the action.

Consumer. Organisms that obtain their nutrition by feeding on other organisms.

Contact inhibition. The cessation of cell growth in vitro when the cells make contact with one another.

Convergence. Superficial similarities between organisms that live in the same kind of environment. For example, the streamlined form of a porpoise and a shark.

Coronary heart disease. Changes in the coronary arteries that decrease the flow of blood to the heart.

Cornea. A transparent portion of the sclera at the front of the eyeball.

Corpus callosum. Large, compact band of nerve fibers that connects the cerebral hemispheres.

Corpus luteum. The remains of a ruptured follicle in the ovary. Produces progesterone for a short period of time.

Counter-current exchange. Liquids flowing in opposite directions in such a manner that they can exchange heat, gases, or ions in a very efficient manner.

Cowper's glands. A pair of small glands in males that opens into the urethra and secrete a small amount of semen.

Cranial nerves. Twelve pairs of nerves that extend from the underside of the brain, mostly to various structures in the head region.

Creationism. The idea that the earth and all species as we now see them were created by a supernatural power as described in the book of Genesis.

Cretinism. Mental retardation and dwarfism in children due to loss of the thyroid gland.

Cro-Magnon man. Our immediate evolutionary predecessors that probably replaced Neandertal man.

Crossing over. Exchange of pieces between two entwined homologous chromosomes.

Cryptic coloration. Color patterns that help animals resemble their environment.

Crystalline lens. Transparent lens-shaped mass of tissue that focuses images on the retina.

Cultural evolution. The changes that have taken place in our cultural and social advancement during the past 10,000 years.

Cushing's disease. A condition caused by excess glucocorticoid hormone from the adrenal glands. Characterized by an obese torso, thin arms and legs, rounded face, thin hair, diabetes, and high blood pressure.

Cyclic AMP. A chemical produced in cells in response to the enzyme adenyl cyclase. Sometimes called the second messenger.

Cystic fibrosis. Disease in which the mucus is too thick to function properly. A hereditary disease due to recessive autosomal genes. The most common hereditary disease among Caucasians.

Cytochrome c. One of the enzymes involved in the cytochrome system.

Cytochrome system. A series of enzymes involved in forming ATP in cellular respiration.

Cytokinesis. Division of the cytoplasm near the end of cell division.

Cytoplasm. That portion of the cell outside the nucleus.

Cytosine. An organic base unit in a nucleic acid; part of the chemical code.

Daughter cell. A cell produced as a consequence of cell division.

DDT. A chlorinated hydrocarbon that was formerly used as an insecticide.

Decomposer. Organism that breaks down the organic matter in dead plants and animals. Mostly bacteria and fungi.

Deficiency disease. Symptoms that arise when an important nutrient is lacking in the diet.

Dendrite. Portion of a neuron that receives impulses from other neurons.

Dermis. The deeper layers of skin cells.

Dialysis. The removal or addition of substances from a fluid by using osmosis.

Diaphragm. Rubber cap that is placed over the cervix to prevent sperm from entering the uterus.

Diastole. Relaxation of the ventricles of the heart.

Differentiation. Specialization of a cell for a specific funcion.

Diffusion. The movement of molecules from areas where they are highly concentrated to areas of lesser concentration.

Digestion. An enzyme-controlled chemical reaction in which organic molecules are broken down, with water molecules attached to the broken bonds.

Diploid. Two sets of chromosomes, the normal condition found in the body cells of eukaryotes.

Disparlure. A synthetic form of the gypsy moth sex attractant.

Distal convoluted tubule. Portion of a nephron in which selective reabsorption takes place.

Diuretic. Substance that increases the flow of urine from the kidneys.

Divergence. Refers to groups of organisms becoming different from one another during their evolutionary history.

Dizygotic twins. Twins that develop from two zygotes in the same uterus.

DNA: deoxyribose nucleic acid. A large, chemically coded molecule found in all cells; functions as genetic material and as controller of cellular activities.

DNAase. An enzyme that breaks down DNA.

Double sugar. A sugar such as sucrose made of two simple sugar units.

Down's disease. A condition caused by an extra body chromosome. Symptoms include mental retardation, slanting eyes, thick flattened tongue, abnormal palm prints, and various other anatomical abnormalities.

Drosophila. A genus of small fruit flies used extensively in genetics studies.

Dryopithecus. Fossil thought to be ancestral to the great apes.

Duodenum. First portion of the small intestine.

Dysmenorrhea. Painful menstrual period.

Ecological pyramid. A graphic means of showing the numbers, mass, or energy contents of the producers and consumers in an ecosystem.

Ecosystem. A unit of plants and animals interacting with each other and their surroundings, and self-sufficient when provided with an energy source.

Ectoderm. Layer of cells in the early embryo that forms structures such as the nervous system and epidermis of the skin.

Edema. A swelling produced by water accumulating in the intercellular spaces.

Ejaculation. Expulsion of semen through the urethra by rhythmical contraction of the ejaculatory duct and urethra.

Ejaculatory duct. Tube that transports sperm from the seminal vesicles to the urethra.

Electroencephalogram. A recording of electrical activity at the surface of the brain.

Electrolyte. An atom or small molecule in an ionic form (electrically charged) found in body fluids. Examples are calcium (Ca^{++}) and carbonate (HCO_3^-).

Element. One of the fundamental substances that constitutes all matter.

Embolus. A portion of a blood clot that breaks free within the circulatory system and may plug up a small blood vessel.

Embryo. Organism in its early stages of development. In human beings, the stages of development until the eighth week at which time it is called a fetus.

Emphysema. Lung disease in which alveoli lose their elasticity.

Energy. Capacity for performing work.

Endocrine gland. A specialized mass of tissue that produces a hormone.

Endoderm. Layer of cells in the early embryo that forms the gut and digestive organs such as the liver.

Endometrium. The vascularized tissue that forms a lining in the uterus and is shed each month during mensis.

Endoplasmic reticulum. A system of flattened, membrane-enclosed spaces in the cytoplasm of cells.

Endorphin. Substance in the brain and pituitary gland that suppresses pain.

Engram. A neural pathway in the brain.

Enkephalin. Neurotransmitter associated with opiate receptors in the brain.

Enzyme. A special type of protein that serves as a catalyst for a chemical reaction in a cell. Virtually all chemical reactions in biological systems are controlled by specific enzymes.

Epidermis. The outer layers of skin cells.

Epididymis. A collecting tube that transports sperm from the testes into the vas deferens.

Epiglottis. Fold of tissue that momentarily seals the larynx when an individual swallows.

Epithelium. A type of tissue that covers surfaces, lines hollow organs, and forms glands.

Epitope. A specific pattern of molecules on the surface of a cell. It is the cell's identification pattern.

Erection. Enlargement of the penis caused by blood flowing into spongy tissue within it.

Esophagus. Tube extending from the pharynx to the stomach.

Essential amino acid. An amino acid that an organism cannot manufacture and therefore must obtain in its diet.

Estradiol. Estrogen hormone that causes the uterine lining to grow thicker.

Estrogens. Collective term for a group of similar female ovarian hormones.

Estrous. Time during which a female is receptive to mating.

Ethics. A discipline of philosophy that deals with what is right or good in relation to human conduct.

Eukaryote. Type of cell containing a membrane-bound nucleus, chromosomes, numerous membranous organelles, and showing reproduction by mitosis.

Eustachian tube. A membranous tube that connects the middle ear and the pharynx.

Eutrophication. Enrichment of a body of water by phosphates and nitrates until plant growth and decomposition become damaging to the ecosystem.

Evolution. Change through time.

Experiment. A procedure used in the scientific method to determine whether cause-and-effect relationships exist between two or more variables.

Fallopian tube. Membranous tube that extends from the

ovary to the uterus. It is the site of fertilization in many organisms and transports the egg into the uterus.

Fat. One of the major kinds of organic substances. A fat consists of glycerol to which three fatty acids are attached. Fats are often used as an energy source by virtue of their high proportion of hydrogen atoms. Fats are a major type of storage material and are found in all cell membranes.

Fat soluble vitamin. A vitamin that is stored in fatty tissues of the body: A, D, E, or K.

Fatty acid. A chain of carbon atoms, usually 16 or 18, to which are attached hydrogen atoms and, at the end, a carboxyl group (—COOH).

Feedback. Information sent into a control system for appropriate action.

Fertilization. Union of an egg and a sperm nucleus.

Fetus. Term applied to the developing human embryo from the eighth week to birth.

Fibrin clot. A mat of tiny fibers that forms at an injury site to prevent excessive bleeding.

Fibrinogen. A fibrous protein that plays a major role in forming blood clots.

Fibrous protein. A category of proteins in which the polypeptide chains are in long ropelike forms, often intertwined.

Filter feeder. An organism that obtains food by straining it out of its environment.

Flagellum. A long slender organelle made of microtubules and used by cells for motility.

Fluid feeder. An organism that obtains food by sucking juices from animals or from plants.

Food chain. A producer-consumer chain of organisms.

Food web. A diagram that attempts to relate all of the food chains in a community.

Fovea. Small area in the retina where most of the rods are located. It is the area of sharpest vision.

Fossil. Any evidence of life that lived in the past.

Frame-shift mutation. Additions or deletions of base pairs in a DNA molecule.

FSH. Follicle Stimulating Hormone produced by the pituitary gland.

Gamete. Egg or sperm.

Ganglion. A cluster of nerve cells. See **Solar plexus** as an example.

Gastrin. A hormone produced by cells in the stomach wall. Its action is to stimulate other stomach-wall cells to make mucus, hydrochloric acid, and pepsinogen.

Gastrulation. Formation of the germinal layers in the early embryo.

Gene. Unit of heredity; a series of nucleotides in a DNA molecule that codes for a polypeptide chain.

Gene mapping. Determining the location of genes on specific chromosomes.

Genetic engineering. Altering an organism's genes to produce a desired phenotype.

Genetics. A branch of biology that specializes in the study of heredity.

Germinal layers, "germ" layers. Three layers of cells in the early embryo that form various parts of the body. Ectoderm, endoderm, and mesoderm.

Gigantism. Abnormal body growth due to an excess amount of the growth hormone, somatotropin.

Glaucoma. A buildup of fluid in the front chamber of the eye. Causes blindness if not corrected.

Globular protein. A category of proteins in which the polypeptide chains are tightly folded and coiled to give them a ball-like form.

Globulin. A protein that functions like a generalized antibody in the blood stream.

Glomerulus. Small ball of capillaries located in a Bowman's capsule at the beginning of a nephron.

Glucagon. Hormone produced by the pancreas that causes glycogen to be converted into blood sugar.

Glucose. The common sugar found in most animal bodies and used as a source of energy. It is often called blood sugar.

Glycerol. A three-carbon molecule that forms a portion of the fat molecule.

Glycogen. A long chain storage form of carbohydrate in animals, mostly stored in the liver.

Goiter. An enlargement of the thyroid gland, usually caused by an iodine-deficient diet.

Golgi body. A set of flattened sacs stacked closely in the cytoplasm. Substances made in a cell are often transported to Golgi bodies to be packaged in tiny sacs for use elsewhere.

Graves' disease. A condition caused by excess amounts of thyroid hormone. Symptoms are goiter, protruding eyeballs, nervousness, weight loss, and trembling hands.

Gray matter. Unmyelinated neurons in the central nervous system.

Green Revolution. The development and introduction of spectacular new food plant varieties into tropical countries.

Greenhouse effect. Trapping of infrared radiation on the earth due to the layer of CO_2 in the atmosphere.

Gross production. The amount of organic matter produced by photosynthesis in an ecosystem in a specified time period.

Growth phase one. The growth shown by a daughter cell immediately following cell division.

Growth phase two. The growth that takes place in a cell after it duplicates its DNA.

Guanine. An organic base unit in a nucleic acid. Part of the chemical code.

Hair cell. A specialized cell found in considerable numbers in the inner ear. It converts vibrations into nerve impulses.

Haploid. One set of chromosomes in a cell in contrast to the diploid or double set of chromosomes.

Hemodialysis. The process of passing blood through a dialysis (artificial kidney) apparatus.

Hemoglobin. A complex protein adapted for transporting oxygen in red blood cells.

Hemophilia. Disease in which the blood does not clot properly. It is inherited as a mutant gene on the X chromosome.

Hemorrhoid. An enlarged vein in the anal region.

Henle's loop. Portion of a nephron where selective reabsorption of salt takes place.

Hepatic portal system. Set of veins that transports blood from the small intestines to the liver.

Herbivore. An organism that is specialized for using plants for food.

Hernia. Protrusion of a portion of a body organ, such as a loop of the intestine, through a weakened area of the body wall.

Heterotroph. An organism that must obtain its energy supply from other organisms.

Hindgut fermenter. A herbivore in which digestion of plant matter (fermentation) takes place in the large intestine.

Histamine. Substance released from damaged cells that causes capillary walls to become porous to white blood cells.

Homeostasis. The control of internal features of the body by feedback systems. Self-regulation of the internal environment.

Hominidae. The family of Anthropoids that contains man.

Homo erectus. Mid-Pleistocene cave dwellers, fire users, tool makers, and hunters. Similar to modern human beings in size; distributed over Europe, Asia, and Africa.

Homo habilis. A fossil skull found by Richard Leakey and considered by him to be an early ancestor of our own species.

Homo sapiens. The scientific name for all contemporary human beings.

Homologous chromosome. One of a pair of identical chromosomes.

Homologous structures. Features that have the same embryological origin in organisms but not necessarily the same function.

Hormone. A chemical produced by a cell or group of specialized cells that regulates part of the body's internal environment.

Hybridoma. The product of fusing a B cell with a myeloma cancer cell.

Hydra. A small, simply organized fresh water organism with a saclike body, mouth, and tentacles.

Hydrolysis. Chemical reaction in digestion in which nutrients are split into smaller units and water molecules are attached to the broken bonds.

Hydroxyl group. An —OH group found in many organic substances in cells.

Hymen. Thin membrane that often partly covers the vaginal opening in young females.

Hypermetropia. Farsightedness: The image attempts to focus behind the retina.

Hypothalamus. A small but important part of the brain that contains many vital control centers and also produces hormones that activate the anterior pituitary gland.

Hypothesis. A tentative explanation that relates a series of observations.

Hysterectomy. Surgical removal of the uterus.

Ileum. A portion of the small intestine.

Immune system. Protective system in the blood, consisting of lymphocytes and the antibodies they produce.

Immunotherapy. A treatment procedure designed to help the body's immune system reject tumors.

Impotency. Inability to have an erection of the penis.

Independent assortment. Each homologous pair of chromosomes arranges itself on the metaphase plate in meiosis independently of how the other pairs are oriented.

Industrial melanism. An increase in the dark-colored forms of organisms in response to increased industrialization (more soot) in an area.

Inferior vena cava. A major vein that drains blood from the lower portion of the body to the heart.

Inguinal canal. A passageway through the lower abdominal wall in which the sperm duct is located.

Inner cell mass. Disc of cells that develops into the mammalian embryo.

Insulin. Hormone produced by the pancreas that causes glucose to be removed from the blood stream.

Interacting genes. Genes that interact with one another to produce an intermediate condition, wavy hair, for example.

Interferon. A protein produced by several kinds of white blood cells that helps fight virus infections.

Interphase. All stages in a cell's life cycle when the cell is not dividing.

Intrauterine device, IUD. Object placed in the uterus as a birth control device. Usually a plastic coated wire of various shapes.

Intuition. An insight or hunch.

In vitro fertilization. Fertilization of an egg in an artificial environment such as in laboratory glassware.

Ion. An electrically charged atom or compound.

Ionizing radiation. Rapidly moving particles or energy waves that create a path of ions as they pass through matter.

Iris. Layer of tissue that covers the front of the lens except for a small opening through which light passes. Muscles in the iris regulate the opening to restrict or let in more light.

Irritability. The ability of an organism to react to stimuli.

Jejunum. A portion of the small intestine.

Karyotype. A set of chromosomes that has been photographed and rearranged in matching pairs.

Keratin. Protein material that fills the outer dead layers of skin cells.

Kidney. An organ that selectively removes a variety of substances from the blood by means of tiny units called nephrons.

Kilocalorie. A unit of measurement of energy. It is the amount of heat required to change the temperature of 1000 grams of water one degree, from 14.5 to 15.5°C. Often used in reference to the energy content of foods.

Kinins. A chemical found in blood and tissue fluid that causes capillary walls to become porous to white blood cells.

Klinefelter's syndrome. Condition manifested by sparse body hair, enlarged breasts, and underdeveloped male genitalia. Caused by the inheritance of an extra X chromosome (XXY).

Kreb's cycle. A series of energy-yielding reactions in cell respiration in which citrate molecules are systematically stripped of carbon dioxides and hydrogens.

Kwashiorkor. A protein-deficiency disease in children. Its symptoms include puffy facial features, swollen abdomen, spindly limbs, blotchy skin, and apathetic behavior.

Labia, major and minor. Liplike folds of tissue that constitute the vulva or external genitalia of females.

Labor. Contractions of the uterus that open the cervix and expel the fetus. Usually described in three stages.

Lacteal. A tiny lymph tubule found in a villus in the intestinal wall.

Lamarckianism. Concept that traits acquired from the environment are inherited.

Laparoscope. Surgical instrument used to seal off the Fallopian tubes.

Larynx. The voice box and entrance into the windpipe.

Laterite soil. A type of soil found in many tropical areas of the world.

Law of dominance. A state in which one member of a pair of allelic genes is expressed.

Leucocyte. A white blood cell.

Leukemia. Cancer of the bone marrow which produces abnormal numbers of white blood cells.

L.H. Luteinizing Hormone from the pituitary gland; causes ovulation.

Life sciences. All of the various fields in biology that deal with living matter and organisms.

Limbic system. A portion of the cerebrum that functions with the thalamus and hypothalamus in controlling emotional behavior.

Linked genes. Genes located on the same chromosome.

Lipase. An enzyme that digests lipids (fats).

Lipid. A diverse group of organic substances that includes fats, cholesterol, and phospholipids.

Luteinizing hormone. Hormone from the pituitary gland that causes ovulation.

Lymph. Thin, watery, intercellular fluid that drains into lymph vessels.

Lymph node. A mass of tissue found in larger lymph vessels that filters the lymph and produces lymphocyte and monocyte blood cells.

Lymphatic system. System of thin-walled tubes that transports lymph to the venous system.

Lymphocyte. A type of white blood cell that has important functions in the body's immune system.

Lymphoma. Type of blood cancer in which abnormal numbers of lymphocytes are formed by the spleen and lymph nodes. Second most common type of human cancer (5.4 percent).

Lysosome. Tiny sac containing enzymes and formed from Golgi bodies. The enzymes are capable of breaking down a variety of organic materials.

Macromolecule. An extremely large molecule such as a protein or DNA.

Macronutrient. A mineral found in amounts equal to one gram or more per kilogram of food material or tissue.

Malthusian doctrine. Concept that human population growth will eventually outstrip its food supply, leading to famine, disease, and war.

Mammal. A vertebrate that has hair, suckles, its young with milk, and is warm blooded.

Medulla oblongata. First part of the brain stem containing all nerve tracts from the spinal cord and nerve cell centers for many basic body functions such as breathing, heartbeat, swallowing, and sneezing.

Meiosis. A special and complex type of cell division in which a cell's chromosome number (and hereditary material) is reduced by one half.

Melanin. Pigment produced by melanocyte cells in the epidermis of the skin.

Melanocyte. A cell in the epidermis of the skin that produces the pigment called melanin.

Membrane, cell or plasma. A thin boundary structure made of a complex of lipids and proteins.

Membrane potential. See **Resting potential.**

Menopause. Cessation of ovulation and menstruation, usually during mid-forties.

Menstrual cycle. Series of changes that occur in the lining of the uterus during the female reproductive cycle.

Mesoderm. Mass of cells in the early embryo that forms bones, muscles, and the blood system.

mRNA. Messenger RNA: a complementary copy of a segment of a DNA molecule that provides a pattern for making a polypeptide chain in the cytoplasm of a cell.

Metaphase. A stage in cell division in which chromosomes become arranged at the middle of the spindle.

Microfiber. Long, extremely slender organelles that are especially abundant in muscle cells.

Micronutrient. A mineral found in microgram amounts per kilogram of food material or tissue.

Microtome. A tool for cutting extremely thin slices of tissue.

Microtubule. An extremely small, long tubular body found in cells. Microtubules are the building units of cilia, flagella, and centrioles.

Midbrain. Portion of the brain containing centers for visual and auditory reflexes.

Mimicry. The use of color patterns by one species in which the mimic resembles another species, the model. The mimic is helped because the model has its own protective devices such as poison glands and is seldom molested by predators.

Minamata disease. A disease of the nervous system caused by ingesting methyl mercury.

Mineral. An element or a compound found in the earth and formed by nonliving processes.

Mineral cycle. The movement of an element from the earth or atmosphere into living things and back into the earth. The carbon cycle is an example.

Minipill. Birth control pill containing progestin only.

Mitochondria. Relatively large membranous body containing numerous internal plates. Most of the events of cellular respiration take place within mitochondria.

Mitosis. Division of the nucleus by means of a series of stages that keep the chromosome number unchanged in cells.

Molecule. The smallest particle of an element or compound that retains the properties of that substance.

Monoclonal antibody. A quantity of one type of antibody.

Monohybrid inheritance. Transmission of one pair of genes.

Monozygotic twins. Twins that develop from one zygote that separates into two cell masses.

Mons veneris. A mound of fatty tissue located over the pubic bone in females.

Moral-ethical problem. The dilemma that arises when the action (the morality) taken to solve a problem differs from the action that ought to have been taken if based on a particular set of values.

Motor neuron. A nerve cell that conducts impulses from the central nervous system to an effector.

Mucin. Slimy glycoprotein material found in saliva, nasal chambers, and throughout the respiratory tract. An important protective substance and lubricant in the body.

Multiple alleles. Condition in which more than two forms (alleles) of a gene exist in the gene pool but an organism can inherit only two of the forms. ABO blood types are an example.

Muscular dystrophy. Hereditary disease due to a gene carried on an X chromosome; thus, a sex-linked gene.

Mutagen. A substance that causes cancer.

Mutation. Change in the hereditary material of an organism and hence passed on to its offspring.

Mutualism. An interaction in which both organisms obtain benefit.

Myelin sheath. Fatty white sheath around the axons of most vertebrate nerve cells. Made of Schwann cell membranes.

Myoglobin. A protein similar to hemoglobin and found in muscle tissue.

Myoneural junction. Synapse between a motor fiber and a muscle fiber.

Myopia. Nearsightedness: the image falls ahead of the retina and hence is blurred.

Natural history. A study of any subject found in nature.

Natural killer cell. Lymphoid cells that are especially efficient in attacking cancer cells.

Natural selection. Nature selects the best adapted for survival as the fittest. The best fit therefore pass their traits on to the next generation.

Neandertal man. Prehistoric human beings that inhabited caves in Europe and the Middle East about 35,000 years ago.

Nekton. Aquatic organism with well-developed locomotor abilities.

Neodarwinism. The modern concept of natural selection utilizing knowledge about genetics, gene pools, and mutation.

Nephron. The filtering unit of the kidney consisting of Bowman's capsule, proximal convoluted tubule, Henle's loop, distal convoluted tubule, and collecting duct.

Nerve. A bundle of thousands of neurons.

Nerve cell body. The nucleus of a neuron.

Nerve impulse. A self propagating wave of depolarization moving along a nerve fiber. It involves changes in electrical potentials in the nerve cell membrane but is not an electrical current.

Net production. The amount of organic matter produced by photosynthesis in an ecosystem less the amount needed by the photosynthesizers for their own cellular respiration.

Neuron. A cell modified for transmitting nerve impulses.

Neuropeptide. One of a group of peptides (chains of amino acids) found in the nervous system and whose functions are still being studied.

Neurotransmitter. A chemical released from the end of an axon into the synaptic cleft when an action potential arrives. It changes the potential across the membrane of the next nerve fiber.

Neutrophil. A very abundant type of white blood cell.

Nitrogen cycle. A biogeochemical cycle in which atmospheric nitrogen is converted by nitrogen-fixing bacteria into nitrogenous compounds that plants and animals can use. Waste products and decomposition of dead plants and animals also play a role. Denitrifying bacteria add nitrogen back into the atmosphere.

Nodes of Ranvier. Interruptions in the myelin sheath of axons.

Nucleolus (_nucleoli_ plural). A dense mass of granular material found in the nucleus.

Nucleotide. A basic unit of a DNA molecule consisting of a nitrogenous base, a deoxyribose sugar, and a phosphate group.

Nutrient. Any substance taken into the body that has a useful function. Included are proteins, lipids, carbohydrates, vitamins, minerals, and water.

Nutritional marasmus. Wasting of the body due to starvation.

Obesity. A condition in which an individual is more than 20 percent heavier than is recommended for the person's weight, sex, and body build.

One gene-one polypeptide hypothesis. That each kind of polypeptide chain in a cell is made under the direction of a different gene.

Ontogeny recapitulates phylogeny. The development (ontogeny) of an organism repeats the evolutionary history (phylogeny) of its group.

Omnivore. An organism that feeds on both animal and plant matter.

Oocyte. Potential egg cell in the ovary.

Oogenesis. Development of an egg in a follicle in the ovary.

Opiate receptor. Specialized receptor sites for opiate chemicals found in certain brain cells.

Open-type circulatory system. A blood system in which the blood flows into spaces around the organs and is not confined within capillaries.

Order Primates. A group (Order) of mammals distinguished by advanced brain structure, stereoscopic vision, good eyesight, poor sense of smell, flattened nails and claws, prehensile hands and feet, an opposable thumb, and a well-developed collar bone. Includes lemurs, tarsiers, monkeys, apes, and human beings.

Organ. A group of associated tissues that functions as an integrated unit.

Organ of Corti. The special sensory cells for hearing, housed in a canal in the inner ear.

Organelle. A discrete structure within a cell, a mitochondrion, for example.

Organic phosphate. A chemical found in some types of very potent and dangerous pesticides.

Organic substance. A chemical constructed around a chain of carbon atoms.

Orgasm. Intensely pleasurable physical and emotional event. Accompanied by muscle spasms and increases in heartbeat, blood pressure, and breathing rate.

Osmoregulation. The passage of water molecules and salts through cell membranes.

Osmosis. Diffusion of water molecules through a semipermeable membrane.

Otoliths. Tiny calcium carbonate crystals that rest on the hair cells in a sacculus and a utriculus.

Ovarian cycle. Series of events within the ovary that constitute part of the female reproductive cycle.

Ovary. The female reproductive organ, producing ova and hormones.

Ovulation. Rupture of an egg through the surface of the ovary.

Ozone. Layer of O_3 in the upper atmosphere.

Pacemaker. The sinoatrial node in the heart that sets the pace of heartbeat.

Pap test. Test for cancer of the uterus, named after Dr. Papanicolaou.

Parasitism. An interaction in which one member lives upon or in another organism and also uses it for a food supply.

Parasympathetic nerves. A portion of the autonomic nervous system serving most internal organs.

Parthenogenesis. Development of an unfertilized egg.

PCM. Protein-calorie malnutrition. General malnutrition (starvation) symptoms resulting from lack of sufficient carbohydrates, fats, and proteins in the diet.

Pedigree. Listing of the ancestral background of an organism.

Penis. Male organ used for copulation and the elimination of urine. Enlarges to become erect during sexual excitement.

Pepsinogen. An inactive form of the digestive enzyme pepsin.

Pepsin. Digestive enzyme that begins the breakdown of protein molecules in the stomach.

Peripheral nerves. Nerves extending out into the body from the brain and spinal cord.

Peristalsis. Waves of muscular contraction that pass along a digestive tube and thereby push small masses of food through it.

Peroxisome. Small sac filled with granules and enzymes called oxidases that destroy hydrogen peroxide.

Phagocytosis. The ability of some cells to surround and ingest particles, and sometimes other cells.

Pharynx. Back portion of the mouth that leads into the esophagus and into the larynx.

Pheromone. Chemical released by an animal into its surrounding that influences the behavior of other members of the same species.

Photochemical smog. A mixture of chemical byproducts, often harmful, formed when sunlight passes through a mixture of air pollutants including hydrocarbons and nitrogen oxides.

Photoperiodism. Physiological responses to seasonal changes in the proportion of daylight to darkness in a 24 hour time period.

Physiological adaptation. A change in the functioning of some part of the body that better enables a species to survive.

Physiology. The study of the functions of organs and parts of the body.

Pituitary. Endocrine gland at the base of the brain that produces hormones with many regulatory activities in the body.

Placenta. An organ that transports nutrients, gases, and waste products between the embryo and its mother's uterus.

Plankton. Aquatic life with limited or no means of locomotion. Usually very small.

Plasma. The liquid portion, mostly water, that comprises about 55 percent of the blood.

Platelets. Tiny fragile bodies that circulate with the blood and release clotting agents if damaged.

Pneumonia. Disease of the lungs that causes the alveoli to fill with fluid.

Point mutation. Loss or repetition of a single nucleotide unit in a DNA molecule. Albinism is an example.

Polar body. Mass of chromosomes thrown off from an egg during oogenesis.

Polygenic trait. A trait controlled by many genes.

Polymer. A compound made of repeating chemical units.

Polypeptide. A chain of amino acids that may help to constitute a protein, or may have other functions in a cell.

Polyunsaturated fatty acid. A fatty acid containing two or more sets of double bonds between its carbon atoms.

Pons. A portion of the brain located just above the medulla. It contains nerve cell centers involved in hearing, taste, and eyeball movement.

Predation. The use of one organism by another for food.

Predator. An organism that uses another organism for food.

Prey. An organism used for food by another organism.

Primary consumers (herbivores). Plant eaters.

Primary producers. Photosynthetic organisms such as algae and all higher green plants.

Primary sex characteristics. The reproductive organs present at birth in both sexes.

Primates. An order of mammals distinguished by advanced brain structure, stereoscopic vision, good eyesight, poor sense of smell, flattened nails and claws, prehensile hands and feet, an opposable thumb, and a well-developed collar bone. Includes monkeys, apes, and humans.

Productivity. Total amount of organic matter formed by the producers in a community.

Progesterone. Hormone from the corpus luteum that helps maintain the endometrium (lining) of the uterus.

Progestin. A chemical produced synthetically that functions like progesterone.

Prokaryote. Simple type of cell found in bacteria and blue green algae. They do not have distinct nuclei, DNA is not in the form of a chromosome, and no membranous organelles are present.

Prophase. The initial stage of cell division when the chromatin forms into chromosomes.

Proprioception. The sense of awareness of the position of various body parts.

Prostate gland. Walnut-sized gland in males found beneath the bladder and containing the united ejaculatory ducts and urethra. Produces seminal fluid.

Protein. Large organic substance that often forms a major structural material in cells. Proteins are made of amino acids linked in long chains and in complex shapes.

Protein-calorie malnutrition (PCM). Starvation symptoms resulting from lack of sufficient carbohydrates, fats, and proteins in the diet.

Proteinase. An enzyme that digests proteins.

Proximal convoluted tubule. Portion of a nephron in which massive reabsorption takes place.

Pulmonary circulation. Movement of blood from the heart to the lungs and back to the heart.

Pulmonary veins. Vessels that transport blood from the lungs into the left atrium of the heart.

Punnett square. A set of boxes used in finding all possible gametic combinations when solving genetics problems.

Pupil. Opening through the iris that allows light to pass into the eyeball.

Pyruvate. A three-carbon organic acid important in living systems.

Quantitative inheritance. Inheritance involving the simultaneous action of many genes in an accumulative or additive fashion. Height, weight, and body size are inherited in this way.

Radiation sickness. Symptoms that arise from excessive exposure to radiation: nausea, loss of hair, internal bleeding, and ulceration of the intestinal wall are typical.

Radioactive dating. A method of determining the age of rocks by using the radioactive decay rate of isotopes in the rock. Carbon-14 is a widely used isotope for this method.

Radioisotope. The radioactive form of a chemical element. Radioisotopes are useful in biological studies because they can be detected or traced with appropriate instruments.

Ramapithecus. Fossil primate with human-like teeth from the Miocene period; possibly the beginning of man's ancestry 10 million years ago.

Receptors. Nerve cells specialized for detecting changes in the environment.

Recessive trait. One that is expressed only when both recessive genes are present.

Recombinant DNA. DNA that contains genes from another organism.

Recombination. The consequence of fertilization; uniting of genes from a male and a female organism.

Recommended Dietary Allowance (RDA). The amount of any specific type of food considered necessary daily to maintain a human being of a specific age category.

Red corpuscle. A red blood cell.

Reflex. A functional unit in the nervous system consisting of a receptor, sensory neuron, association neuron, motor neuron, and effector.

Renal artery. A branch from the abdominal aorta that transports blood into the kidney.

Replication. The duplication of a DNA molecule.

Repolarization. A process that changes the potential on a nerve fiber back to its resting condition after an impulse has passed.

Resolving power. Ability of a microscope to distinguish objects lying near one another.

Respiratory gases. Oxygen and carbon dioxide.

Respirometer. A device for measuring the rate at which oxygen is used and carbon dioxide is produced as an individual breathes.

Resting potential. The electrical charge difference between the inside and outside of a cell membrane.

Reticular activating system. Center of arousal and wakefulness. Located in the brain stem.

Retina. Complex layers of nerve cells containing light receptors that line the back of the eyeball.

Rh blood factor. Blood antigen present in about 85 percent of the population. First detected in Rhesus monkeys, hence its name.

Rhodopsin. Light sensitive pigment found in rods in the retina.

Rhythm method of birth control. Avoiding intercourse around the time of ovulation.

Ribosomes. Smaller granular bodies that form the sites where proteins are made in the cell. Usually extremely abundant and attached to endoplasmic reticulum membranes.

RNA (ribonucleic acid). See mRNA and tRNA.

Rods. Specialized light sensitive cells in the retina. Rods are especially sensitive to dim light.

Ruminant. A herbivore in which the lower esophagus and stomach are enlarged into compartments for fermentation.

Saccule. Organ in the inner ear that contains otoliths resting on hair cells. Detects changes in head movement.

Sarcoma. A solid type tumor found in muscle, bone, and other connective tissues. Relatively uncommon.

Saturated fatty acid. Fatty acids that cannot take on any more hydrogen atoms. They are mostly animal fats.

Schwann cell. Specialized cell whose cell membrane forms a sheath called myelin around an axon.

Scientific law. A theory that has been repeatedly verified.

Scientific method. An approach used by scientists for obtaining new knowledge. Its success and reliability depend on repeatability by others.

Sclera. Tough outer coat of connective tissue around the eyeball. The "white" of the eye.

Scrotum. Sac containing the testes, located near the base of the penis.

Second messenger concept. Theory to explain how a hormone brings about specific responses in target cells. The hormone (first messenger) causes cells to produce cyclic AMP (second messenger) that in turn affects a specific cellular response.

Secondary consumers (carnivores). Forms that feed on herbivores.

Secondary sex characteristics. Features of the body that appear at puberty such as pubic and armpit hair, change in voice, and a change in general body proportions.

Segregation. Separation of homologous chromosomes in the first part of meiosis.

Self-regulation (homeostasis). The control of internal features of the body by feedback systems.

Semen. Seminal fluid containing sperm cells.

Semicircular canals. Fluid-filled passageways that function to provide a sense of equilibrium.

Seminal vesicle. Gland that adds seminal fluid to sperm cells as they enter the ejaculatory duct.

Sensory neuron. Nerve cells that conduct impulses into the nervous system from sense organs.

Sex chromosome. Chromosome concerned with sex determination. Termed X or Y chromosomes in many organisms.

Sex indicator. See **Barr body.**

Sex linked gene. Gene located on an X chromosome.

Sexual dimorphism. Traits that distinguish males from females.

Sexual reproduction. The union of specialized sex cells (egg and sperm) to produce a new organism.

Sickle cell anemia. Hereditary disease caused by a defective hemoglobin molecule.

Simple sugar. A small carbohydrate made of from three to seven carbon atoms.

Sinoatrial node. Mass of tissue in the wall of the right atrium that helps coordinate the contractions of the heart chambers. It is sometimes called the "pacemaker."

Social dominance. An activity found in flocks and herds where members become arranged in a social hierarchy or pecking order. It helps to decrease intragroup strife.

Solar plexus. Cluster of autonomic nerve ganglia located in the abdomen.

Speciation. Formation of new species, thought to require isolation of a portion of a population from the parent group for a great time period.

Species. A group of organisms that is isolated reproductively from other similar groups.

Spermatid. A haploid cell prior to undergoing specialization into a mature spermatozoan.

Spermatocyte. Cell in the testes that is destined to become a sperm cell.

Spermatogenesis. Series of events including meiosis and specialization that a cell undergoes in order to become a spermatozoan.

Spermatozoan. Cell that has undergone meiosis and become modified into a motile sperm cell.

Spinal cord. A compact mass of motor, sensory, and association neurons extending from the brain stem into the vertebral column. It is sheathed in protective membranes and bathed in spinal fluid.

Spinal nerves. Thirty-one pairs of nerves that extend from the spinal cord into skeletal muscles and sense organs in the skin. A spinal nerve is composed of sensory and motor neurons.

Spindle. A structure made of microtubules that functions in separating chromosomes during cell division.

Starch. A storage form of carbohydrate in plants. It consists of numerous linked simple sugar units.

Sterile male technique. Releasing sterile males in a population as a population control technique.

Stimulus. A change in the environment.

Structural adaptation. An alteration in form or anatomy that better enables a species to survive.

Structural protein. A protein used in the structure of various parts of the body including cell membranes, connective tissues, and protective tissues such as skin, and bones.

Sucrose. Table sugar made commercially from sugar cane and sugar beets, also common in many other plant juices.

Suction curettage. Suction device used for removing an embryo or fetus in an abortion.

Supercoil Theory. The theory that a chromosome consists of DNA coated with nucleoproteins and wrapped into a complex coil.

Superior vena cava. A major vein that drains blood from the upper part of the body into the heart.

Survivorship curve. A graph of the percentage of survivors of a group after a period of time.

Symbiosis. A close association (living together) between two dissimilar organisms.

Sympathetic nerves. A portion of the autonomic nervous system serving most internal organs.

Synapse. Junction between two nerve cells.

Synaptic cleft. The tiny fluid-filled space between a synaptic knob and a postsynaptic fiber.

Synaptic transmission. Continuation of an impulse across a synaptic cleft by means of the diffusion of a neural transmitter chemical such as acetylcholine.

Synaptic transmitter. Chemical such as acetylcholine produced into a synaptic cleft from the ends of axons.

Synthesis phase. A stage in the life cycle of a cell during which its DNA is duplicated.

Systemic circulation. The transport of blood from the left ventricle out to the major organs of the body and back to the right atrium.

Systole. Contraction of the ventricles of the heart.

T cell. Lymphocyte that has passed through the thymus gland and is specialized in destroying foreign cells that get into the body.

Tapetum. Layer of light-reflecting cells behind the retina (not found in humans).

Target tissue. A tissue or group of cells specifically affected by a hormone.

Telophase. The final stage of cell division leading to the formation of daughter cells.

Temperature inversion. Unusual condition in which a layer of cool air is held at ground level by an overlying layer of warm air.

Territoriality. The establishment of a circumscribed area by an organism, usually a male, who defends it against other males and courts females therein.

Testis. The male reproductive organ, producing sperm and the hormone testosterone.

Thalamus. A small region of the brain containing sensory tracts going to the cerebrum.

Theory. A statement that summarizes or relates a series of observations.

Thrombus. A clot that forms within a blood vessel.

Thymine. An organic base unit in a nucleic acid; part of the chemical code.

Thyroid. An endocrine gland located in the neck. Its hormones help regulate general growth and development, plus metabolism.

Tissue. A group of specialized cells that performs a specific function.

Tissue culture. Growing cells or small bits of tissues in chemical media in glassware.

Trace element. See micronutrient.

Tracheae. A network of tiny air tubes used by many arthropods as a respiratory system.

Transcription. The copying of the DNA triplet code into ribonucleic acid (RNA).

Translation. The formation of a polypeptide chain based on the triplet codes in a strand of mRNA.

Transmission genetics. A study of the mechanisms by which hereditary traits are passed from one generation to the next.

Triglyceride. The chemical term for a fat (a glycerol molecule combined with three fatty acids; hence the prefix *tri-*).

Triplet code. A sequence of three consecutive organic bases in a nucleotide chain.

tRNA, Transfer Ribonucleic Acid. A small "cloverleaf" shaped string of nucleotides that transports an amino acid and is coded to match a triplet code on messenger RNA.

Unsaturated fatty acid. A fatty acid capable of accepting more hydrogens by chemical means.

Uracil. A nitrogen-containing unit in a RNA molecule.

Urethra. Tube that extends from the urinary bladder to the vulva in females or through the penis of males.

Uterus. Muscular organ in pelvic region in which a fertilized egg can implant and grow.

Utricle. Organ in the inner ear that contains otoliths resting on hair cells. Detects changes in head movement.

Vagina. A muscular tube that connects the uterus with the vulva. It functions in sexual intercourse and as the birth canal.

Value. A belief.

Vas deferens. Small tube that transports sperm from the epididymis up into the pelvic cavity into the seminal vesicles.

Vasectomy. Cutting the sperm ducts as a method of birth control.

Vasoconstriction. Constriction of the arterioles leading into capillary beds so that less blood enters the beds.

Vasodilation. Dilation of the arterioles leading into the capillary beds so that more blood enters the beds.

Veins. Blood vessels that transport blood toward the heart. They contain only small amounts of smooth muscle tissue and have valves to prevent backflow of blood.

Vena cava. A large vein that receives blood from most of the body and empties it into the right atrium of the heart.

Ventricle, brain. One of four small passageways through the brain. Contains cerebrospinal fluid.

Ventrical, heart. Lower chamber of the heart. Right ventricle pumps blood to the lungs. Left ventricle pumps blood to the body.

Venule. A tiny vessel that receives blood after it passes through a capillary bed.

Vertebrate. An animal that has a backbone, a concentration of sense organs in the head region, and a protective cranium around the brain.

Villi. Microscopically tiny projections (villus singular).

Vitamins. Chemicals required in small amounts by animals in order to maintain most of their essential chemical functions.

Vitreous humor. Jellylike material found in the back chamber of the eyeball.

Vulva. External genitalia of the female consisting of the labia, clitoris, and openings of the urethra and vagina.

Water cycle. Water evaporates from the oceans, lakes, and so on, circulates in the atmosphere, and returns eventually as rain or snow.

Water soluble vitamin. One of a group of vitamins that is soluble in water in the body: the B vitamins and vitamin C.

Watson-Crick model. The structure of DNA as proposed by James Watson and Francis Crick.

White matter. Myelinated neurons in the central nervous system.

X chromosome (sex chromosome). Females contain two X chromosomes in every cell.

Y chromosome. Sex chromosome; determines maleness.

Yolk sac. A sac that contains nutrient material for the embryo in most organisms. The yolk sac in mammals does not contain yolk and persists for only a short period of time.

Zona pellucida. Mass of follicle cells around an egg cell.

Zygote. A fertilized egg; the consequence of the union of an egg and a sperm.

photo credits

Chapter 1

Opener Jeff Jacobson/Archive Pictures.
Figure 1-1 Courtesy of French Government Tourist Office.
Figure 1-2 American Museum of Natural History.
Figure 1-3 George Bellerose/Stock, Boston.
Figure 1-4 Courtesy of the Institute for Scientific Information.
Figure 1-7 Top left, Courtesy of The Memorial Sloan-Kettering Cancer Center; top right, and bottom left, Grant Heilman; bottom right, Andrew Rokoczy.
Figure 1-10 George Whiteley/Photo Researchers.

Chapter 2

Opener Alan Carey/The Image Works.
Figure 2-7 (a), Russ Kinne/Photo Researchers; (b & c), A. W. Ambler/National Audubon Society-Photo Researchers; (d), Gordon Smith/National Audubon Society-Photo Researchers.
Figure 2-8 Left, Bernard P. Wolfe/Photo Researchers; right, Tom McHugh/Photo Researchers.
Figure 2-12 Top, Teri Leigh Stratford; bottom, courtesy of Yerkes Regional Primate Research Center, Emory University.
Figure 2-13 Courtesy of Hewlett-Packard.
Figure 2-14 Miguel Castro/Photo Researchers.

Chapter 3

Opener Courtesy of Keith R. Porter.
Figure 3-2 Rare Book Division, The New York Public Library, Astor, Lenox and Tilden Foundations.
Figure 3-3 Courtesy of Susumo Ito, Ph.D. Professor of Anatomy, Harvard Medical School.
Figure 3-6 Courtesy of New Brunswick Scientific Co., Inc., New Jersey.
Figure 3-7 (a), Courtesy of Carl Zeiss; (b), courtesy of George E. Palade.
Figure 3-11 Left, endoplasmic reticulum, courtesy of Don Fawcett; ribosomes, courtesy of George E. Palade; golgi, courtesy of David Friend; mitochondrion, courtesy of K. R. Porter.
Figures 3-12 & 3-13 Courtesy of Dr. D. W. Fawcett, Harvard Medical School.
Figure 3-14 (a), Courtesy of Charlotte Memorial Hospital/American Cancer Society; (b), courtesy of American Cancer Society.
Figure 3-17 (a), Courtesy of Dr. Jerome Gross, Massachusetts General Hospital, Boston; (b), Courtesy of D. W. Fawcett.
Figure 3-19 John Wiley & Sons, Picture Research File.

Chapter 4

Opener Owen Franken/Stock, Boston.
Figure 4-3 (a), top and bottom, Courtesy of World Health Organization.
Figure 4-4 Both, Courtesy of Food and Agriculture Organization.
Figure 4-5 Courtesy of Rockefeller Foundation.
Figure 4-7 Teri Leigh Stratford.
Figure 4-8 Paul Almasy/WHO.
Figures 4-9, 4-10, 4-11 & 4-12 Teri Leigh Stratford.
Page 68 Both, Drawing by Drucker © The New Yorker Magazine.

Chapter 5

Opener Mark Antman/The Image Works.
Figure 5-2 Courtesy of D. Fulton, Ohio State University.
Figure 5-7 (a), Dr. William C. Roberts, Dr. L. Maximilian Buja, Dr. Bernadine Healy Bulkley, National Heart and Lung Institute, National Institute of Health; (b), courtesy of American Heart Association; (c), Dr. William C. Roberts, Dr. L. Maximilian Buja, Dr. Bernadine Healy Bulkley, National Heart and Lung Institute, National Institute of Health.
Figure 5-10 Dick Raphael for *Sports Illustrated* © Time, Inc.
Page 87 Sidney Harris.

Chapter 6

Opener Jen & Des Bartlett/Photo Researchers.
Figure 6-5 Teri Leigh Stratford.
Figure 6-6a Farber & Papa, Johnson & Johnson.
Figure 6-7 (a), George Holton/Photo Researchers; (b), Mark Boulton/Photo Researchers; (c), Karl H. Maslowski/Photo Researchers.
Figure 6-8 Robert Hermes/National Audubon Society-Photo Researchers.
Figure 6-9 Jen & Des Bartlett/Photo Researchers.
Figure 6-10 A. W. Ambler/National Audubon Society-Photo Researchers.

Chapter 7

Opener   Diego Goldberg/Sygma.
Figure 7-3   Courtesy of Dr. Thomas C. Hayes, Lawrence Berkeley Laboratory, University of California, Berkeley.
Figure 7-4   Courtesy of L. W. Diggs, M.D. From L. W. Diggs, D. Strum, A. Bell, *Morphology of Human Blood Cells*, 4th edition, 1978, Abbott Laboratories.
Figure 7-13   Courtesy of Dr. J. A. Kylston, Duke University Medical Center.
Figure 7-14   (a), George Porter/National Audubon Society-Photo Researchers; (b), Runk-Schoenberger/Grant Heilman.

Chapter 8

Opener   Joel Gordon.
Figure 8-1   (a and b), Wide World.
Figure 8-2   (a and b), Courtesy of Ralph G. Somes, Jr., University of Connecticut, Storrs.
Figure 8-3   Armed Forces Institute of Pathology.
Figure 8-7   Wide World.
Figure 8-8   From N. Coshing, *The Pituitary Body and Its Disorders*, Philadelphia, J. P. Lippincott, 1912.
Figure 8-9   (a and b), Lester V. Bergman.
Figure 8-10   Courtesy of Massachusetts General Hospital.

Chapter 9

Opener   Courtesy of Edwin R. Lewis, Thomas E. Everhart, Yehoshua Y. Zeevi, University of California, Berkeley.
Figure 9-7   Courtesy of Terence H. Williams.

Chapter 10

Opener   David Powers/Stock, Boston.
Figure 10-4   (a, b and c), Courtesy of Anita Anderson, Department of Ophthalmology, Columbia University.
Figure 10-5   Edwin R. Lewis, Yehoshua Y. Zeevi, and Frank S. Werblin, University of California, Berkeley
Figure 10-13   (a), Courtesy of Neuropathology, New York State Psychiatric Institute, State of New York, Department of Mental Hygiene.

Chapter 11

Opener   Mimi Forsyth/Monkmeyer.
Figure 11-10   Christa Armstrong/Rapho-Photo Researchers.

Chapter 12

Opener   Cary Wolinsky/Stock, Boston.
Figure 12-5   Courtesy of Dr. Nathaniel E. Rodman, Medical Center, West Virginia University.
Figure 12-9   United Press International.

Chapter 13

Opener   Barbara Alper/Stock, Boston.
Figure 13-1   Courtesy of Jordache ®. Campaign created by Hick & Gareist Advertising, Inc., Creative Director, Aric Frons.
Figure 13-2   CCM: General Biological Supply, Inc., Chicago.
Figure 13-5   (a and b), From *Scanning Electron Microscopy in Biology*, Kessel & Sheh, Springer-Verlag 1974.
Figures 13-10 & 13-11   Courtesy of Dr. R. J. Blandau, Ph.D, School of Medicine, University of Washington, Seattle.
Figure 13-17   Teri Leigh Stratford.
Figure 13-18   Top, Teri Leigh Stratford.

Chapter 14

Opener   Rick Winsor/Woodfin Camp.
Figure 14-1   Courtesy of Dr. Pat Colarco, School of Medicine, University of California, San Francisco.
Figure 14-2   Carnegie Institution of Washington.
Figure 14-4   (b), Courtesy of Dr. Ronan O'Rahilly, Carnegie Institution of Washington, Department of Embryology, Davis Division.
Figure 14-8   (b), Courtesy of Dr. Ronan O'Rahilly, Carnegie Institution of Washington, Department of Embryology, Davis Division.
Figure 14-10   Joe Baker.
Figure 14-11   Courtesy of Dr. Ronan O'Rahilly, Carnegie Institution of Washington, Department of Embryology, Davis Division.
Figure 14-17   Left, Gabor Demjen/Stock, Boston; center, Thomas Hopker/Woodfin Camp; right, Timothy Eagan/Woodfin Camp.

Chapter 15

Opener   Courtesy of Gunther F. Bahr, Armed Forces Institute of Pathology.
Figure 15-10   (c), Ada L. Olins and Donald E. Olins, Oak Ridge Graduate School of Biomedical Sciences, Oak Ridge National Laboratory, Tennessee.
Figure 15-12   Courtesy of New York University Cytogenetics Laboratory.
Figure 15-13   Top, Norman R. Lightfoot/Photo Researchers; bottom, Harvey Stein.
Figure 15-14   Courtesy of Dr. Stanley N. Cohen, Stanford University.

Chapter 16

Opener   Michal Heron/Woodfin Camp.
Figure 16-1   Joel Gordon.

Figure 16-3 Courtesy of Australian News and Information Bureau.

Figure 16-4 (a), Grant Heilman; (b), courtesy William P. Nye; (c and d), Carolina Biological Supply Company.

Figure 16-5 Top, Francois Morel, California Institute of Technology, Pasadena; bottom, courtesy Springer-Verlag, Publishers, from *Corpuscles* by Marcel Bessis.

Figure 16-6 Murray L. Barr, M.D.

Figure 16-7 Victor A. McKusick, Johns Hopkins University School of Medicine, Baltimore.

Figure 16-8 Marjorie Pickens.

Figure 16-9 Maxwell Coplan/DPI.

Figure 16-10 Michael Weisbrot and Family.

Figure 16-11 Photo Communications.

Figure 16-13 Left, Teri Leigh Stratford; right, Kathy Bendo.

Figure 16-14 Courtesy of Australian News Information Bureau.

Figure 16-16 Peter Menzel.

Chapter 17

Opener Courtesy of American Museum of Natural History.

Figure 17-4 Courtesy of Smithsonian Institution.

Figures 17-5 & 17-6 Courtesy of American Museum of Natural History.

Figure 17-8 Courtesy of J. William Schopf, University of California, Los Angeles.

Figure 17-14 Courtesy of Dr. H. B. D. Kettlewell.

Figure 17-15 Walter Chandoha.

Chapter 18

Opener Gideon Nelson.

Figure 18-1 (a), Mark Boulton/National Audubon Society-Photo Researchers; (b), George S. Harrison/Grant Heilman; (c), Jen & Des Bartlett/Photo Researchers; (d), Leonard Lee Rue III/Bruce Coleman; (e), Runk-Schoenberger/Grant Heilman; (f), Grant Heilman.

Figure 18-2 Top and bottom, Grant Heilman.

Figure 18-3 (a and b), Tom McHugh/Photo Researchers.

Figure 18-4 (a and b), Jack Dermid/National Audubon Society-Photo Researchers.

Figure 18-5 Eric Hosking.

Figure 18-7 Top, Courtesy H. Douglas Pratt.

Figure 18-11 Top, Courtesy of American Museum of Natural History; bottom, Maurice Wilson, British Museum of Natural History.

Figure 18-13 Top, Courtesy of American Museum of Natural History.

Figure 18-14 Courtesy of American Museum of Natural History.

Figure 18-15 Top, National Maritime Museum.

Figure 18-16 Michael Weisbrot and family.

Chapter 19

Opener Keith Gunnar/Photo Researchers.

Figure 19-2 Stephen Collins/National Audubon Society-Photo Researchers.

Figure 19-6 The Nitrogen Company, Inc., Milwaukee.

Figure 19-12 R. S. Virdee/Grant Heilman.

Figure 19-13 Ylla/Rapho-Photo Researchers.

Figure 19-14 Left, Grant Heilman; right, Des Bartlett/Photo Researchers.

Figure 19-16 Courtesy of American Museum of Natural History.

Figure 19-18 Douglas Faulkner.

Chapter 20

Opener George Bellerose/Stock, Boston.

Figure 20-2 Top left, Tom McHugh/Photo Researchers; top center, Steve McCutcheon; top right, Eric Hosking/Bruce Coleman; center and bottom left, Jen & Des Bartlett/Photo Researchers; bottom center and bottom right, Leonard Lee Rue III/Bruce Coleman.

Figure 20-3 Steve McCutcheon/Photo Researchers.

Figure 20-4 Top left and center, Steve McCutcheon/Alaska Pictorial Service; top right, Leonard Lee Rue III/Bruce Coleman; center, Weyerhauser Company; bottom left, G. Ronald Austing/National Audubon Society-Photo Researchers; bottom center, Harry Engles/National Audubon Society-Photo Researchers; bottom right, Jen & Des Bartlett/Photo Researchers.

Figure 20-5 Top left, Elinor S. Beckwith/DPI; top right and center left, Alvin E. Staffan/National Audubon Society-Photo Researchers; center, Karl H. Maslowski/National Audubon Society-Photo Researchers; center right, Leonard Lee Rue III/Photo Researchers; bottom left, A. W. Ambler/National Audubon Society-Photo Researchers; bottom center, Grant Heilman; bottom right, Leonard Lee Rue III/National Audubon Society-Photo Researchers.

Figure 20-6 Top left, Karl H. Maslowski/National Audubon Society-Photo Researchers; top center, Joseph Van Wormer/National Audubon Society-Photo Researchers; top right, E. R. Kalmbach/U.S. Fish and Wildlife Service; center, U.S. Forest Service; bottom left, Jen & Des Bartlett/Photo Researchers; bottom right, Grant Heilman.

Figure 20-7 Smithsonian Institution/National Anthropological Archives.

Figure 20-8 Top left, Shelly Grossman/Woodfin Camp; top right, Tom McHugh/Photo Researchers; bottom right, Grant Heilman; center left and center, Grant Heilman; center right, Verna R. Johnston/Photo Researchers; bottom left, Hal Harrison/Grant Heilman; bottom center, Verna R. Johnston/Photo Researchers; bottom right, Tom McHugh/Steinhart Aquarium-Photo Researchers.

Figure 20-9   Mary Stuart Lang.

Figure 20-10   Top, Carl Frank/Photo Researchers; center left, Fritz Polking/Bruce Coleman; center, L. and D. Klein/Photo Researchers; center right, Jen & Des Bartlett/Photo Researchers; bottom left, Jen & Des Bartlett/Photo Researchers; bottom center, A. W. Ambler/National Audubon Society-Photo Researchers; bottom right, Tom McHugh/Photo Researchers.

Figure 20-11   Paul Almasy.

Figure 20-12   Jack Dermid/National Audubon Society-Photo Researchers.

Figure 20-13   Gideon Nelson.

index

Italicized numbers indicate illustrations.